Grundlehren der mathematischen Wissenschaften 282

A Series of Comprehensive Studies in Mathematics

Editors

M. Artin S.S. Chern J.M. Fröhlich E. Heinz
H. Hironaka F. Hirzebruch L. Hörmander
S. MacLane W. Magnus C.C. Moore J.K. Moser
M. Nagata W. Schmidt D.S. Scott Ya.G. Sinai
J. Tits B.L. van der Waerden M. Waldschmidt
S. Watanabe

Managing Editors

M. Berger B. Eckmann S.R.S. Varadhan

Pierre Lelong   Lawrence Gruman

Entire Functions of Several Complex Variables

Springer-Verlag
Berlin Heidelberg NewYork Tokyo

Professor Dr. Pierre Lelong
Université Paris VI
4, Place Jussieu, Tour 45–46
75230 Paris Cedex 05
France

Dr. Lawrence Gruman
UER de mathématiques
Université de Provence
3 place Victor Hugo
13331 Marseille
France

Mathematics Subject Classification (1980): 32A15

ISBN 3-540-15296-2
Springer-Verlag Berlin Heidelberg New York Tokyo
ISBN 0-387-15296-2
Springer-Verlag New York Heidelberg Berlin Tokyo

Library of Congress Cataloging in Publication Data
Lelong, Pierre.
Entire functions of several complex variables.
(Grundlehren der mathematischen Wissenschaften; 282)
Bibliography: p.
Includes index.
1. Functions, Entire. 2. Functions of several complex variables. I. Gruman, Lawrence, 1942–.
II. Title. III. Series.
QA353.E5L44 1986 515.914 85-25028
ISBN 0-387-15296-2 (U.S.)

Typesetting, printing and bookbinding: Universitätsdruckerei H. Stürtz AG, Würzburg
2141/3140-543210

Introduction

I – Entire functions of several complex variables constitute an important and original chapter in complex analysis. The study is often motivated by certain applications to specific problems in other areas of mathematics: partial differential equations via the Fourier-Laplace transformation and convolution operators, analytic number theory and problems of transcendence, or approximation theory, just to name a few.

What is important for these applications is to find solutions which satisfy certain growth conditions. The specific problem defines inherently a growth scale, and one seeks a solution of the problem which satisfies certain growth conditions on this scale, and sometimes solutions of minimal asymptotic growth or optimal solutions in some sense.

For one complex variable the study of solutions with growth conditions forms the core of the classical theory of entire functions and, historically, the relationship between the number of zeros of an entire function $f(z)$ of one complex variable and the growth of $|f|$ (or equivalently $\log|f|$) was the first example of a systematic study of growth conditions in a general setting.

Problems with growth conditions on the solutions demand much more precise information than existence theorems. The correspondence between two scales of growth can be interpreted often as a correspondence between families of bounded sets in certain Fréchet spaces. However, for applications it is of utmost importance to develop precise and explicit representations of the solutions.

If we pass from $\mathbb{C}$ to $\mathbb{C}^n$, new problems such as problems of value distribution for holomorphic mappings from $\mathbb{C}^n$ to C^m arise. On the other hand, new techniques are often needed for classical problems to obtain solutions and representations of the solutions. Zeros of entire functions f are no longer isolated points; a measure of the zero set is obtained by the representation of the divisor X_f of f (and more generally of analytic subvarieties) by closed and positive currents, a class of generalized differential forms.

Paradoxally, it is the non-holomorphic objects, the "soft" objects (objets souples in French, see [C]) of complex analysis, principally plurisubharmonic functions and positive closed currents, which are adapted to problems with growth conditions, giving global representations in $\mathbb{C}^n$. Very often properties of the classical (i.e. holomorphic) objects will be derived from properties obtained for the soft objects. Plurisubharmonic functions were introduced in 1942 by K. Oka and P. Lelong. They occur in a natural way from the beginning of this book. Indicators of growth for a class of entire functions f are obtained as upper bounds for $\log|f|$; for $\log|f|$. To solve Cousin's Second Problem, i.e. to find (with growth conditions) an entire function f with given zeros X in $\mathbb{C}^n$, we solve first the general equation $i\partial\bar{\partial}V=\theta$ for a closed and positive current θ; if $\theta=[X]$, the current of integration on X, we then obtain f by $V=\log|f|$. Properties of plurisubharmonic functions appear again in a remarkable (and unexpected) result of H. Skoda (1972): there exists a representation for the analytic subvarieties Y in $\mathbb{C}^n$ of dimension $p(0\leqq p\leqq n-1)$ as the zero set $Y=F^{-1}(0)$ of an entire mapping $F=(f_1,\ldots,f_{n+1})$ such that $\|F\|$ is controlled by the growth of the area of Y. Plurisubharmonic functions obtained from potentials seem well adapted to the construction of global representations in $\mathbb{C}^n$; the method avoids the delicate study of ideals of holomorphic functions vanishing on Y and satisfying growth conditions.

The same methods using the soft object's properties of the current $(i\partial\bar{\partial}V)^p$ and the Monge-Ampère equation for plurisubharmonic functions V are employed for recent results obtained in value distribution theory of holomorphic mappings $\mathbb{C}^n\to\mathbb{C}^m$ or $X\to Y$, two analytic subvarieties in $\mathbb{C}^n$.

II - Before summarizing the content of this book, we would like to make some remarks.

a) We have not sought to give an exhaustive treatment of the subject (problems for $n>1$ are too numerous for a single book). We have tried to introduce the reader to the central problems of current research in this area, essentially that which had led to general methods or new technics. Applications appear only in Chapter 6 (to analytic number theory) and in Chapters 8 and 9 (to functional analysis).

b) On the other hand, we have tried to make the book self-contained. Some knowledge in the theory for one complex variable is required of the reader, as well as on integration, the calculus of differential forms and the theory of distributions. A list of books where the reader can find general results not developped here is given before the bibliography (such references are given by a capital roman letter).

The proofs of complementary results appear in three appendices: Appendix I for general properties of plurisubharmonic functions, Appendix II for the technic of proximate orders Appendix III for the $\bar{\partial}$ resolution for (0, 1) forms with L^2-estimates by Hörmander's method.

c) The importance of analytic representations, particulary for some applications, has made it necessary to give certain calculations *in extenso*. The authors are aware of the technical aspect of some developments given in the book. We recommand that the reader first read over the proof in order to assimilate the general idea before immersing himself in the details of the calculations.

d) The literature on the subject of entire functions is enormous. The bibliography, without pretending to be exhaustive, gives an overview of those areas of current interest. Each chapter has a short historical note which is an attempt to explain the origin of the given results.

III - Chapter 1 gives the basic definitions of the growth scales in $\mathbb{C}^n$, the notion of order and type, the indicator of growth and proximate orders. These classical notions extend trivially to plurisubharmonic functions and to entire functions in $\mathbb{C}^n$. In Chapter 2, we introduce the reader to the fundamental properties of positive differential forms and of positive and closed currents. Chapter 3 studies the solution with growth conditions of the equation $i\partial\bar{\partial}V=\theta$ for θ a positive closed current of type (1, 1) in $\mathbb{C}^n$, from which we deduce for $V=\log|f|$ the solution with growth conditions in $\mathbb{C}^n$ of Cousin's Second Problem and the representation of entire functions with a given zero set. The result for an entire function of finite order in $\mathbb{C}^n$ gives an extension of classical results of J. Hadamard and E. Lindelöf for $n=1$. Chapter 4 studies the class of entire functions f of regular growth. Certain results are given here for the first time. The importance of this study, which is based on the preceeding chapters, is in the numerous applications (Fourier transforms, differential systems) and the possibility of associating the regular growth of $\log|f|$ with the regular distribution of the zero set of f.

Chapter 5 studies the problems of entire maps $F: \mathbb{C}^n \to \mathbb{C}^m$. The first portion is devoted to the development of a representation of an analytic subvariety Y of $\mathbb{C}^n$ as the zero set of an entire map $F: \mathbb{C}^n \to \mathbb{C}^{n+1}$, that is $Y = F^{-1}(0)$, $F=(f_1, \ldots, f_{n+1})$, with control of the growth of the function $\|F\|$. The second part studies the growth of the fibers $F^{-1}(a) \cap B(0, r)$, where $B(0, r)=\{z: \|z\| < r\}$, when $r \to +\infty$. The third part studies the relationship between the growth of the area of an analytic set in $\mathbb{C}^n$ and its trace on linear subspaces of $\mathbb{C}^n$. The cases of slow growth and algebraic growth are also studied.

Chapter 6 gives an example of an application of the methods of the preceeding chapters to a problem in number theory. We show that the set of points of $\mathbb{C}^n$ where certain families of meromorphic functions of finite order take on algebraic values is contained in an algebraic subvariety of $\mathbb{C}^n$ whose degree can be bounded: this famous result of E. Bombieri (1970) gave a very deep and unexpected application of the theory of closed positive currents t and of the number $\nu_t(x)$ (a kind of multiplicity for x on the support of t) to number theory, via a classical method of Siegel and L^2-estimates for the $\hat{\partial}$ operator. The same idea was also fundamental some time later in Siu's Theorem about the structure of closed positive currents.

Chapter 7 establishes the theory of the indicator of growth theorem for entire functions of finite order in $\mathbb{C}^n$: every plurisubharmonic function positively homogeneous of order ρ is the (regularized) indicator of growth function of an entire function of order ρ.

Chapters 8 and 9 concern applications of entire functions to classes of linear operators. Indeed the space $\hat{\mathscr{D}}(\Omega)$ of the Fourier transforms of the distributions defined in a bounded domain Ω of $\mathbb{C}^n$ is a subspace of the space $\mathscr{H}(\mathbb{C}^n)$ of the entire functions in $\mathbb{C}^n$, and many problems characterize classes of distributions in Ω by growth properties of the image in $\mathscr{H}(\mathbb{C}^n)$. This method leads to analytic functionals. The analytic functionals are the elements of the dual space of the space $\mathscr{H}(\Omega)$ of holomorphic functions in Ω, equipped with the topology of uniform convergence on compact subsets of Ω. Chapter 8 gives a study of the Fourier-Borel transform and of the Laplace transform in order to obtain properties for analytic functionals and their supports.

Chapter 9 gives a general treatment of convolution operators in linear spaces of entire functions. New results in particular for the functions of order $\rho < 1$ are given as consequences of the techniques developped in preceeding sections of the book.

We use the following system of notations for references: a statement (theorem, lemma, proposition, definition etc.) is given two numbers, the first indicating the chapter in which it is found and the second indicating its position in that chapter. Thus, Theorem 8.23 refers to the 23-rd statement in Chapter 8. Figures within parentheses refer to equations in the text, for instance (4, 18) refers to the eighteenth equation in Chapter 4. Roman numerals I, II, III, refer to the three appendices which are at the end of the book.

The authors thank K. Diederich, R. Gay, H. Skoda and M. Waldschmidt who read different parts of the manuscript and whose remarks were fruitful for many improvements. They are indebted to Mireille Geurts for typing the manuscript, which was no small nor easy task. And we should like to thank the editors of Springer-Verlag for their excellent and rapid preparation of the book.

Paris and Marseille,
January 1986

P. Lelong, Université Paris VI
L. Gruman, C.N.R.S. Marseille

Table of Contents

Chapter 1. Measures of Growth 1

§1. Preliminaries 1
§2. Subharmonic and Plurisubharmonic Functions 2
§3. Norms on $\mathbb{C}^n$ and Order of Growth 5
§4. Minimal Growth: Liouville's Theorem and Generalizations . . . 6
§5. Entire Functions of Finite Order 8
§6. Proximate Orders 14
§7. Regularizations 18
§8. Indicator of Growth Functions 20
§9. Exceptional Sets for Growth Conditions 24
Historical Notes 28

Chapter 2. Local Metric Properties of Zero Sets
and Positive Closed Currents 30

§1. Positive Currents 30
§2. Exterior Product 35
§3. Positive Closed Currents 37
§4. Positive Closed Currents of Degree 1 40
§5. Analytic Varieties and Currents of Integration 48
Historical Notes 58

Chapter 3. The Relationship Between the Growth
of an Entire Function and the Growth of its Zero Set 59

§1. Positive Closed Currents of Degree 1 Associated
with a Positive Divisor 60
§2. Indicators of Growth of Cousin Data in $\mathbb{C}^n$ 63
§3. Canonical Potentials in $\mathbb{R}^m$ 64
§4. The Canonical Representation of Entire Functions of Finite Order 67
§5. Solution of the $\partial\bar{\partial}$ Equation 73
§6. The Case of a Cousin Data 77
§7. Slowly Increasing Cousin Data: the Genus $q=0$;
the Algebraic Case 79
§8. The Case of Integral Order: Extension of a Theorem of Lindelöf . 82

§9. Trace of a Cousin Data on Complex Lines 86
§10. The Case of a Cousin Data of Infinite Order 89
Historical Notes 94

Chapter 4. Functions of Regular Growth 95

§1. General Properties of Functions of Regular growth 96
§2. Distribution of the Zeros of Functions of Regular Growth . . . 106
Historical Notes 114

Chapter 5. Holomorphic Mappings from $\mathbb{C}^n$ to $\mathbb{C}^m$ 116

§1. Representation of an Analytic Variety Y in $\mathbb{C}^n$ as $F^{-1}(0)$ 117
§2. Local Potentials and the Defect of Plurisubharmonicity 118
§3. Global Potentials 123
§4. Construction of a System F of Entire Functions such that $Y=F^{-1}(0)$ 126
§5. The Case of Slow Growth 130
§6. The Algebraic Case 133
§7. The Pseudo Algebraic Case 136
§8. Counterexamples to Uniform Upper Bounds 136
§9. An Upper Bound for the Area of $F^{-1}(a)$ for a Holomorphic Map 138
§10. Upper and Lower Bounds for the Trace of an Analytic Variety on Complex Planes 143
Historical Notes 154

Chapter 6. Application of Entire Functions in Number Theory 155

§1. Preliminaries from Number Theory 155
§2. A Schwarz Lemma 160
§3. Statement and Proof of the Main Theorem 162
Historical Notes 165

Chapter 7. The Indicator of Growth Theorem 167

Historical Notes 176

Chapter 8. Analytic Functionals 177

§1. Convex Sets and the Fourier-Borel Transform 178
§2. The Projective Indicator 178
§3. The Projective Laplace Transform 183
§4. The Case of M a Complex Submanifold of $\mathbb{C}^n$ 185
§5. The Generalized Laplace Transform and Indicator Function . . . 186
§6. Support for Analytic Functionals 188
§7. Unique Supports for Domains in $\mathbb{C}^n$ 191
§8. Unique Convex Supports 195
Historical Notes 200

Chapter 9. Convolution Operators on Linear Spaces of Entire Functions 201

§1. Linear Topological Spaces of Entire Functions 201
§2. Theorems of Division 207
§3. Applications of Convolution Operators in the Spaces $E_p^{\rho(r)}$ and E^0 210
§4. Supplementary Results for Proximate Orders with $\rho > 1$ 212
§5. The Case $\rho = 1$ 217
§6. More on Functions of Order Less than One 225
§7. Convolution Operators in $\mathbb{C}^n$ 227
Historical Notes 229

Appendix I. Subharmonic and Plurisubharmonic Functions 230

Appendix II. The Existence of Proximate Orders 242

Appendix III. Solution of the $\bar{\partial}$-Equation with Growth Conditions . . 245

Bibliography 254

Index 269

Chapter 1. Measures of Growth

§ 1. Preliminaries

We will let $\mathbb{C}$ represent the field of complex numbers and $\mathbb{R}$ the subfield of real numbers. Let $z=(z_1, \ldots, z_n)$ be an element of $\mathbb{C}^n$ and $\mathbb{R}^{2n}$, the underlying space of real coordinates. The transformations from the complex to the real coordinates are given by $z_k=x_k+iy_k$, $\bar{z}_k=x_k-iy_k$ and $x_k=\frac{z_k+\bar{z}_k}{2}$, $y_k=\frac{z_k-\bar{z}_k}{2i}$. Unless specified to the contrary, we equip $\mathbb{C}^n$ with the Euclidian metric of $\mathbb{R}^{2n}$:

$$ds^2=\sum_{k=1}^{n}(dx_k^2+dy_k^2)=\sum_{k=1}^{n} dz_k \cdot d\bar{z}_k \tag{1,1}$$

and we choose for $\mathbb{C}^n$ the volume form

$$\tau=\bigwedge_{k=1}^{n}(dx_k \wedge dy_k)=(i/2)^n dz_1 \wedge d\bar{z}_1 \wedge \ldots \wedge dz_n \wedge d\bar{z}_n.$$

By a domain Ω, we shall always mean an open connected set. We let $d_\Omega(z)$, the distance to the boundary, be defined for $z\in\Omega$ by $d_\Omega(z)=\inf_{z'\notin\Omega}\|z-z'\|$ (where $\| \ \|$ represents the Euclidean norm) and set $d_\Omega(z)=+\infty$ if $\Omega=\mathbb{C}^n$. Let $\alpha=(\alpha_1, \ldots, \alpha_n)$ be a multi-indice of non-negative integers. We then define $|\alpha|$ by $|\alpha|=\sum_{i=1}^{n}\alpha_i$, the differential operator D^α by $D^\alpha=\frac{\partial^{|\alpha|}}{\partial z_1^{\alpha_1}\ldots\partial z_n^{\alpha_n}}$, and z^α by $z^\alpha=z_1^{\alpha_1}\ldots z_n^{\alpha_n}$.

We let $\mathscr{C}^k(\Omega)$ be the set of functions defined on Ω all of whose derivatives up to order $|\alpha|\leqq k$ are continuous and $\mathscr{C}^\infty(\Omega)$ the set of functions whose derivatives of all orders are continuous. By $\mathscr{C}_0^k(\Omega)$ (resp. $\mathscr{C}_0^\infty(\Omega)$) we will mean the subset of $\mathscr{C}_0^k(\Omega)$ (resp. $\mathscr{C}_0^\infty(\Omega)$) composed of those functions whose support in Ω is compact. We let ∂ and $\bar{\partial}$ be the exterior differential operators defined by

$$\partial=\sum_{k=1}^{n}\frac{\partial}{\partial z_k}dz_k, \qquad \bar{\partial}=\sum_{k=1}^{n}\frac{\partial}{\partial \bar{z}_k}d\bar{z}_k$$

and set

$$d=\partial+\bar{\partial}=\sum_{k=1}^{n}\left(\frac{\partial}{\partial x_k}dx_k+\frac{\partial}{\partial y_k}dy_k\right).$$

A function $f\colon \Omega\subset\mathbb{C}^n\to\mathbb{C}$ is said to be holomorphic in Ω if $f\in\mathscr{C}^1(\Omega)$ and $\bar\partial f=0$. In particular, this means that $\frac{\partial f}{\partial \bar z_k}=0$ for $1\leqq k\leqq n$. The domain $\Delta(z',r)=[z\colon |z'_k-z_k|<r_k,\ r_k>0,\ 1\leqq k\leqq n]$ is called the *polydisc of center* z', of radii r_k. For $\Delta(z',r)\Subset\Omega$, and f holomorphic in Ω, the iteration of the Cauchy Integral Formula for one complex variable gives for $z\in\Delta(z',r)$ the *integral representation:*

$$(1,2)\qquad f(z)=(2\pi)^{-n}\int_0^{2\pi}\dots\int_0^{2\pi}\frac{f(z'_1+r_1e^{i\theta},\dots,z'_n+r_ne^{i\theta})}{\prod_{k=1}^{n}(z'_k+r_ke^{i\theta_k}-z_k)}\,d\theta_1\dots d\theta_n.$$

As for $n=1$, we deduce from (1, 2) a Taylor series expansion:

$$f(z)=\sum_{(\alpha)}C_\alpha(z-z')^\alpha,\qquad \alpha=(\alpha_1,\dots,\alpha_n),$$

which converges uniformly for $|z'_k-z_k|\leqq r'_k<r_k$. Then we obtain a Taylor series expansion on each compact polydisc of Ω. We designate by $\mathscr{H}(\Omega)$ the family of functions holomorphic in Ω. By an *entire function*, we shall mean an element of $\mathscr{H}(\mathbb{C}^n)$. Thus, an entire function $f(z)$ has a Taylor series expansion $f(z'+z)=\sum_m\sum_{|\alpha|=m}P_\alpha(z')z^\alpha$ which, for every point z', converges uniformly in z on compact subsets of $\mathbb{C}^n$. We say that $\sum_{|\alpha|=m}P_\alpha(z')z^\alpha$ is the homogeneous polynomial of degree m in the Taylor series expansion of $f(z)$ at the point z'.

§2. Subharmonic and Plurisubharmonic Functions

In our study of entire functions f of several complex variables, we shall be interested in the asymptotic growth of $|f|$, $f\in\mathscr{H}(\mathbb{C}^n)$, or equivalently by the asymptotic growth of $\log|f|$. Suppose for instance that $\varphi(t)$ is an increasing function of t for $t\geqq 0$ such that $\limsup_{t\to\infty}\frac{\varphi(tu)}{\varphi(t)}<\infty$, for $u\geqq 0$ and $\lim_{t\to\infty}\varphi(t)=\infty$, and consider in $\mathscr{H}(\mathbb{C}^n)$ the subclass M_φ defined by the condition

$$\log|f(z)|\leqq\varphi(\|z\|)+C(f).$$

Then the function

$$\chi_f(z)=\limsup_{t\to\infty}\frac{\log|f(tz)|}{\varphi(t)}$$

measures an asymptotic growth with respect to the weight factor $\varphi(t)$ on the real lines through the origin. Thus we are led to consider expressions of the form $\limsup_{i\in I} C_i\log|f_i|$, $f_i\in\mathscr{H}(\mathbb{C}^n)$, $C_i\in\mathbb{R}^+$. This leads us to study filtered families included in a larger class of functions, the plurisubharmonic functions introduced by K. Oka and P. Lelong. This family is closed under the operation of taking the smallest upper semi-continuous majorant of a filtered family uniformly bounded above on compact subsets (in fact, one can show that the functions $C\log|f|$, $f\in\mathscr{H}(\mathbb{C}^n)$, $C\in\mathbb{R}^+$, generate locally the plurisubharmonic functions under this operation, but we shall not need this property). The plurisubharmonic functions play the same role for $n>1$ complex variables that the subharmonic function play in complex analysis of one complex variable. Moreover, in $\mathbb{C}^n$, for $n\geqq 2$, the growth of entire functions has properties which can be compared to the classical properties (pseudo-convexity, sometimes R^n-convexity) of domains of holomorphy. The use of plurisubharmonic functions and the systematic exploitation of their properties will lead to most of the results in this book.

We begin by recalling important definitions. The proofs of the properties that we shall need can be found in Appendix I (referred to by App. I).

Definition 1.1. Let $\Omega\subset\mathbb{R}^m$ be a domain. A real valued function $\varphi(-\infty\leqq\varphi(x)<+\infty)$ is said to be subharmonic in Ω if

a) φ is upper semi-continuous and $\varphi(x)\not\equiv-\infty$ in Ω,

b) $\varphi(x)\leqq\lambda(x,r,\varphi)\equiv\omega_m^{-1}\int\varphi(x+r\alpha)d\omega_m(\alpha)$ for $r<d_\Omega(x)$ where ω_m is the Lebesgue measure of the unit sphere S^{m-1} in $\mathbb{R}^m$ and λ is the mean value of φ on S^{m-1} relative to the Haar measure $d\omega_m(\alpha)$.

Definition 1.2. Let $\Omega\subset\mathbb{C}^n$ be a domain. A real valued function $\varphi(-\infty\leqq\varphi(z)<+\infty)$ is said to be *plurisubharmonic* in Ω if it has property (a) above and in addition b$_2$) $\varphi(z)\leqq l(z,w,r,\varphi)=\frac{1}{2\pi}\int_0^{2\pi}\varphi(z+wre^{i\theta})d\theta$ for all w, r such that $z+uw\subset\Omega$ for $u\in\mathbb{C}$, $|u|\leqq r$.

In the sequel we denote by $D(z,w,r)$ the compact disc

$$\{z'\in\mathbb{C}^n\colon z'=z+uw,\ u\in\mathbb{C},\ |u|\leqq r\}.$$

We shall let $S(\Omega)$ designate the set of subharmonic functions defined on a domain $\Omega\subset\mathbb{R}^m$ and by $\mathrm{PSH}(\Omega)$ the set of plurisubharmonic functions defined on a domain $\Omega\subset\mathbb{C}^n$. We recall some classical properties of the sets $S(\Omega)$ and $\mathrm{PSH}(\Omega)$ (we refer the reader to Appendix I for the proofs):

i) if τ_m is the volume of the unit ball in $\mathbb{R}^m$ and $\varphi\in S(\Omega)$, then $\varphi(x)\leqq\tau_m^{-1}r^{-m}\int_{\|x'\|\leqq r}\varphi(x+x')d\tau(x')=A(x,r,\varphi)$ for $r<d_\Omega(x)$ (cf. Remark after Definition I.1);

ii) $S(\Omega)\subset L^1_{\mathrm{loc}}(\Omega)$, the family of locally Lebesgue integrable functions, and $\mathrm{PSH}(\Omega)\subset S(\Omega)$ for $\Omega\subset\mathbb{C}^n$ (Proposition I.9).

iii) the set $\{x\in\Omega\colon\varphi(x)=-\infty,\ \varphi\in S(\Omega)\}$ is of Lebesgue measure zero in Ω (Corollary I.12);

iv) for $\varphi\in S(\Omega)$ and $x\in\Omega$ either $\varphi(x)<\sup_\Omega\varphi(x)$ or φ is a constant (Proposition I.13);

v) if $\varphi \in \mathrm{PSH}(\Omega)$ and $D(z,w,r) \subset \Omega$, then the two functions $r \to l(z,w,r,\varphi)$ and $r \to \sup_{z \in D(z,w,r)} \varphi(z)$ are increasing convex functions of $\log r$, and as a consequence $\lambda(z,r,\varphi)$ and $M(z,r,\varphi) = \sup_{\|z'-z\| \leqq r} \varphi(z')$ are increasing convex functions of $\log r$ (Proposition I.17);

vi) if $F: \Omega \to \Omega'$ is a holomorphic homeomorphism (analytic isomorphism) of Ω onto Ω', then the map $T: \varphi \in \mathrm{PSH}(\Omega) \to \varphi \circ F^{-1} \in \mathrm{PSH}(\Omega')$ is a bijection (i.e. the class of plurisubharmonic functions is invariant with respect to holomorphic homeomorphisms and is thus an object of the analytic structure; this is false for the larger class $S(\Omega)$).

The preceeding remarks shall play a crucial role, since for $f \in \mathscr{H}(\mathbb{C}^n)$, the functions $\log|f|$ form a subset $V(\mathbb{C}^n)$ of $\mathrm{PSH}(\mathbb{C}^n)$. But the class $\mathrm{PSH}(\mathbb{C}^n)$ also contains certain functions which do not belong to $V(\mathbb{C}^n)$ (such as convex functions in the space of the variables $(\log r_1, \ldots, \log r_n)$) (cf. App. I). What is more, it is fruitful to introduce certain measure theoretic concepts in the class $\mathrm{PSH}(\mathbb{C}^n)$.

On $\mathrm{PSH}(\Omega)$ and $S(\Omega)$, we consider the topology $L^1_{\mathrm{loc}}(\Omega)$ defined by the seminorms $N_K(\varphi) = \int_K |\varphi(z)|\,d\tau(z)$ where $d\tau$ is the Lebesgue measure and K is compact in Ω. Actually, it is sufficient to consider the semi-norms $N_i(\varphi) = \int_{B_i} |\varphi(z)|\,d\tau(z)$ where B_i runs over a countable family of compact balls which cover Ω (in $L^1_{\mathrm{loc}}(\Omega)$, we do not distinguish between two functions which are equal almost everywhere). We note that with this topology, $L^1_{\mathrm{loc}}(\Omega)$ is a Fréchet space; $S(\Omega)$ and $\mathrm{PSH}(\Omega)$ are convex cones in $L^1_{\mathrm{loc}}(\Omega)$, closed for this topology (see App. I).

Theorem 1.3. *A subset $M \subset S(\Omega)$ is bounded in $L^1_{\mathrm{loc}}(\Omega)$ if and only if the elements of M have a common upper bound on every compact subset of Ω and if M does not contain a sequence which converges uniformly to $-\infty$ on each compact set $K \subset \Omega$.*

Proof. If M is bounded in $L^1_{\mathrm{loc}}(\Omega)$, then it does not contain a sequence φ_k which tends uniformly to $-\infty$ on any compact ball B, since the integrals $\int_B |\varphi_k|\,d\tau$ are uniformly bounded. Let K be a compact subset of Ω and define K' by $K' = \bigcup_{x \in K} \bar{B}(x, \frac{1}{2}\delta_K)$ where $\delta_K = \inf_{x \in K} [d_\Omega(x), 1]$.

Then K' is compact in Ω and for $x \in K$ and $\varphi \in S(\Omega)$ we have

$$\varphi(x) \leqq A(x, \tfrac{1}{2}\delta_K, \varphi) \leqq \tau_m^{-1} 2^m \delta_K^{-m} \int_{K'} |\varphi(x)|\,d\tau(x).$$

Thus, if M is bounded, the elements of M have a common upper bound on every compact subset of Ω.

Conversely, let $M \subset S(\Omega)$ and suppose that M is uniformly bounded on every compact subset of Ω. If there exists a semi-norm N_i such that $\{N_i(\varphi), \varphi \in M\}$, is not bounded, we can find a sequence $\varphi_k \in M$ such that $N_i(\varphi_k) \to \infty$, and since the φ_k are uniformly bounded above on B_i, $\lim_{k\to\infty} \int_{B_i} \varphi_k d\tau = -\infty$. Let $\delta > 0$ be the distance of $B_i = B(x_i, r_i)$ to $\complement\Omega$, $\delta = \inf_{x' \in B_i} d_\Omega(x')$, and let α be such that $0 < 2\alpha < \delta$. For $x \in B(x_i, \alpha)$, we have

$$B_i \subset B(x_i, r_i + \alpha) \Subset B'_i = B(x_i, r_i + 2\alpha) \Subset \Omega.$$

If σ is an upper bound for φ_k in B'_i, then for $x \in B(x_i, \alpha)$, we obtain by the Mean Value Property for subharmonic functions:

$$\begin{aligned}\varphi_k(x) - \sigma &\leqq A(x, r_i + \alpha, \varphi_k - \sigma) \leqq [\tau_m (r_i + \alpha)^m]^{-1} \int_{B_i} [\varphi_k(x) - \sigma] d\tau_m \\ &\leqq \tau_m^{-1} (r_i + \alpha)^{-m} [C\sigma + \int_{B_i} \varphi_k(x') d\tau_m(x')]\end{aligned}$$

which proves the uniform convergence of the sequence φ_k to $-\infty$ on $B(x_i, \alpha)$.

Let $\hat{\Omega}$ be the largest open subset of Ω such that $\{\varphi_k\}$ converges to $-\infty$ uniformly on every compact subset of $\hat{\Omega}$. Since $B(x_i, \alpha) \subset \hat{\Omega}$, we know that $\hat{\Omega}$ is not empty. Moreover, if x' is a limit point of $\hat{\Omega}$ in Ω, there exists a ball $B(x', \rho) \Subset \Omega$ such that $B(x', \rho) \cap \hat{\Omega}$ contains a set $K \Subset \Omega$ of positive Lebesgue measure. Then $\lim_{k\to\infty} \int_{B(x', \rho)} \varphi_k d\tau = -\infty$, and by the above reasoning, $\varphi_k(x) \to -\infty$ uniformly on $B(x', \alpha)$ for $\alpha > 0$ such that $\rho + \alpha < d_\Omega(x')$. Thus $x' \in \hat{\Omega}$ so $\hat{\Omega}$ is closed. Since $\hat{\Omega}$ is open, closed, and is a domain, $\hat{\Omega} = \Omega$. □

§3. Norms on $\mathbb{C}^n$ and Order of Growth

Let $p(z)$ be a real valued function on $\mathbb{C}^n$. We say that $p(z)$ is *subadditive* if $p(z+z') \leqq p(z) + p(z')$; we say that $p(z)$ is *positively homogeneous* of order ρ if $p(tz) = t^\rho p(z)$ for $t \geqq 0$; we say that $p(z)$ is *complex homogeneous* of order ρ if $p(uz) = |u|^\rho p(z)$, $u \in \mathbb{C}$. If $p(z)$ is subadditive and $p(tz) = |t| p(z)$ for $t \in \mathbb{R}$ (resp. $p(\lambda z) = |\lambda| p(z)$ for $\lambda \in \mathbb{C}$), we say that $p(z)$ is a *real (resp. complex) semi-norm.* If, in addition, $p(z) = 0$ if and only if $z = 0$, then $p(z)$ is a *real* (resp. *complex*) norm.

If $p(z)$ is a norm, we define the p-ball of center z and radius r by $B_p(z, r) = \{z' : p(z - z') < r\}$, and if the norm is not specified, it will be assumed to be the Euclidean norm $\|z\|$. We recall that if p and q are two norms on $\mathbb{C}^n$, then each determines the unique separated vector space topology on $\mathbb{C}^n$, and there exist positive finite constants C_1 and C_2 such that

$$0 < C_1 \leqq \frac{p(z)}{q(z)} \leqq C_2. \tag{1,3}$$

Given a function $a(z): \mathbb{C}^n \to \mathbb{R}^+ = \{r \in \mathbb{R} : r > 0\}$, we consider

$$M_{a,p}(r) = \sup_{p(z) \leqq r} a(z). \tag{1,4}$$

It then follows from (1,3) that there exist constants C and C', $0 < C < C' < \infty$, depending only on $p(z)$ and $q(z)$, such that for every real valued function $a(z)$

$$M_{a,q}(Cr) \leqq M_{a,p}(r) \leqq M_{a,q}(C'r). \tag{1,5}$$

The functions we shall consider will often be plurisubharmonic, and in this case we have:

Proposition 1.4. *If $\varphi(z)$ is plurisubharmonic in $\mathbb{C}^n$, then*

a) $m_\varphi(z, z', r) = \sup_{|u| \leqq r} \varphi(z + uz')$ *is identically* $-\infty$ *or an increasing convex function of* $\log r$;

b) *if $p(z)$ is a complex norm, then $M_{\varphi,p}(r)$ is an increasing convex function of* $\log r$.

Proof. For a): $\varphi(z+uz') = -\infty$ for all $u \in \mathbb{C}$ or $\varphi(z+uz')$ is a subharmonic function of the variable $u = \alpha + i\beta$ in $\mathbb{C} = \mathbb{R}^2$ (cf. Remark 2 after Definition I.2).

For b): Consider $M_{\alpha,p}(r) = \sup_{z \in p^{-1}(1)} [\sup_{|u| \leqq r} \varphi(uz)]$ and remark that $\sup_{|u| \leqq r} \varphi(uz)$ is an increasing convex function of $\log r$ or identically $-\infty$, but is not identically $-\infty$ for all z. □

§4. Minimal Growth: Liouville's Theorem and Generalizations

The existence of a minimal growth for a non-constant function $\varphi \in \mathrm{PSH}(\mathbb{C}^n)$ is just a consequence of the convexity properties of Proposition 1.4 and formula (1,5).

Theorem 1.5. i) *Let $p(z)$ be a norm and $\varphi(z)$ a plurisubharmonic function in $\mathbb{C}^n$. Then $C = \lim_{r \to \infty} \frac{M_{\varphi,p}(r)}{\log r}$ and $C(z, z') = \lim_{r \to \infty} \frac{M_\varphi(z, z', r)}{\log r}$ exist, either finite or infinite, with the following properties:*

a) $C \geqq 0$; *moreover $C(z, z') \geqq 0$ with the possible exception $C(z, z') = -\infty$ in which case $\varphi(z + uz') \equiv -\infty$ for $u \in \mathbb{C}$.*

b) $C(z, uz') \equiv C(z, z')$ *for every $u \in \mathbb{C}$, $u \neq 0$.*

ii) *if* $p(z)$ *is a norm on* $\mathbb{C}^n$ *and* $\varphi(z+uz')\not\equiv -\infty$, $u\in\mathbb{C}$, *then* $\dfrac{\partial}{\partial \log r} M_{\varphi,p}(r)$ *and* $\dfrac{\partial}{\partial \log r} m_\varphi(z,z',r)$ *exist except perhaps for a countable set of* r *and*

$$\lim_{r\to\infty}\frac{\partial}{\partial \log r} M_{\varphi,p}(r)=\lim_{r\to\infty}\frac{M_{\varphi,p}(r)}{\log r};$$

$$\lim_{r\to\infty}\frac{\partial m_\varphi(z,z',r)}{\partial \log r}=\lim_{r\to\infty}\frac{m_\varphi(z,z',r)}{\log r}.$$

Proof. There exists $r_0>0$ such that $M_{\varphi,p}(r_0)>-\infty$, and since it is an increasing convex function of $\log r$, $M_{\varphi,p}(r)>-\infty$ for $r\geqq r_0$, which proves that $C\geqq 0$. If $\varphi(z+uz')\equiv-\infty$, then for $r>1$, $(\log r)^{-1}m_\varphi(z,z',r)=-\infty$, hence $C(z,z')=-\infty$. Otherwise, $\varphi(z+uz')$ is an $\mathbb{R}^2$-subharmonic function of u and hence by the above reasoning. $C(z,z')\geqq 0$. From the definition, by an obvious calculation we obtain $C(z,z')=C(z,uz')$ for $u\neq 0$. Part (ii) follows directly from Proposition 1.4. □

Theorem 1.6. *Suppose* $f\in\mathscr{H}(\mathbb{C}^n)$ *and set* $\varphi(z)=\log|f(z)|$. *Let*

$$C(z')=\liminf_{r\to\infty}\frac{m_\varphi(0,z',r)}{\log r}=\lim_{r\to\infty}\frac{m_\varphi(0,z',r)}{\log r}.$$

Then

i) $\eta_\rho=\{z';\,C(z')\leqq\rho\}$ *is a cone and if* η_ρ *is not contained in an algebraic hypersurface defined as the zero set of a homogeneous polynomial of degree* $\rho'\leqq\rho$, *then* f *is a polynomial of degree at most* ρ;

ii) *if* f *is a polynomial of degree* m, *then*

$$C'_m(z')=\liminf_{r\to\infty}\,[m_\varphi(0,z',r)-m\log r]=m_\psi(0,z',1)$$

where $\psi=\log|P_m|$ *and* P_m *is the homogeneous polynomial of maximal degree* m *in* f. *Furthermore for* $z'\in\mathbb{C}^n-\{0\}$ *we have*

$$\{z';\,C'_m(z')=-\infty\}=\{z';\,P_m(z')=0\}=\{z';\,C(z')\neq m\}.$$

Proof. Let $f(z')=\sum_{k=0}^{\infty}P_k(z')$ be the Taylor series expansion of $f(z')$ in terms of homogeneous polynomials. Then for $u\in\mathbb{C}$, $f(uz')=\sum_{k=0}^{\infty}P_k(z')u^k$. It follows from the Cauchy Integral Formula that $P_k(z')=1/2\pi\int_0^{2\pi}f(re^{i\theta}z')r^{-k}e^{-ik\theta}d\theta$, and hence $\log|P_k(z')|\leqq m_\varphi(0,z',r)-k\log r$. Thus, if there exists a sequence $r_\mu(z')\to\infty$ such that $[m_\varphi(0,z',r_\mu(z'))-k\log r_\mu(z')]\to-\infty$, we have $P_k(z')=0$. So if η_ρ is not contained in the zeros of a homogeneous polynomial of degree $\rho'\leqq\rho$, then $P_k(z')\equiv 0$ for $k>\rho$. If $f(z)$ is a polynomial of degree m, then $|f(rz')|\leqq|P_m(z')r^m|+O(r^{m-1})$, from which the second part follows. □

Since $M_{\varphi,p}(r) = \sup\limits_{p(z')\leqq 1} m_\varphi(0,z',r)$, we obtain:

Corollary 1.7. *If* $\liminf\limits_{r\to\infty} \dfrac{M_{\varphi,p}(r)}{\log r} = \rho < +\infty$, *then* f *is a polynomial of degree at most* ρ. *If, in addition,* ρ *is an integer and* $\liminf\limits_{r\to\infty} [M_{\varphi,p}(r) - \rho \log r] = -\infty$, *then degree* $f \leqq \rho - 1$. *In particular, if* $\liminf\limits_{r\to\infty} [M_{\varphi,p}(r) - \log r] = -\infty$, f *is the constant function* $f(0)$.

§5. Entire Functions of Finite Order

We equip the complex vector space $\mathscr{H}(\mathbb{C}^n)$ with the topology of uniform convergence on compact subsets so that it is a topological vector space. The topology can be defined by a sequence of normes $N_m(f) = \sup\limits_{\|z\|\leqq m} |f(z)|$, and thus $\mathscr{H}(\mathbb{C}^n)$ is a Fréchet space. A metric invariant with respect to translations is given by $\delta(f,g) = d(f-g)$ where $d(f) = \sum\limits_{m=1}^{\infty} 2^{-m} \dfrac{N_m(f)}{1+N_m(f)}$. A set $\beta \subset \mathscr{H}(\mathbb{C}^n)$ is bounded if and only if there exists $\psi(r)$ for $r \geqq 0$ such that $0 \leqq \psi(r) < +\infty$ and $M_{\log|f|,p}(r) \leqq \psi(r)$ for $f \in \beta$ and $p(z) = \|z\|$. Thus, the study of the common properties of $f \in \mathscr{H}(\mathbb{C}^n)$ which have a certain rate of growth is in reality the study of the bounded sets of $\mathscr{H}(\mathbb{C}^n)$.

As we have seen in the last section, there exists a minimal scale of growth for the non-constant functions in $\mathscr{H}(\mathbb{C}^n)$. Once past this minimal scale, however, the choice of a growth function $\psi(r)$ becomes somewhat arbitrary, as long as we stay in the family of increasing convex functions of $\log r$. We shall choose as a scale of growth functions $\psi(r)$ of the type σr^k, $\sigma > 0$, $k > 0$. This choice is motivated by the fact that the most familiar (and useful) transcendental entire functions fall into this class as well as the Fourier transforms of measures and distributions with compact support. Thus, the theory has an enormous range of applications from number theory to partial differential equations.

Definition 1.8. The *order* ρ of a positive real valued function $a(z)$ with respect to a norm $p(z)$ is given by

$$\rho = \limsup_{r\to\infty} \frac{\log M_{a,p}(r)}{\log r}.$$

If $\rho < +\infty$, $a(z)$ is said to be of maximal, normal, or minimal type according to whether the positive number

$$\sigma = \limsup_{r\to\infty} \frac{M_{a,p}(r)}{r^\rho}$$

is $+\infty$, finite, or zero, and σ is said to be the type of a with respect to $p(z)$. For $\varphi\in \mathrm{PSH}(\mathbb{C}^n)$ we define the order ρ of φ by using $a(z)=\sup[\varphi(z),0] =\varphi^+(z)$.

If f is an entire function, by an abuse of language, we shall say that f is of order ρ if $\log|f|$ is of order ρ, and we shall denote $M_{\log|f|,p}(r)$ by $M_{f,p}(r)$ for simplicity. An entire function of order at most 1 and finite type is said to be of *exponential type*.

Remark. The order and the nature of the type are not changed by a change in norms or a translation (cf. (1,5)) and thus depend only on the topology of the space $\mathbb{C}^n$.

Examples. i) If μ is a measure with compact support in $\mathbb{C}^n$, then $f(z)=\int_{\mathbb{C}^n} \exp\left(\sum_{k=1}^n z_k\cdot\xi_k\right)d\mu(\xi)$ is of order 1 and of normal or minimal type;

ii) if $P(z)$ is a polynomial, $\exp P(z)$ is of order $\deg P$;

iii) if $f_1(z_1),\ldots,f_n(z_n)$ are entire functions of finite order, and $P(z)$ a polynomial, then $P\circ F$ is an entire function of finite order.

iv) $\cos\sqrt{z_1 z_2}$ in $\mathbb{C}^2$ is an entire function of order $\rho=1$ and type $\sigma=\dfrac{1}{\sqrt{2}}$.

Theorem 1.9. *If* $f(z)=\sum_{q=0}^{\infty} P_q(z)$ *is the expansion of an entire function in homogeneous polynomials and* $C_q=\sup_{p(z)\leq 1}|P_q(z)|$, *then the order* ρ *and for* $\rho>0$ *the type* σ *of* f *with respect to* $p(z)$ *are given by*

a) $$-\frac{1}{\rho}=\limsup_{q\to\infty}\frac{\log C_q}{q\log q}=\limsup_{q\to\infty}\frac{1}{q\log q}[\sup_{p(z)\leq 1}\log|P_q(z)|].$$

b) $$\log\sigma=\limsup_{q\to\infty}\left[\frac{\log C_q}{q\log q}+\frac{1}{\rho}\right]\rho\log q-\log\rho-1.$$

b′) $$\log\sigma e\rho=\limsup_{q\to\infty}\left(\frac{\rho}{q}\log C_q+\log q\right).$$

c) $$\sigma=\limsup_{q\to\infty}\left[\Gamma\left(\frac{q}{\rho}\right)C_q\right]^{1/q}.$$

d) *For* $\rho=1$, $\sigma=\limsup_{q\to\infty}[q!\,C_q]^{1/q}$.

Proof. Let z_q be a point on the unit p-ball for which $|P_q(z_q)|=C_q$. By a rotation, we can assume that $z_q=(x_q,0,\ldots,0)$. If $\tilde f(u)=f(u,0,\ldots,0)$ and $\tilde f(u)=\sum_{m=0}^{\infty} a_m u^m$ is the Taylor series expansion of $\tilde f$ at the origin, then $|a_q x_q^q|=C_q$, and by the Cauchy Integral Formula, $C_q\leqq r^{-q}\exp M_{f,p}(r)$. If

$M_{f,p}(r) \leqq A r^k$ for $r > R(A, k)$, then $C_q \leqq r^{-q} \exp A r^k$. Since

$$\frac{d}{dr}(r^{-q} \exp A r^k) = r^{-q-1}(-q + A k r^k) \exp A r^k,$$

the minimum of this expression occurs when $r = \left(\frac{q}{kA}\right)^{1/k}$ and is equal to $\left(\frac{eAk}{q}\right)^{q/k}$. We then have for q sufficiently large

$$\log C_q \leqq \frac{q}{k}[1 + \log A k - \log q]. \tag{1,6}$$

Thus, given $\varepsilon > 0$, for q sufficiently large, $k \geqq \frac{q \log q}{-\log C_q} - \varepsilon$, so if f is of finite order ρ, $\rho \geqq \limsup_{q \to \infty} \frac{q \log q}{-\log C_q}$ or equivalently, $-\rho^{-1} \geqq \limsup_{q \to \infty} \frac{\log C_q}{q \log q}$. If f is of finite type σ and $A > \sigma$, (1,6) gives (with $k = \rho$): $A e \rho \geqq q C_q^{\rho/q}$ for q sufficiently large (depending on A), so that $\sigma e \rho \geqq \limsup_{q \to \infty} q C_q^{\rho/q}$.

Suppose now that $C_q \leqq \left(\frac{eAk}{q}\right)^{q/k}$ for $q \geqq q_0(A, k)$. If $q > m_r$, where m_r is the largest integer smaller than or equal to $2^k e A k r^k$ and r is sufficiently large, then for $p(z) \leqq r$,

$$|P_q(z)| \leqq \left(\frac{eAk}{q}\right)^{q/k} r^q \leqq \left(\frac{1}{2kr^k}\right)^{q/k} r^q \leqq 2^{-q}$$

Thus $|f(z)| < \sum_{q=0}^{m_r} C_q r^q + 2$ for $p(z) \leqq r$. Let $\mu(r) = \sup_q C_q r^q$. Then $\exp M_{f,p}(r) \leqq (1 + 2^k e A k r^k) \mu(r) + 2$. By our assumption on C_q

$$\mu(r) \leqq \sup_q \left(\frac{eAk}{q}\right)^{q/k} r^q \leqq \sup_x \left(\frac{eAk}{x}\right)^{x/k} r^x.$$

This latter function attains its maximum for $x = A k r^k$, and so $\mu(r) \leqq \exp A r^k$ and $M_{f,p} \leqq A r^k + (k+1) \log r$ for r sufficiently large. This shows that

$$\rho \leqq \limsup_{r \to \infty} \frac{q \log q}{-\log C_q} \quad \text{and} \quad (\sigma e \rho)^{1/\rho} \leqq \limsup_{q \to \infty} (q^{1/\rho} C_q^{1/q}).$$

By Stirling's Formula, we have

$$\log \Gamma(x) = x(\log x - 1) - \tfrac{1}{2} \log x + \tfrac{1}{2} \log 2\pi + O(x^{-1}).$$

Thus, we obtain

$$\lim_{q \to \infty} \frac{[\Gamma(q/\rho)]^{1/q}}{q^{1/\rho}} = (e\rho)^{-1/\rho}.$$

□

Corollary 1.10. *If f is an entire function of order ρ and type σ with respect to a norm $p(z)$, then its restriction $f|_L$ to any linear subspace L is of order at most ρ and type at most σ (with respect to the restriction $p|_L$).*

Remark. If $f(z)=\sum_{\alpha} a_\alpha z^\alpha$ is the Taylor series expansion at the origin of an entire function, then the order ρ can be calculated and the type σ (with respect to a given norm) estimated by the Taylor series coefficients a_α (as was previously shown, the order ρ is independant of the norm, but not the type σ). We consider the classical situations $p(z)=\|z\|$ and $p(z)=\sup_{1\leqq k\leqq n}|z_k|$. We compare $m_q=\sup_{|\alpha|=q}|a_\alpha|$ and $C_q=\sup_{p(z)\leqq 1}|P_q(z)|$

a) $p(z)=\|z\|$. The unit ball contains the disc $|z_k|\leqq\frac{1}{\sqrt{n}}$, $k=1,\dots,n$. By the Cauchy Integral Formula, we have

$$|a_\alpha|\leqq C_q(\sqrt{n})^q \quad \text{for } |\alpha|=q \text{ so } m_q\leqq n^{q/2}C_q.$$

Conversely, $C_q\leqq s(n,q)m_q$, where $s(n,q)$ is the number of monomials in P_q, and $s(n,q)\leqq(q+1)^n$. Then

$$\log m_q-\frac{q}{2}\log n\leqq\log C_q\leqq\log m_q+n\log(q+1).$$

From Theorem 1.9 (i), we deduce that $-\frac{1}{\rho}=\limsup_{q\to\infty}\frac{\log m_q}{q\log q}$ and from (ii) that

$$n^{-\rho/2}(e\rho)^{-1}B'\leqq\sigma\leqq(e\rho)^{-1}B' \quad\text{with}\quad B'=\limsup_{q\to\infty}\big(q\sup_{|\alpha|=q}|a_\alpha|^{\rho/q}\big).$$

b) $p(z)=\sup_{1\leqq k\leqq n}|z_k|$. In the same way, we obtain

$$\log m_q\leqq\log C_q\leqq\log m_q+n\log(q+1).$$

Therefore, ρ and σ are given by (i), (ii), and (iii) of Theorem 1.9 where we replace C_q by $m_q=\sup_{|\alpha|=q}|a_\alpha|$.

The order ρ defined above is sometimes called the *total order* of $f(z_1,\dots,z_n)\in\mathscr{H}(\mathbb{C}^n)$. We are sometimes led to study the growth of an entire function $f\in\mathscr{H}(\mathbb{C}^n\times\mathbb{C}^p)$ where the variables $z\in\mathbb{C}^n$ and $\xi\in\mathbb{C}^p$ play different roles. This is the case, for instance, in the study of Fourier transforms of linear differential and pseudo-differential operators. This leads us to define the growth and the order with respect to each variable z and ξ separately.

Let $f\in\mathscr{H}(\mathbb{C}^n)$ where $\mathbb{C}^n=\mathbb{C}\times\dots\times\mathbb{C}$. We consider the function $M_{f,p}(r)=\sup_{|z_k|\leqq r}\log|f(z_1,\dots,z_n)|$ with respect to the norm $p(z)=\sup_k|z_k|$ and write $\hat{M}_f(r)$ instead of $M_{f,p}(r)$. In addition, we let $\tilde{M}_f(r_1,\dots,r_n)=\sup_{|z_k|\leqq r_k}\log|f(z_1,\dots,z_n)|$.

Definition 1.11. Given $f\in\mathscr{H}(\mathbb{C}^n)$, we define the *indicator of growth of f with respect to the variable z_k, $M_f^{(k)}(r)$*, by

$$M_f^{(k)}(r)=\tilde{M}_f(1,\dots,1,r,1,\dots,1) \tag{1,7}$$

(r_j is replaced by 1 in $\tilde{M}_f(r_1, \ldots, r_n)$ if $j \neq k$ and $r_k = r$). We say that f is of *order* ρ_k *in* z_k if $M_f^{(k)}(r)$ is of order ρ_k.

Remark. If f is of finite order ρ, then $\rho_k \leqq \rho$, for we have $M_f^{(k)}(r) \leqq M_{f,p}(r)$ for $r \geqq 1$.

Reciprocally, we shall see that there exists an upper bound for the total order ρ in terms of the ρ_k. We note that for $|z_k| = r_k$

$$\tilde{M}_f(r_1, \ldots, r_n) = \sup_{|\alpha_k| \leqq 1} \log |f(\alpha_1 z_1, \ldots, \alpha_n z_n)| \tag{1,8}$$

is a convex function of each $u_k = \log |z_k|$ with finite values for $-\infty < a_k \leqq u_k \leqq b_k < +\infty$. It follows from the continuity of $\log |f|$ in a neighborhood of every point z for which $f(z) \neq 0$ that $\psi(u_1, \ldots, u_n) = \tilde{M}_f(r_1, \ldots, r_n)$ is continuous for $u = (u_1, \ldots, u_n) \in \mathbb{R}^n$. By (1,8), it is a plurisubharmonic function of $z = (z_1, \ldots, z_n)$ which depends only on $|z_k|$, so $\psi(u)$ is a convex function of u (cf. Proposition I.25). Hence $\tilde{M}_f(r_1, \ldots, r_n)$ is a continuous convex function of the variables $\log r_k = u_k$ which is increasing in each r_k.

Let $(u_1, \ldots, u_n) \in (\mathbb{R}_+)^n$. The convexity of $\psi(u)$ then gives

$$\psi(C_1 u_1, \ldots, C_n u_n) \leqq \sum_{k=1}^{n} C_k \psi(0, \ldots, 0, u_k, 0, \ldots, 0)$$

for every vector $C = (C_1, \ldots, C_n)$ such that $0 \leqq C_k \leqq 1$ and $\sum_{k=1}^{n} C_k = 1$. We rewrite this

$$\tilde{M}_f(r_1, \ldots, r_n) \leqq \sum_{k=1}^{n} C_k M_f^{(k)}(r'_k) \quad \text{with } r'_k = r_k^{1/C_k}. \tag{1,9}$$

From formula (1.9), we can immediately make two observations:

i) if the growth in z_k is of finite order for every k, then f is of finite order;

ii) the order in z_k could as well be defined using any polydisc $|z_j| \leqq r_j^0$, $|z_n| \leqq r$, r_j^0 fixed, in place of the polydisc $|z_j| \leqq 1$, $|z_k| \leqq r$.

Theorem 1.12. *We have*

$$\hat{M}_f(r) = \tilde{M}_f(r, \ldots, r) \leqq \sum_{k=1}^{n} \gamma_k^{-1} M_f^{(k)}(r^{\gamma_k}) \tag{1,10}$$

for every system of numbers $(\gamma_1, \ldots, \gamma_n)$ *with* $\gamma_k > 0$ *and* $\sum_{k=1}^{n} \gamma_k^{-1} = 1$.

Proof. We let $r_k = r$ in (1,9). Then $r'_k = r^{\gamma_k}$. □

Corollary 1.13. *Let* $f \in \mathscr{H}(\mathbb{C}^n)$. If f has order ρ_k with respect to the variable z_k, $k = 1, \ldots, n$, *then* f *is of finite order* ρ *and* $\sup_{k=1,\ldots,n} \rho_k \leqq \rho \leqq \sum_{k=1}^{n} \rho_j$.

Proof. The first inequality is a result of the inequality $M_f^{(k)}(r) \leqq \hat{M}_f(r)$ for $r \geqq 1$. To prove the second, we note that if ρ_k is finite, then for every $\rho_k' > \rho_k$, there exists A_k such that $M_f^{(k)}(r) \leqq A_k r^{\rho_k'}$, $1 \leqq k \leqq n$, so if we set $\lambda = \sum_{k=1}^{n} \rho_k'$ and $\gamma_k = \lambda \rho_k^{-1}$, then from Theorem 1.12 we obtain

$$\hat{M}_f(r) = M_f(r, \ldots, r) \leqq \lambda^{-1} \left(\sum_{k=1}^{n} \rho_k A_k \right) r^{\sigma}.$$

□

We shall now apply (1,8) in another context by letting the numbers $u_1, \ldots, u_{n-1}$ remain fixed as u_n goes to infinity and choosing the C_k to be variable functions of u_n.

Theorem 1.14. *Let $f \in \mathscr{H}(\mathbb{C}^n)$ and let $\tau_1, \ldots, \tau_{n-1}$ be fixed positive numbers. Then there exists a positive function $\varepsilon(r)$ and $r_0 > 1$ such that $\varepsilon(r)$ goes to zero when r tends to infinity and for $r > r_0$,*

$$\tilde{M}_f(\tau_1, \ldots, \tau_{n-1}, r) \leqq \tilde{M}_f(1, \ldots, 1, r^{1+\varepsilon(r)}) = M_f^{(n)}(r^{1+\varepsilon(r)}) \tag{1,11}$$

The function $\varepsilon(r)$ and r_0 depend on the τ_j and f.

Proof. From (1,9), we obtain

$$\tilde{M}_f(\tau_1, \ldots, \tau_{n-1}, r) \leqq \sum_{k=1}^{n-1} C_k M_f^{(k)}(\tau_k^{1/C_k}) + C_n M_f^{(n)}(r^{1/C_n}).$$

We choose functions $C_k(r)$ with $C_k(r) \geqq 0$, $\lim_{r \to \infty} C_k(r) = 0$, $\lim_{r \to \infty} C_n(r) = 1$ and $\sum_{k=1}^{n} C_k(r) = 1$. Then

$$\tilde{M}_f(\tau_1, \ldots, \tau_{n-1}, r) \leqq M_f^{(n)}(r^{1/C_n(r)}) + \sum_{k=1}^{n-1} C_k(r) [M_f^{(k)}(\tau_k^{1/C_k(r)}) - M_f^{(n)}(r^{1/C_n(r)})].$$

We make the $(n-1)$ brackets on the right negative by choosing the $C_k = C_k(r)$ (as functions of the τ_k) so that $M_f^{(k)}(\tau_k^{1/C_k}) = M_f^{(n)}(r)$. Define $\alpha(r)$ by

$$\alpha(r) = \sup \{ \lambda : M_f^{(k)}(\tau_k^{\lambda}) \leqq M_f^{(n)}(r), \ 1 \leqq k \leqq n-1 \}$$

Then we obtain $\alpha(r) > n-1$ for $r > r_0$, where r_0 is large enough so that $M^{(n)}(r_0) > \sup_k M_f^{(k)}(\tau_j^{1/n-1})$. For $r > r_0$, $\alpha(r)$ is a continuous increasing function of r and $\lim_{r \to \infty} \alpha(r) = +\infty$. Let $\varepsilon(r) = \left[\frac{\alpha(r)}{n-1} - 1 \right]^{-1}$. Then $\varepsilon(r)$ decreases to zero and

$$\alpha(r) \leqq \inf_{k=1, \ldots, n-1} C_k(r)^{-1} \leqq (n-1) \left[\sum_{k=1}^{n-1} C_k(r) \right]^{-1}$$

hence $\varepsilon(r) \geqq [C_n(r)]^{-1} - 1$. □

Remark. Theorems 1.12 and 1.14 can be applied in the same way to an entire function $f \in \mathscr{H}(E)$, $E = E_1 \times \ldots \times E_n$ with $E_k = \mathbb{C}^{q_k}$. Let π_k be the projection $\pi_k: E \to E_k$ and p_k a norm on E_k. We choose as norm on E the number $p(z) = \sup_k p_k \circ \pi_k(z)$. For $f \in \mathscr{H}(E)$, we set

$$\tilde{M}_f(r) = \sup_{p(z) \leqq r} \log |f(z)| \quad \text{and} \quad \tilde{M}(r_1, \ldots, r_n) = \sup_{p_k \circ \pi_k(z) \leqq r_j} \log |f(z)|.$$

If we let $M_f^{(k)}(r)$ be the indicator of growth on E_j defined from $\tilde{M}(r_1, \ldots, r_n)$ by letting $r_j = 1$ if $k \neq j$ and $r_k = r$, the conclusions of the two theorems remain valid.

§6. Proximate Orders

The growth scale of order and type has been further refined by Valiron. He introduced intermediate functions of comparison, called proximate orders, which make it unnecessary to consider functions of minimal or maximal type. For entire functions of $n \geqq 2$ complex variables the advantage of this method stems from the fact that only functions of normal type will have non-trivial indicator functions (see below) which give the growth in different directions.

Definition 1.15. A proximate order $\rho(r)$ (for the order $\rho \geqq 0$) is a function $\rho(r) \geqq 0$ defined for $r \in \mathbb{R}^+$ such that

i) $\lim_{r \to \infty} \rho(r) = \rho$ and ii) $\lim_{r \to \infty} \rho'(r) r \log r = 0$.

Example. We define by induction the function $\log_j r = \log(\log_{j-1} r)$. Then the functions

$$r^{\rho(r)} = r^{\rho} \log_1^{\beta_1} r \ldots \log_m^{\beta_m} r \qquad \beta_j > 0,\ j = 1, \ldots, m$$

are proximate orders (of course, we have to modify the function in a neighborhood of the origin).

Definition 1.16. If $\varphi(z)$ is a real valued function in $\mathbb{C}^n$, its *type* σ with respect to the norm $p(z)$ and the proximate order $\rho(r)$ is given by $\sigma = \limsup_{r \to \infty} \dfrac{M_{\varphi, p}(r)}{r^{\rho(r)}}$; φ is said to be of minimal, normal, or maximal type (with respect to $\rho(r)$) depending on whether $\sigma = 0$, $0 < \sigma < +\infty$, or $\sigma = +\infty$.

We note that for a given φ, the number σ depends upon $\rho(r)$ as well as p and φ.

Definition 1.17. A function $L(r)$ defined on $r>0$ is said to be *slowly increasing* if for every compact interval I of $(0,+\infty)$, for every $\varepsilon>0$, there exists r_0 such that for $r>r_0$, $\left|\frac{L(kr)}{L(r)}-1\right|<\varepsilon$ for every $k\in I$.

Theorem 1.18. *If $\rho(r)$ is a proximate order, then $L(r)=r^{\rho(r)-\rho}$ is a slowly increasing function.*

Proof. Let $\varepsilon>0$ be given. By definition

$$\log\left(\frac{L(kr)}{L(r)}\right)=[\rho(kr)-\rho]\log k+[\rho(kr)-\rho(r)]\log r.$$

We first assume that $0<a\leqq k<1$. By the Mean Value Theorem,

$$\left|\frac{\rho(r)-\rho(kr)}{r-kr}\right|=|\rho'(\xi)|$$

for some ξ, $kr\leqq\xi\leqq r$ so by (ii) of Definition 1.15, there exists R_1 such that $|\rho'(\xi)|\leqq\frac{\varepsilon}{3(a^{-1}-1)\xi\log\xi}$ for $\xi>aR_1$. Thus, for r sufficiently large, we have the following bounds:

$$(\log r)|\rho(kr)-\rho(r)|<\frac{\varepsilon(\log r)}{3(a^{-1}-1)}\frac{(r-kr)}{kr\log kr}$$

$$\leqq\frac{\varepsilon}{3(a^{-1}-1)}\frac{(a^{-1}-1)}{\log ar}\log r<\varepsilon/2.$$

Furthermore, by (ii) of Definition 1.15:

$$|\rho(kr)-\rho|<\varepsilon 2^{-1}(-\log a)^{-1}$$

and hence

$$\left|\log\left(\frac{L(kr)}{L(r)}\right)\right|<\varepsilon.$$

If $1<k\leqq b$, we reason in the same way to conclude that $\left|\log\left(\frac{L(kr)}{L(r)}\right)\right|<\varepsilon$ for $r>r_0$ and all $k\in[a,b]$. □

Proposition 1.19. *If $\rho(r)$ is a proximate order with $\rho>0$, then there exists R such that $r^{\rho(r)}$ is strictly increasing for $r>R$.*

Proof. $\frac{d}{dr}(r^{\rho(r)})=\rho(r)r^{\rho(r)-1}+r^{\rho(r)}\rho'(r)\log r$. By (i) of Definition 1.15, for $r>R_1$, $\rho(r)>\rho/2$ and by (ii) of Definition 1.15, for $r>R_2$, $|\rho'(r)r\log r|<\rho/4$, so for $r>\sup(R_1,R_2)$:

$$\frac{d}{dr}(r^{\rho(r)})>\rho/4 \quad r^{\rho(r)-1}>0.$$ □

Note. Since in the study of the asymptotic properties of entire functions we are only interested in their properties for r sufficiently large, we can always change $\rho(r)$ on a bounded set without affecting the asymptotic properties we study. Thus, for $\rho>0$, we can always assume that $r^{\rho(r)}$ is everywhere strictly increasing on the set $r>0$.

Proposition 1.20. *Given $\varepsilon>0$, there exists an $R(\varepsilon)$ such that*

$$(1-\varepsilon)k^{\rho}r^{\rho(r)}<(kr)^{\rho(kr)}<(1+\varepsilon)k^{\rho}r^{\rho(r)}$$

uniformly for $0<a\leqq k\leqq b<+\infty$ for $r>R(\varepsilon)$.

Proof. By Theorem 1.18, for r large enough

$$(1-\varepsilon)\leqq\frac{(kr)^{\rho(kr)}r^{\rho}}{(kr)^{\rho}r^{\rho(r)}}\leqq(1+\varepsilon). \qquad \square$$

Definition 1.21. A proximate order $\rho(r)$ is a *strong proximate order* if $\rho(r)$ is twice continuously differentiable for $r>0$ and $\lim_{r\to\infty}\rho''(r)r^2\log r=0$.

Since we are interested in convexity properties, the following is essential:

Proposition 1.22. *If $\rho(r)$ is a strong proximate order, $\rho>0$, then $r^{\rho(r)}$ is a convex function of $\log r$ for r large. If $\rho>1$, then $r^{\rho(r)}$ is a convex function of r for r large.*

Proof. By Proposition 1.19, $r^{\rho(r)}$ is an increasing function of r. A simple calculation shows that

$$\frac{d}{d(\log r)}r^{\rho(r)}=\rho(r)r^{\rho(r)}+r^{\rho(r)}\rho'(r)r\log r,$$

which is positive for r sufficiently large by (i) and (ii) of Definition 1.15. Furthermore,

$$\begin{aligned}\frac{d^2}{d(\log r)^2}r^{\rho(r)}&=r\{\rho'(r)r^{\rho(r)}+\rho(r)^2r^{\rho(r)-1}\\&\quad+\rho(r)r^{\rho(r)}\rho'(r)\log r+\rho''(r)r^{\rho(r)}\log r\\&\quad+\rho'(r)r^{\rho(r)-1}+\rho(r)\rho'(r)r^{\rho(r)-1}\log r+[\rho'(r)r\log r]^2r^{\rho(r)}\}\\&>\frac{\rho^2}{2}r^{\rho(r)}\qquad\text{for } r \text{ sufficiently large}\end{aligned}$$

by (i) and (ii) of Definition 1.15 and Definition 1.21. Similarly, a simple calculation shows that

$$\frac{d^2 r^{\rho(r)}}{dr^2} = \rho'(r) r^{\rho(r)-1} + \rho(r)(\rho(r)-1) r^{\rho(r)-2}$$
$$+ \rho(r)\, r^{\rho(r)-1}\, \rho'(r) \log r + \rho''(r)\, r^{\rho(r)} \log r + \rho'(r)\, r^{\rho(r)-1}$$
$$+ \rho(r) \rho'(r)(\log r) r^{\rho(r)-1} + [\rho'(r) \log r]^2 r^{\rho(r)}$$
$$> \frac{\rho(\rho-1)}{2} r^{\rho(r)-1} \quad \text{for } r \text{ sufficiently large.} \qquad \square$$

A fundamental result that we shall need (for the proof see Appendix II) is that *for any positive continuous increasing function $a(r)$ of finite order ρ there exists a (strong) proximate order with respect to which $a(r)$ is of normal type.* We apply this result to $M_{f,p}(r)$ for $f \in \mathscr{H}(\mathbb{C}^n)$.

In Theorem 1.9, we obtained a formula for the type of an entire function of finite order ρ in terms of its Taylor series expansion in homogeneous polynomials. A similar formula exists for proximate orders. Since by Proposition 1.19, $r^{\rho(r)}$ is an increasing function for $r > 0$, if $\rho > 0$ the equation $t = r^{\rho(r)}$ admits a unique solution for $t > 0$. We will denote by $r = \varphi(t)$ this solution; $\varphi(t)$ is just the inverse function of $r^{\rho(r)}$. Of course, $\varphi(t)$ depends on $\rho(r)$, but the proximate order in question will be clear from the context, so we will not note this dependence.

Theorem 1.23. *Let $f(z) = \sum_q P_q(z)$ be the Taylor series expansion of the entire function $f(z)$ of finite order $\rho > 0$ and of proximate order $\rho(r)$, and let $C_q = \sup_{p(z) \leq 1} |P_q(z)|$. Then the type σ of $f(z)$ with respect to the norm $p(z)$ and to the proximate order $\rho(r)$ is given by*

$$\frac{1}{\rho} \log \sigma = \limsup_{q \to \infty} \left[\frac{1}{q} \log C_q + \log \varphi(q) \right] - \frac{1}{\rho} - \frac{\log \rho}{\rho}, \qquad \rho > 0.$$

The function $r = \varphi(t)$ is the inverse function of $t = r^{\rho(r)}$.

Proof. 1) We first show that $\lim_{t \to \infty} \frac{\varphi(kt)}{\varphi(t)} = k^{1/\rho}$, $0 < k < +\infty$. Since $\log t = \rho(r) \log r$, $\frac{d(\log t)}{d(\log r)} = \rho'(r)\, r \log r + \rho(r)$, which tends to ρ when r tends to $+\infty$ by Definition 1.15. Furthermore,

$$\frac{d(\log t)}{d(\log r)} = \frac{\frac{d}{dt}(\log t)}{\frac{d}{dt}(\log r)} = \frac{\frac{d}{dt}(\log t)}{\frac{d}{dt}(\log \varphi(t))},$$

so for r sufficiently large,

$$\left(\frac{1}{\rho} - \varepsilon\right) \frac{d}{dt}(\log t(r)) < \frac{d}{dt} \log \varphi(t(r)) < \left(\frac{1}{\rho} + \varepsilon\right) \frac{d}{dt}(\log t(r)).$$

If we integrate from t to kt, we obtain

$$\left(\frac{1}{\rho}-\varepsilon\right)\log k<\log\left(\frac{\varphi(kt)}{\varphi(t)}\right)<\left(\frac{1}{\rho}+\varepsilon\right)\log k \quad \text{so} \lim_{t\to\infty}\frac{\varphi(kt)}{\varphi(t)}=k^{1/\rho}.$$

2) Let $\tilde{\sigma}>\sigma$. Then, as in the proof of Theorem 1.9, it follows from the Cauchy Integral Formula that $\log C_q<\tilde{\sigma}\, r^{\rho(r)}-q\log r$ for r large, and if r_q is the solution of the equation $q=\tilde{\sigma}\,\rho\, r^{\rho(r)}$, then for q large we have $\log C_q<\frac{q}{\rho}-q\log\varphi\left(\frac{q}{\rho\,\tilde{\sigma}}\right)$. From this, it follows that

$$\log[\varphi(q)\, C_q^{1/q}]<\frac{1}{\rho}+\log\frac{\varphi(q)}{\varphi\left(\frac{q}{\rho\,\tilde{\sigma}}\right)},$$

and hence $\limsup_{q\to\infty}(\varphi(q)\, C_q^{1/q})\leqq(\tilde{\sigma}\,\rho\, e)^{1/\rho}$ by (1). This is true for all $\tilde{\sigma}>\sigma$; thus we have proved: $\limsup_{q\to\infty}(\varphi(q)\, C_q^{1/q})\leqq(\sigma\,\rho\, e)^{1/\rho}$.

3) Let $\hat{\sigma}$ be defined by the formula $\limsup_{q\to\infty}\varphi(q)\, C_q^{1/q}=(\hat{\sigma} e)^{1/\rho}$. We shall show that $\hat{\sigma}\geqq\sigma$. Suppose $\hat{\sigma}<\sigma$ and choose σ' and σ'' such that $\hat{\sigma}<\sigma'<\sigma''<\sigma$. For q sufficiently large, we deduce

$$C_q<\left\{\frac{(\sigma'\rho\, e)^{1/\rho}}{\varphi(q)}\right\}^q\leqq\left\{\frac{e^{1/\rho}}{\varphi\left(\frac{q}{\sigma''\rho}\right)}\right\}^q$$

by (1) above.

Thus, there exists q_0 such that for $q\geqq q_0$,

$$C_q r^q\leqq\left\{\frac{e^{1/\rho}}{\varphi\left(\frac{q}{\sigma''\rho}\right)}\right\}^q r^q. \tag{1,12}$$

Let $\mu(r)=\sup_q C_q r^q$ and let q_r be the greatest integer less than or equal to $\sigma''\rho r^{\rho(r)}$. Then $\varphi(r^{\rho(r)})=r$ and $\lim_{r\to\infty}\frac{\varphi(r^{\rho(r)}-1)}{\varphi(r^{\rho(r)})}=1$ by (1). It follows from (1,12) that $\mu(r)<(1+\varepsilon)\exp\sigma'' r^{\rho(r)}$ for r sufficiently large. Choose $\rho_1>\rho$. Then for $q>2^\rho\sigma''^{\rho_1}r^{\rho_1}$, we have the bound $C_q r^q<2^{-q}$ by (1,12) for r sufficiently large (since $\varphi(r^{\rho_1})>\varphi(r^{\rho(r)})=r$ for r large). Thus, for r sufficiently large, $M_{f,p}(r)<(2+2^\rho\sigma^{1/\rho}r^{\rho_1})\exp\sigma'' r^{\rho(r)}$ and hence $\sigma\leqq\sigma''$, which is a contradiction. □

§ 7. Regularizations

If $\{\varphi_i\}_{i=1}^m$ is a finite family of subharmonic (resp. plurisubharmonic) functions defined in a domain Ω, then $\psi(z)=\sup_{1\leqq i\leqq m}\varphi_i(z)$ is subharmonic (resp.

plurisubharmonic). However, if the family is infinite, even when it is uniformly bounded on compact subsets, the supremum is not in general upper semi-continuous and so is not subharmonic (resp. plurisubharmonic). We seek to remedy this situation by finding the smallest subharmonic (resp. plurisubharmonic) majorant of a family $\{\varphi_i\}_{i\in I}$ of subharmonic (resp. plurisubharmonic) functions defined in a domain Ω.

Definition 1.24. Let $\varphi(x)$ be a function defined in a domain $\Omega\subset\mathbb{R}^m$ and locally bounded from above. Then $\varphi^\star(x)$, the *regularization* of $\varphi(x)$, is defined by $\varphi^\star(x)=\limsup\limits_{x'\subset\Omega\to x}\varphi(x')$.

Obviously, $\varphi^\star(x)$ is the smallest upper semi-continuous majorant of $\varphi(x)$.

Lemma 1.25. *Let Ω be a domain in $\mathbb{C}^n$. Let φ be an upper semi-continuous function locally bounded from above in Ω. Let $w\in\mathbb{C}^n$, $w\neq 0$, be fixed* and suppose that for some $z_0\in\Omega$, $D(z_0,w,1)=\{z_0+uw\in\Omega\colon |u|\leqq 1\}\subset\Omega$. *Then* $g(z)=\frac{1}{2\pi}\int\limits_0^{2\pi}\varphi(z+we^{i\theta})d\theta$ *is an upper semi-continuous function of z for* $z\in\Omega'=\{z\in\Omega\colon D(z,w,1)\subset\Omega\}$.

Proof. Ω' is open and non-empty, since $z_0\in\Omega'$, so g is defined in a neighborhood of $z\in\Omega'$. Let $z_m\to z$ in Ω'. Then the functions $\varphi(z_m+we^{i\theta})$ are uniformly bounded in m and θ. Fatou's Lemma then gives

$$\begin{aligned}\limsup_{z_m\to z} g(z_m)&=\limsup_{z_m\to z}\frac{1}{2\pi}\int_0^{2\pi}\varphi(z_m+we^{i\theta})d\theta\\&\leqq 1/2\pi\int_0^{2\pi}\limsup_{z_m\to z}\varphi(z_m+we^{i\theta})d\theta\\&\leqq 1/2\pi\int_0^{2\pi}\varphi(z+we^{i\theta})d\theta=g(z).\end{aligned}$$
□

Theorem 1.26. *Let $\{\varphi_i\}_{i\in I}$ be a family of plurisubharmonic functions defined on a domain $\Omega\subset\mathbb{C}^n$ and locally bounded from above and $\psi(z)=[\sup\limits_i\varphi_i(z)]$. Then $\psi^\star\in\mathrm{PSH}(\Omega)$.*

Proof. Let $D(z,w,1)\subset\Omega$. Then

$$\varphi_i(z)\leqq\frac{1}{2\pi}\int_0^{2\pi}\varphi_i(z+we^{i\theta})d\theta\leqq\frac{1}{2\pi}\int_0^{2\pi}\psi^\star(z+we^{i\theta})d\theta,$$

since $\varphi_i\leqq\sup\limits_i\varphi_i\leqq\psi^\star$. Thus $\varphi(z)\leqq\frac{1}{2\pi}\int\limits_0^{2\pi}\psi^\star(z+we^{i\theta})d\theta$. We take the regularization of both sides and apply Lemma 1.25. □

Theorem 1.27. *Let $\{\varphi_i\}_{i\in I}$ be a family of plurisubharmonic functions defined in $\Omega\subset\mathbb{C}^n$ and locally bounded above uniformly in I. Suppose that the family I is* *an ordered filter with the property that the filter of sections* $S_i=\{j\in I;\ j\geqq i\}$ *has a countable basis* S_m. *Then if* $\psi(z)=\limsup_{i\in I}\varphi_i(z)=\lim_{m\to\infty}[\sup_{i\in S_m}\varphi_i(z)]$, $\psi^\star(z)\in \mathrm{PSH}(\Omega)$ *or* $\psi^\star\equiv-\infty$.

Proof. Let $\psi_m(z)=\sup_{i\in S_m}\varphi_i(z)$ and let $\int_{*0}^{2\pi}$ represent the lower Lebesgue integral on the boundary of the disc. Then

$$\psi(z)=\lim_{m\to\infty}\psi_m(z)\leqq\lim_{m\to\infty}\frac{1}{2\pi}\int_{*0}^{2\pi}\psi_m(z+we^{i\theta})d\theta.$$

We apply Fatou's Lemma to the lower Lebesgue integral to obtain

$$\begin{aligned}\psi(z)&\leqq\frac{1}{2\pi}\int_{*0}^{2\pi}\limsup_{m\to\infty}\psi_m(z+we^{i\theta})d\theta\\ &\leqq\frac{1}{2\pi}\int_{*0}^{2\pi}\psi(z+we^{i\theta})d\theta\leqq\frac{1}{2\pi}\int_{0}^{2\pi}\psi^\star(z+we^{i\theta})d\theta.\end{aligned}$$

We now take the regularization of both sides and apply Lemma 1.25 to obtain $\psi^\star(z)\leqq\frac{1}{2\pi}\int_0^{2\pi}\psi^\star(z+we^{i\theta})d\theta$. Thus, if $\psi^\star(z)\not\equiv-\infty$, it is plurisubharmonic. □

In exactly the same way, we have the following result:

Theorem 1.28. *Let* $\{\varphi_i\}_{i\in I}$ *be a family of subharmonic functions defined in* $\Omega\subset\mathbb{R}^m$ *and locally bounded above uniformly in I. Suppose that the family I is an ordered filter with the property that the filter of sections* $S_i=\{j\in I,\ j\geqq i\}$ *has a countable basis* S_n. *Then if* $\varphi(x)=\limsup_{i\in I}\varphi_i(x)=\lim_{n\to\infty}[\sup_{x\in S_n}\varphi_i(x)]$, $\varphi^\star(x)\in S(\Omega)$ *or* $\varphi^\star(x)\equiv-\infty$.

Remark. If the family $\{\varphi_i\}_{i\in I}$, $\varphi_i\in\mathrm{PSH}(\Omega)$, is locally bounded above uniformly, then $\psi(z)=\sup_{i\in I}\varphi_i(z)$ is in fact integrable in θ over the set $\{z+we^{i\theta}: 0\leqq\theta\leqq2\pi\}$ for every disc $D(z,w,1)\subset\Omega$ which is not contained in $\bigcap_{i\in I}\{z\in\Omega:\varphi_i(z)=-\infty\}$. The set $\{\theta:\psi(z+we^{i\theta})<\hat{\psi}^\star(z+we^{i\theta})\}$, where $\hat{\psi}^\star$ is the regularization of ψ on the complex line $z+uw$ with respect to u, is of linear Lebesgue measure zero, so in fact $\int_{*0}^{2\pi}\psi(z+we^{i\theta})d\theta=\int_{*0}^{2\pi}\hat{\psi}^\star(z+we^{i\theta})d\theta$. A similar property holds for the set $S(\Omega)$ and the boundaries of balls in Ω.

§8. Indicator of Growth Functions

Let $\varphi(x)$ be a subharmonic function of finite order ρ and normal type with respect to the proximate order $\rho(r)$. We let $h_r(x,x',\varphi)=\limsup_{\substack{t>0\\ t\to+\infty}}\frac{\varphi(tx+x')}{t^{\rho(t)}}$. If

$\varphi(z)$ is in addition a plurisubharmonic function in $\mathbb{C}^n = \mathbb{R}^{2n}$, we define

$$h_c(z, z', \varphi) = \limsup_{\substack{u\in\mathbb{C}\\ |u|\to\infty}} \frac{\varphi(uz+z')}{|u|^{\rho(|u|)}}.$$

Their regularizations will be noted by

$$h_r^\star(x, x', \varphi) = \limsup_{(y,y')\to(x,x')} h_r(y, y', \varphi);$$

$$h_c^\star(z, z', \varphi) = \limsup_{(w,w')\to(z,z')} h_c(w, w', \varphi).$$

The function $h_r(x, x', \varphi)$ measures the growth of φ along the positive rays emanating from x' and $h_c(z, z', \varphi)$ measures the growth of φ along complex lines emanating from z'. If φ is of normal type with respect to the proximate order $\rho(r)$. then there exists C_φ and A_φ such that $\varphi(x) \leqq A_\varphi \|x\|^{\rho(\|x\|)} + C_\varphi$, hence $\varphi(rx+x') \leqq A_\varphi \|rx+x'\|^{\rho(\|rx+x'\|)} + C_\varphi$, and it follows from Theorem 1.18 that $h_r(x, x', \varphi)$ and $h_c(z, z', \varphi)$ are locally bounded from above in $\mathbb{R}^m \times \mathbb{R}^m$ and $\mathbb{C}^n \times \mathbb{C}^n$ respectively. Thus the function $h_r^\star(x, x', \varphi)$ is subharmonic if φ is subharmonic in $\mathbb{R}^m$ and the functions $h_r^\star(z, z', \varphi)$ and $h_c^\star(z, z', \varphi)$ are plurisubharmonic if φ is plurisubharmonic in $\mathbb{C}^n$.

Definition 1.29. If $\varphi(x)$ is a subharmonic function of order ρ and normal type with respect to the proximate order $\rho(r)$, then we call $h_r^\star(x, x', \varphi)$ its *radial indicator of growth function* with respect to center x'. If $\varphi(z) \in \mathrm{PSH}(\mathbb{C}^n) \subset S(\mathbb{R}^{2n})$, we call $h_c^\star(z, z', \varphi)$ *its circled indicator of growth function* with respect to center z'.

Remark 1. Our principal interest will be the case when $\varphi = \log|f|$ for $f \in \mathscr{H}(\mathbb{C}^n)$ an entire function of order ρ. In this case, we will say that $h_r^\star(z, z', \varphi)$ and $h_c^\star(z, z', \varphi)$ are the radial and circled indicator functions of f.

Remark 2. The dependence on the function φ will usually be clear from the context, and so will not always be noted.

Proposition 1.30. *For $x' \in \mathbb{R}^m$ fixed, the functions $h_r(x, x', \varphi)$ and $h_r^\star(x, x', \varphi)$ are positively homogeneous of order ρ. For $z' \in \mathbb{C}^n$, the functions $h_c(z, z', \varphi)$ and $h_c^\star(z, z', \varphi)$ are complex homogeneous of order ρ* (i.e. $h_r(tx, x', u) = t^\rho h_r(x, x', u)$, $t \geqq 0$ *and* $h_c^\star(uz, z', \varphi) = |u|^\rho h(z, z', \varphi) u \in \mathbb{C}$).

Proof. We shall prove only the case of the radial indicator, as the proof for the case of the circled indicator is practically identical (cf. Proposition 1.34). From Theorem 1.12, if $L(r) = r^{\rho(r)-\rho}$, then for t fixed, $t > 0$, $\lim\limits_{r\to\infty} \frac{L(tr)}{L(r)} = 1$. Thus

$$h_r(tx, x', \varphi) = \limsup_{r\to\infty} \frac{\varphi(rtx+x')}{r^{\rho(r)}} = \limsup_{r\to\infty} \frac{\varphi(rtx+x')}{(rt)^{\rho(rt)}} \cdot \frac{(rt)^{\rho(rt)}}{r^{\rho(r)}}$$

$$= \limsup_{rt\to\infty} \frac{\varphi(rtx+x')}{(rt)^{\rho(rt)}} \frac{(rt)^{\rho(rt)-\rho}}{r^{\rho(r)-\rho}} \cdot t^\rho = t^\rho h_r(x, x', \varphi)$$

and

$$\limsup_{(y,y')\to(tx,x')} h_r(y,y',\varphi) = t^\rho \limsup_{(y,y')\to(tx,x')} h_r\left(\frac{y}{t}, y', \varphi\right)$$

$$= t^\rho \limsup_{(\tilde{y},y')\to(x,x')} h_r(\tilde{y},y',\varphi) = t^\rho h_r^\star(x,x',\varphi). \qquad \square$$

Theorem 1.31 (Hartog's Lemma). *Let $v_t(x)$, $t>0$, be a family of subharmonic functions uniformly bounded above on compact subsets in the domain $\Omega \subset \mathbb{R}^m$. Suppose that for a compact set K in Ω there exists a constant C such that $w(x) = [\limsup_{t\to\infty} v_t(x)]^\star \leqq C$ on K. Then for every $\varepsilon>0$, there exists T_ε such that $v_t(x) \leqq C+\varepsilon$ for $t \geqq T_\varepsilon$ and $x\in K$.*

Proof. We replace Ω by an open neighborhood $\hat{\Omega}$ of K relatively compact in Ω such that $w(x) \leqq C + \frac{\varepsilon}{3}$ in $\hat{\Omega}$. Since $v_t(x)$ is bounded above in $\hat{\Omega}$, by subtracting a constant, we may assume that $v_t(x)<0$.

Choose r so small that $B(x,3r)\subset\hat{\Omega}$ for $x\in K$. Then $v_t(x) \leqq A(x,r,v_t)$, and by Fatou's Lemma, $\limsup_{t\to\infty} A(x,r,v_t) \leqq C+\varepsilon/3$. Thus, for $x\in K$, there exists T_x such that $A(x,r,v_t) \leqq C+\varepsilon/2$ for $t>T_x$. Since $v_t<0$, if $\|x'-x\|<\delta<r$,

$$\tau_m (r+\delta)^m v_t(x') \leqq \int_{\|x''-x'\|<r+\delta} v_t(x'')\,d\tau(x'')$$

$$\leqq \int_{\|x''-x\|<r} v_t(x'')\,d\tau(x''),$$

and $v_t(x') \leqq C+\varepsilon$ if $\delta<\delta_x$ and $t>T_x$. Since K is compact, we can cover K by a finite number of balls $B(x_i,\delta_{x_i})$, and if we choose $t > \sup_i T_{x_i}$, the conclusion of the Theorem is valid. $\square$

Corollary 1.32. *If, with the above hypotheses, $g(x)$ is a continuous function on K such that $[\limsup_{t\to\infty} v_t(x)]^\star \leqq g(x)$ on K, then there exist T_ε such that $v_t(x) \leqq g(x)+\varepsilon$ for $t\geqq T_\varepsilon$.*

Proof. Let $\varepsilon>0$ be given. Since K is compact, g is uniformly continuous on K, so there exists δ such that $|g(x')-g(x)|<\varepsilon/4$ for $|x-x'|<\delta$. For $x\in K$, there exists T_x such that $v_t(x')<g(x)+\varepsilon/2$ for $t>T_x$ and $|x'-x|<\delta$ by Theorem 1.31. Since K is compact, we can choose a finite number of balls $B(x_i,\delta)$ which cover K. Then for $t>\sup_i T_{x_i}$, $v_t(x')<g(x')+\varepsilon$. $\square$

Theorem 1.33. *The functions $h_r^\star(x,x',\varphi)$ and $h_r^\star(z,z',\varphi)$ are independent of the center x' or z' (this property holds for a proximate order $\rho(r)$ or the usual order ρ).*

Proof. We prove only the property for $h_r^\star(x,x',\varphi)$. The proof for the circled indicator is similar (and in fact can be deduced directly from Theorem 1.34).

Let $x_0 \in \mathbb{R}^m$ and define $\tilde{h}(x) = \limsup_{x'' \to x} h_r(x'', x_0, \varphi)$, which is a subharmonic function of x. Suppose $x \neq 0$. Given $\varepsilon > 0$, there exists $\delta > 0$ such that $h_r(x'', x_0, \varphi) \leqq \tilde{h}(x) + \varepsilon/2$ for $\|x'' - x\| < \delta$ by the upper semi-continuity of $\tilde{h}$, and so by Theorem 1.31, there exists R_ε such that for $r > R_\varepsilon$ and $\|x'' - x\| < \delta$, $\frac{\varphi(rx'' + x_0)}{r^{\rho(r)}} \leqq \tilde{h}(x) + \varepsilon$. Let x_1 be an arbitrary point in $\mathbb{R}^n$ and suppose $\|y - x_1\| < 1$. Then if $\|x'' - x\| \leqq \delta/2$ and r is sufficiently large (depending on x_0), $\tilde{x} = \frac{(y - x_0)}{r} + x''$ satisfies $\|\tilde{x} - x\| \leqq \delta$. Thus we see that

$$\frac{\varphi(rx'' + y)}{r^{\rho(r)}} = \frac{\varphi(r\tilde{x} + x_0)}{r^{\rho(r)}} \leqq \tilde{h}(x) + \varepsilon$$

and hence $h_r(x', y, \varphi) \leqq \tilde{h}(x) + \varepsilon$ for $\|x' - x\| < \delta/2$ and $\|y - x_1\| \leqq 1$ or $h_r^\star(x, x_1, \varphi) \leqq \tilde{h}(x) + \varepsilon$. Since ε was arbitrary, $h_r^\star(x, x_1, \varphi) \leqq h^\star(x) = h_r^\star(x, x_0, \varphi)$. Since x_0 and x_1 were arbitrary, the result now follows. □

Remark. Theorem 1.33 remains true if φ is only supposed to be subharmonic in an open cone Γ and if there exist A_φ and C_φ such that $\varphi(x) \leqq A_\varphi r^{\rho(r)} + C_\varphi$ for $\|x\| \leqq r$ and $x \in \Gamma$. The indicator function is then defined and subharmonic in Γ; it is positively homogeneous of order ρ; if Γ is convex, $x \in \Gamma$, and $\tilde{\varphi}^\star(x', x) = \left[\limsup_{r \to \infty} \frac{\varphi(rx' + x)}{r^{\rho(r)}}\right]$, then $\tilde{\varphi}^\star(x) = \varphi^\star(x)$.

Theorem 1.34. *Let $\varphi(z)$ be a plurisubharmonic function of finite order ρ and normal type with respect to the proximate order $\rho(r)$. Then* $h_c^\star(z, \varphi) = \sup_{0 \leqq \theta \leqq 2\pi} h_r^\star(ze^{i\theta}, \varphi)$.

Proof. It follows from the definition of the two functions that $h_c^\star(z, \varphi) \geqq h_r^\star(z e^{i\theta}, \varphi)$ for all θ. Suppose that $\sup_{0 \leqq \theta \leqq 2\pi} h_r^\star(z_0 e^{i\theta}, \varphi) = b < a = h_c^\star(z_0, \varphi)$ for $z_0 \neq 0$. The family of plurisubharmonic functions $v_t(z) = \frac{\varphi(tz)}{t^{\rho(t)}}$ is locally bounded above uniformly in t. If $\varepsilon > 0$ is so small that $b + 2\varepsilon < a$, then $\omega = \left\{z' \colon h_r^\star(z') < (b + \varepsilon)\left(\frac{\|z'\|}{\|z_0\|}\right)^\rho\right\}$ is an open neighborhood of the set $S_{z_0} = \{z_0 e^{i\theta} \colon 0 \leqq \theta \leqq 2\pi\}$, which is compact. Let $\omega_1 \Subset \omega$ be an open neighborhood of S_{z_0} and let $\bar{\omega}_2 = \omega_1 \cap \{z \colon \|z\| = \|z_0\|\}$. If $z \in \bar{\omega}_2$, $\limsup_{t \to \infty} v_t(z) \leqq h_r^\star(z) \leqq (b + \varepsilon)\left(\frac{\|z\|}{\|z_0\|}\right)^\rho$. Thus, by Theorem 1.31 there exists T_0 such that $v_t(z) \leqq b + 2\varepsilon$ for all $z \in \bar{\omega}_2$ and $t > T_0$. But then $h_c(z) \leqq (b + 2\varepsilon)$ for all $z \in \bar{\omega}_2$ and so $h_c^\star(z_0) \leqq b + 2\varepsilon < a$ which is a contradiction. □

We end this section with another application of Hartogs Lemma (Theorem 1.31):

Proposition 1.35. *Let $f \in \mathscr{H}(\mathbb{C}^n)$ be an entire function of order ρ, and let $\rho(z)$ be the order of the function $f(uz)$ as an element of $\mathscr{H}(\mathbb{C})$. Then $\rho = \sup_z \rho(z)$.*

Proof. It follows from Corollary 1.10 that $\rho \geqq \sup_z \rho(z)$. Suppose that there exists $\varepsilon > 0$ such that $\rho(z) \leqq \rho - \varepsilon$. If $\rho' = \rho - \varepsilon/2$ and $v_r(z) = \dfrac{\log|f(rz)|}{r^{\rho'}}$, then $\limsup_{r\to\infty} v_r(z) \leqq 0$ and so for $\|z\| = 1$, by Theorem 1.31, there exists R_0 such that for $r > R_0$, $v_r(z) \leqq 1$. Thus, $M_{f,p}(r) \leqq C_f r^{\rho'}$ and $\rho \leqq \rho'$, which is a contradiction. □

§9. Exceptional Sets for Growth Conditions

Our purpose here is to classify those complex lines in $\mathbb{C}^n$ on which the growth of an entire function differs from its global growth. A natural way of describing these exceptional sets is in terms of the pluripolar sets. We recall the definition:

Definition 1.36. Let $\Omega \subset \mathbb{C}^n$ be a domain. A set $E \subset \Omega$ is said to be *pluripolar* in Ω if there exists $\varphi \in \mathrm{PSH}(\Omega)$ such that $E \subset \{z\colon \varphi(z) = -\infty\}$.

Proposition 1.37. *Let Ω be a domain in $\mathbb{C}^n$. Then a countable union of pluripolar sets in Ω is pluripolar in Ω.*

Proof. Let $A_q' \subset A_q = \{z \in \Omega\colon \varphi_q(z) = -\infty,\ \varphi_q \in \mathrm{PSH}(\Omega)\}$ and let $E = \bigcup_{q=1}^{\infty} A_q'$. Let Ω_q be an exhaustion of Ω, that is $\Omega_q \Subset \Omega_{q+1}$ and $\bigcup_{q=1}^{\infty} \Omega_q = \Omega$. Since measure$(A_q) = 0$, there exists $\xi \notin \bigcup_{q=1}^{\infty} A_q$. Let $M_q = \sup_{\Omega_q} \varphi_q$ and set $S_m(z) = \sum_{q=1}^{m} C_q[\varphi_q(z) - M_q]$, where the $C_q > 0$ are chosen so that $\sum_{q=1}^{\infty} C_q|\varphi_q(\xi) - M_q| < +\infty$. Then $S_m \in \mathrm{PSH}(\Omega)$ and S_m decreases on Ω_s for $m \geqq s$. Furthermore, $\lim_{m\to\infty} S_m(\xi) > -\infty$. Thus, $S(z) = \lim_{m\to\infty} S_m(z) \in \mathrm{PSH}(\Omega_q)$ for every q (Proposition I.3). Hence $S(z) \in \mathrm{PSH}(\Omega)$ (Corollary I.20) and $E \subset \{z \in \Omega\colon S(z) = -\infty\}$. □

Proposition 1.38. *Let $\varphi \in \mathrm{PSH}(\mathbb{C}^n)$ be bounded. Then $\varphi \equiv \varphi(0)$ is a constant.*

Proof. By Proposition 1.17, $M_\varphi(r) = \sup_{\|z\| \leqq r} \varphi(z)$ is an increasing convex function of $\log r$. Thus, if $M_\varphi(r)$ is bounded, it is constant, and it now follows from the Maximum Principle that φ is constant. □

We recall a basic result, the Inverse Function Theorem for plurisubharmonic functions (cf. Theorem I.28). Let $\Omega \subset \mathbb{C}^{n-1}$ be a domain and

$\varphi \in \mathrm{PSH}(\Omega \times \mathbb{C})$. Set $M_\varphi(z,r) = \sup_{|u| \leqq r} \varphi(z,u)$, which is either constant or an increasing convex function of $\log r$ for fixed z. For $z \in \Omega$, we let $\delta(z,m) = \{\sup r : r > 0,\ M(z,r) < m\}$, which is defined for $m > \varphi(z,0)$. Then $\varphi(z,m) = -\log \delta(z,m)$ is plurisubharmonic on every connected component of Ω_q for $m > q$, where $\Omega_q = \{z \in \Omega : M(z,1) < q\}$, or is the constant $-\infty$ in Ω_q; furthermore $\varphi(z,m) \leqq 0$ in Ω_q for $m > q$.

Proposition 1.39. *Let $\{\varphi_q\}$ be a sequence of plurisubharmonic functions uniformly bounded above on compact subsets in a domain $\Omega \subset C^n$, with $\limsup_{q \to \infty} \varphi_q \leqq 0$ and suppose that there exists $\xi \in \Omega$ such that $\limsup_{q \to \infty} \varphi_q(\xi) = 0$. Then $A = \{z \in \Omega : \limsup_{q \to \infty} \varphi_q(z) < 0\}$ is pluripolar in Ω.*

Proof. Let Ω_q be an exhaustion of Ω, that is $\Omega_q \Subset \Omega_{q+1}$, $\bigcup_{q=1}^{\infty} \Omega_q = \Omega$. By Theorem 1.31, there exists T_q such that $\varphi_l \leqq q^{-2}$ for $l \geqq T_q$. Thus, we can choose a subsequence $\{\varphi'_q\}$ such that $\varphi'_q \leqq q^{-2}$ on Ω_q and $\sum_{q=1}^{\infty} |\varphi'_q(\xi)| < +\infty$. Let $S_m(z) = \sum_{q=1}^{m} [\varphi'_q(z) - q^{-2}] \in \mathrm{PSH}(\Omega)$. Then $\lim_{m \to \infty} S_m(\xi) > -\infty$, and hence $S(z) = \lim_{m \to \infty} S_m(z) \in \mathrm{PSH}(\Omega)$ by Proposition I.3. For $z \in A$, $S(z) = -\infty$. □

Proposition 1.40. *Let $\Omega \subset \mathbb{C}^{n-1}$ and $\varphi \in \mathrm{PSH}(\Omega \times \mathbb{C})$, $\varphi \geqq 1$. Let $\rho(z')$ be the order of $\varphi_{z'}(u) = \varphi(z',u)$. Then for $\Omega' \Subset \Omega$ a domain, there exists a sequence of negative plurisubharmonic functions $\{\psi_q\}$ on Ω' such that $-[\rho(z')]^{-1} = \limsup_{q \to \infty} \psi_q(z')$.*

Proof. By definition, $\rho(z') = \limsup_{r \to \infty} \dfrac{\log[\sup M(z',r), 1]}{\log r}$. Let $M_0 > \sup_{z \in \Omega'} M(z',1)$ and $m > \sup(M_0, 1)$. Choose r_m so that $M(z', r_m) = m$. Then

$$\rho(z') = \limsup_{m \to \infty} (\log m)(\log \delta(z',m))^{-1}$$

and hence

$$-[\rho(z')]^{-1} = \limsup_{m \to \infty} (\log m)^{-1}[-\log \delta(z',m)]. \qquad \square$$

Theorem 1.41. *Let $\Omega \subset \mathbb{C}^{n-1}$ and $\varphi \in \mathrm{PSH}(\Omega \times \mathbb{C})$. For $z = (z',u)$, $z' \in \Omega$, $u = z_n$, let $\rho(z')$ be the order of $u \to \varphi(z',u)$. Then if for some $\Omega' \Subset \Omega$, $\rho(z)$ is finite for $z \in M$ a non pluripolar set in Ω', $\rho(z')$ is bounded on each compact set in Ω and $\rho^\star(z') = \limsup_{z'' \to z'} \rho(z'') \in \mathrm{PSH}(\Omega)$. Then we say that $\varphi(z',u)$ is of finite order with respect to the variable u.*

Proof. Let $\bar{\Omega}'$ be compact in Ω and Ω'' a subset of Ω such that $\bar{\Omega}' \subset \Omega'' \Subset \Omega$; suppose $m > \sup[1, \sup_{z' \in \Omega'', |u| \leqq 1} \varphi(z',u)]$, and consider

$$\psi(z',m) = (\log m)^{-1}[-\log \delta(z',m)] < 0.$$

Then for $g(z')=\limsup_{m\to\infty}\psi(z',m)$, and $g^\star(z')=\limsup_{z''\to z'} g(z'')$, there exist two possibilities:

1) $g^\star(z')\equiv 0$ in Ω''; then there exists $z'_0\in\Omega'$ such that $g(z'_0)=g^\star(z'_0)=0$ (App. I. 27). By Proposition 1.39, the set

$$E=\{z'\in\Omega'' : g(z')<0]=\{z'\in\Omega'' : \rho(z')<\infty\}$$

is pluripolar in Ω'', hence in $\Omega'\subset\Omega''$, contradicting the hypothesis.

2) $g^\star\not\equiv 0$. Then if $g^\star\equiv-\infty$, $g(z')=g^\star(z')\equiv-\infty$ and $\rho(z')\equiv 0$ in Ω''. If $g^\star\in\mathrm{PSH}(\Omega'')$ and $g^\star\not\equiv 0$, by the condition $g^\star(z')\leqq 0$ and the Maximum Principle, $g^\star(z')<0$ has a strictly negative bound on every compact subset of Ω''; then $\rho^\star(z')=-[g^\star(z')]^{-1}$ has a finite bound on K, and $\rho^\star(z')$ is locally plurisubharmonic in Ω, thus $\rho^\star(z')\in\mathrm{PSH}(\Omega)$. □

Corollary 1.42. *Suppose* $\varphi\in\mathrm{PSH}(\mathbb{C}^n)$ *is of finite total order* ρ. *Let* $z=(z',u)$, $z'\in\mathbb{C}^{n-1}$, $u\in\mathbb{C}$. *Then the order* $\rho(z')$ *of* $u\to\varphi(z',u)$ *with respect to* u *is a constant* $\rho_0<\rho$ *except on a pluripolar set in* $\mathbb{C}^{n-1}$, *and* $\rho(z')\leqq\rho_0$.

Proof. By considering $\sup(\varphi,2)$, if necessary, we may assume $\varphi\geqq 2$. Since $\rho^\star(z')\leqq\rho$, $\rho^\star(z')$ is a constant $\rho^\star$ by Proposition 1.38. As before, we write $-\dfrac{1}{\rho(z')}=\limsup_{m\to\infty}\psi(z',m)$, where $\psi(z',m)$ is defined and negative for $\|z'\|\leqq p$, and $m>M_p=\sup_{\|z'\|\leqq p,\,|u|\leqq 1}\varphi(z',u)\geqq 2$. Now we replace $\psi(z',m)$ by a sequence $\psi_p(z')\in\mathrm{PSH}(\mathbb{C}^n)$. We remark that for $m_p>M_p$ and $\|z'\|\leqq p$, $\sup\psi(z',m_p)\leqq -\alpha_p<0$. Thus given $m_p>M_p$ we define:

$$\psi_p(z')=\sup\left[\psi(z',m_p),\log\frac{\|z'\|}{p}\right]\qquad\text{for } \|z'\|\leqq p$$

$$\psi_p(z')=\log\frac{\|z'\|}{p}\qquad\text{for } \|z'\|\geqq p.$$

Because $\psi(z',m_p)$ is bounded by $-\alpha_p<0$ on $\|z'\|\leqq p$, and $\log\dfrac{\|z'\|}{p}$ is a continuous function vanishing for $\|z'\|=p$, the function ψ_p is well defined and $\psi_p\in\mathrm{PSH}(\mathbb{C}^{n-1})$. Now there exists z'_0, $\|z'_0\|<1$ such that

$$-\frac{1}{\rho^\star}=-\frac{1}{\rho(z'_0)}=\limsup_{m\to\infty}\psi(z'_0,m).$$

Then we choose for $p=1,2,\ldots$ a sequence $m_p>M_p$ such that $\limsup_{p\to\infty}\left[\psi_p(z'_0)+\dfrac{1}{\rho^\star}\right]=0$ and apply Proposition 1.39 for $\Omega=C^{n-1}$. Thus the set $[z'\in\mathbb{C}^{n-1} : \rho(z')<\rho^\star]$ is pluripolar in $\mathbb{C}^{n-1}$. □

Remark. We will apply Theorem 1.41 to entire functions $F(z_1,z_2,\ldots,z_n)$. We say that F is of finite order with respect to the variable z_n if for

$z'=(z_1, \ldots, z_{n-1}) \in \mathbb{C}^{n-1}$, the order $\rho(z')$ of $z_n \to F(z', z_n)$ is finite for all z'. From Theorem 1.41 it is so if and only if $\rho(z')$ has finite values on a non pluripolar set. Note that F can be of finite order with respect to z_n even if its total order is infinite.

Corollary 1.43. *Let $\varphi \in \mathrm{PSH}(\mathbb{C}^n)$ and let $\rho(z)$ be the order of $\varphi_z(u)$: $u \to \varphi(uz)$. Then $\rho(z)$ is a constant ρ_0 (finite or infinite) except on a pluripolar cone A with vertex at the origin where $\rho(z) < \rho_0$, $A \subset A' = [z: S(z) = -\infty]$, $S(z) \in \mathrm{PSH}(\mathbb{C}^n)$, and $S(\lambda z) = S(z) + \log(\lambda)$ for $\lambda \in \mathbb{C}$.*

Proof. We suppose $\varphi(z) > 1$ and define

$$\delta(z, m) = \{\sup r, r > 0, M_\varphi(rz) = \sup \varphi(uz) < m\}.$$

Suppose $m > m_0 = \sup_{\|z\| \leqq 1} \varphi(z)$. Then

$$-\log \delta(uz, m) = -\log \delta(z, m) + \log|u|$$

and for $m > m_0 > 1$, the functions

$$\psi_m(z) = (\log m)^{-1}[-\log \delta(z, m)] \in \mathrm{PSH}(\mathbb{C}^n)$$

are uniformly bounded from above on each compact set in $\mathbb{C}^n$. Set

$$g(z) = -\frac{1}{\rho(z)} = \limsup_{m \to \infty} \psi_m(z).$$

Then $g(\lambda z) = g(z)$ if $\lambda \neq 0$, $g(z) \leqq 0$, and $g^\star(z) \in \mathrm{PSH}(\mathbb{C}^n)$; hence $g^\star(z) \leqq 0$. Therefore $g^\star$ is a constant c_0 and $\rho^\star(z) = -c_0^{-1}$. The set $A = [z: \rho(z) < \rho^\star]$ $= [z: g(z) < g^\star]$ is a cone with vertex at the origin in $\mathbb{C}^n$. Moreover there exists $\xi \in \mathbb{C}^n$ such that $\limsup_{m \to \infty} [\psi_m(\xi) - c_0] = 0$, so $\rho(\xi) = \rho^\star$.

a) If $c_0 = 0$, there exist a subsequence $m_q > m_0 > 1$, $\lim_{q \to \infty} m_q = +\infty$ such that $a = \sum_q (\log m_q)^{-1} < \infty$ and $\sum |\psi_{m_q}(\xi)| < \infty$. Then $\rho^\star = +\infty$ and

$$S(z) = a^{-1} \sum_q \psi_{m_q}(z)$$

is plurisubharmonic in $\mathbb{C}^n$, $S(z) = -\infty$ if $\rho(z) < \infty$, and $S(uz) = S(z) + \log|u|$ for $u \in C$.

b) If $c_0 < 0$, the order ρ_0 is finite. Let $\Omega_q \Subset \Omega_{q+1}$ be an exhaustive sequence of relatively compact sets in $\mathbb{C}^n$. We can find a sequence $m_q \to +\infty$, $m_q > m_0 > 1$ (see Proposition 1.39) such that

(1) $$\psi_{m_q}(z) - c_0 - \frac{1}{q^2} < 0 \quad \text{for } z \in \Omega_q$$

(2) $$\sum |\psi_{m_q}(\xi) - c_0| < \infty \quad \text{for some } \xi \in \Omega$$

(3) $$a = \sum (\log m_q)^{-1} < \infty.$$

Then $S(z)=\sum_q \left[\psi_{m_q}(z)-c_0-\frac{1}{q^2}\right]\in \mathrm{PSH}(\mathbb{C}^n)$ has the property that $S(uz)=S(z)+\log|u|$ for $u\in\mathbb{C}$, and the set $\rho(z)<\rho_0=c_0^{-1}$ is contained in the pluripolar cone $S(z)=-\infty$. □

Theorem 1.44. *Let Ω be a domain in $\mathbb{C}^{n-1}$ and $\varphi\in\mathrm{PSH}(\Omega\times\mathbb{C})$. Let $\Omega_1\Subset\Omega$ be a domain and A a non-pluripolar subset of Ω_1. Let $\psi(t)$ be an increasing convex function of t for $t\geqq 0$ such that $M_\varphi(z,r)\leqq\psi(\log r)$ for $r>r_0\geqq 1$ and $z\in A$. Then there exists a function $\sigma(z)\in\mathrm{PSH}(\Omega_1)$ with $1\leqq\sigma(z)<+\infty$ such that $M_\varphi(z,r)\leqq\psi(\sigma(z)\log r)$ for $r>r_0$ and $z\in\Omega_1$.*

Proof. Let $m_0=\sup_{z\in\Omega_1} M(z,1)$. The equation $\psi(\log r)=m$ has as solution $\log r=\log\eta(m)$. The equation $M(z,r)=m$ has as solution $r=\delta(z,m)$. Let $\psi(z,m)=\frac{-\log\delta(z,m)}{\log\eta(m)}$ for $m>m_1=\sup(m_0,\psi(0))$. Then $\psi(z,m)\in\mathrm{PSH}(\Omega_1)$ and $\psi(z,m)<0$. Furthermore, $\psi(z,m)\leqq -1$ for $z\in A$. Let

$$\varphi_A(z)=\sup_{\substack{\tau\in\mathrm{PSH}(\Omega_1)\\ \tau\leqq 0\\ \tau\leqq -1 \text{ on } A}}\{\tau(z)\}$$

Since A is non-pluripolar, we have $\varphi_A^\star(z)\not\equiv 0$ and $\varphi_A^\star\in\mathrm{PSH}(\Omega_1)$. Thus, $\varphi_A^\star(z)<0$ for $z\in\Omega_1$ by the Maximum Principle. Let

$$\sigma(z)=-[\varphi_A^\star(z)]^{-1}\in\mathrm{PSH}(\Omega_1).$$

Thus

$$-\log\delta(z,m)\leqq\varphi_A^\star(z)\log\eta(m)\quad\text{or}\quad\sigma(z)\log\delta(z,m)\geqq\log\eta(m);$$

$\delta(z,m)^{\sigma(z)}\geqq\eta(m)$. In terms of r, we obtain for $r=[\eta(m)]^{1/\sigma(z)}$:

$$M_\varphi(z,r)\leqq M_\varphi(z,\delta(z,m))=m=\psi_\eta(\log\eta(m))=\psi(\sigma(z)\log r).$$ □

Of course our primary purpose is to apply these results to $\varphi(z)=\log|f(z)|$, f an entire function in $\mathbb{C}^n$. As remarked before, the indicator functions for the growth of $|f|$ are plurisubharmonic functions (not necessarily continuous); later we shall apply the same technics to the indicator functions of the zeros of f.

Historical Notes

The idea of using intermediate functions in the definition of type is due to Lindelöf, but the use of proximate orders is due to Valiron [1]. The calculation of the order and type in terms of the Taylor series coefficients is classic for $n=1$. For $n\geqq 2$, this as well as variants has been studied in detail by the Russian school (cf. Gold'berg [1]). Relations between the total order and the orders relative to each variable were first given by Borel [1]; the first comparison with respect to the growth on complex lines was made by Sire [1] at the beginning of the century. The modern treatment of the indicator function as given here is primarily due to Lelong [2]. This generalizes the classical Phragmen-Lindelöf indicator function and was first considered by Lelong [2] and by Deny and Lelong [1] and [2] for subharmonic functions. In particular Lelong developped in his early works Hartog's Theorem in $\mathbb{C}^2$ in the context of subharmonic functions and potential theory. After the introduction due to Oka and Lelong [5, 6] of the class of plurisubharmonic functions (1942), the properties of the indicator function were obtained from the general properties of locally bounded families of plurisubharmonic functions; the characterization of the indicator functions for entire functions of finite order in terms of plurisubharmonicity was given by Kiselman [2] and Martineau [4, 5] and will be presented in Chapter 7. The proof given here that $h^\star(x, x', \varphi)$ is independent of the center has the advantage of working in the class of subharmonic indicators defined in cones. The results of § 9 and the Inverse Function Theorem for plurisubharmonic functions (see Appendix I) were given by Lelong [15] for complex topological vector spaces.

Chapter 2. Local Metric Properties of Zero Sets and Positive Closed Currents

§ 1. Positive Currents

A biholomorphic mapping $F\colon \mathbb{C}^n \to \mathbb{C}^n$ induces a mapping of the underlying real coordinates $\mathbb{R}^{2n}$ whose determinant is just $|J(F)|^2$, where $J(F)$ is the Jacobian of F, and this determinant is positive. Thus, if one chooses once and for all a volume form an $\mathbb{C}^n$, this choice determines an orientation in $\mathbb{C}^n$ which is invariant with respect to holomorphic isomorphisms, and its restriction to subspaces L^p, $p<n$, or to complex submanifolds, determines a volume element and hence an orientation. This led to the introduction of a positive differential form in the exterior differential algebra $E_{2n}(dz)$ with involution (given by $dz \to d\bar{z}$) and its generalization, the positive closed current.

We write $\beta = \frac{i}{2}\partial\bar{\partial}\|z\|^2 = \frac{i}{2}\sum_{j=1}^{n} dz_j \wedge d\bar{z}_j$ and set $\beta_p = (p!)^{-1}\beta^p$, which is just the p-dimensional Euclidean volume measure in $\mathbb{C}^n$.

Definition 2.1. A differential form $\varphi(dz)$ with complex-valued coefficients will be said to be positive of degree p in $E_{2n}(dz)$ if

i) it is homogeneous of type (p,p), $0 \leqq p \leqq n$;

ii) for every system of forms $\alpha_1, \ldots, \alpha_{n-p}$ linear in dz_j (that is such that $\alpha_i = \sum_{j=1}^{n} a_{ij} dz_j$, $a_{ij} \in \mathbb{C}$, is a (1,0) form), the product $\varphi \wedge i\alpha_1 \wedge \bar{\alpha}_1 \wedge \ldots \wedge i\alpha_{n-p} \wedge \bar{\alpha}_{n-p} = \psi \cdot \beta_n$ is such that $\psi \geqq 0$.

For Ω a domain in $\mathbb{C}^n$, we let $\Phi_p^+(\Omega)$ be the positive forms of degree p in Ω with continuous coefficients. As a consequence of Definition 2.1, we obtain:

Proposition 2.2. *A C-linear change of wordinates in $E_{2n}(dz)$: $dz'_j = \sum_{k=1}^{n} C_{j,k} dz_k$, $d\bar{z}'_j = \sum_{k=1}^{n} \bar{C}_{jk} d\bar{z}_k$ transforms a positive form into a positive form. As a consequence a biholomorphic map $F\colon \Omega \to \Omega'$ induces a map of $\Phi_p^+(\Omega)$ onto $\Phi_p^+(\Omega')$.*

Proposition 2.2 permits the definition of positive forms on a complex submanifold $M \subset \Omega$: it is those forms which are positive for every choice of local coordinates.

For $p=0$, $\Phi_0^+(\Omega)$ is just the set of positive continuous functions on Ω. For $p=1$, $\varphi\in\Phi_1^+(\Omega)$ if and only if $\varphi=i\sum_{j,k=1}^{n}\varphi_{jk}dz_j\wedge d\bar{z}_k$, where the matrix $[\varphi_{jk}(z)]$ is positive semi-definite for every $z\in\Omega$. If for any p we have

$$\varphi=i\lambda_1\wedge\bar{\lambda}_1\wedge\ldots\wedge i\lambda_p\wedge\bar{\lambda}_p, \tag{2,1}$$

where the λ_p are complex linear in dz_j with coefficients in $\mathscr{C}^0(\Omega)$, then $\varphi\in\Phi_p^+(\Omega)$. Those φ which can be represented as in (2,1) will be said to be *decomposable.*

If L^p is a complex subspace of dimension p, there exists a rotation $g\in U(n)$ given by $u=g(z)$, such that $g(L^p)$ is defined by the equations $u_{p+1}=\ldots=u_n=0$. We define the form $\tau(L^p)\in\Phi_p^+(\mathbb{C}^n)$ by $\tau(L^p)=g^\star\tilde{\beta}$, substituing in

$$\tilde{\beta}=\frac{i}{2}du_1\wedge d\bar{u}_1\wedge\ldots\wedge\frac{i}{2}du_p\wedge d\bar{u}_p$$

the values $du=g(dz)$. The mapping $L^p\to\Phi_p^+(\mathbb{C}^n)$ is well defined, since the use of different coordinates $u=\gamma(v)$ orthonormal on L^p does not modify $\tau(L^p)$.

We note by $\star\varphi$ the adjoint of a monomial φ of type (p,q). If $\varphi=dz_I\wedge d\bar{z}_J$, then

$$\star\varphi=2^{p+q-n}i^n(-1)^{|I|+|J|+np}dz_{j_{q+1}}\wedge\ldots\wedge dz_{j_n}\wedge d\bar{z}_{i_{p+1}}\wedge\ldots\wedge d\bar{z}_{i_n}$$

so that $\star\tau(L^p)$ is the form $\tau(L^{n-p})$ associated with the orthogonal subspace L^{n-p} and $\tau(L^p)\wedge\tau(L^{n-p})=\beta_n$. We have proved:

Proposition 2.3. *For every linear subspace L^p of dimension p, there exists a positive form $\tau(L^p)\in\Phi_p^+(\mathbb{C}^n)$ associated with L^p via the unitary group $U(n)$. Its adjoint $*\tau(L^p)$ is the form associated with the orthogonal subspace L^{n-p}.*

In Definition 2.1, we can suppose that the linear forms $\alpha_1,\ldots,\alpha_{n-p}$ are linearly independant and are coordinates on an L^{n-p} orthogonal to some L^p defined by $\alpha_1=\ldots=\alpha_{n-p}=0$. The product $i\alpha_1\wedge\bar{\alpha}_1\wedge\ldots\wedge i\alpha_{n-p}\wedge\bar{\alpha}_{n-p}$ which figures in ii) of Definition 2.1 is just $C\cdot\tau(L^{n-p})$ for C a positive constant. We thus obtain:

Proposition 2.4. *A homogeneous form φ of type (p,p) is positive if and only if it satisfies*

$$\varphi\wedge\star\tau(L^p)=C_\varphi(z)\,\beta_n \quad \textit{with } C_\varphi(z)\geqq 0 \textit{ for every } L^p. \tag{2,2}$$

We can render condition (2,2) more explicit by calculating $C_\varphi(z)$ for a form $\varphi=\sum_{I,J}\varphi_{I,J}dz_I\wedge d\bar{z}_J$. We use the restriction of φ for those dz_k belonging to an L^p defined by (2,2). Then if h_I is the determinant $\left[\dfrac{\partial z_s}{\partial u_j}\right]_{j\in J}^{s\in I}$

(2,3) $\varphi=k_p[\sum_{I,J}\varphi_{I,J}h_I\bar{h}_J]\,\tau(L^p)$ on L^p, where $k_p=2^p$ if p is even and $k_p=-2^p i$

if p is odd. Then $C_\varphi(z)$ is equal to the coefficient of $\tau(L^p)$ in (2,3). We obtain finally:

Proposition 2.5. *A necessary and sufficient condition for a form φ homogeneous of type (p,p) to be positive is that its restriction to every complex linear subspace L^p of dimension p is the product of the volume element $\tau(L^p)$ by a positive function $C_\varphi(z)$ given by (2,3). If p is even, then $\varphi_{I,I}\geqq 0$ and if p is odd $-i\varphi_{I,I}\geqq 0$.*

Remarks. 1. For $p=1$ and $\varphi=i\sum\varphi_{jk}dz_j\wedge d\bar z_k$, φ is positive if and only if the hermitian form $h=\sum\varphi_{j,k}dz_j\cdot d\bar z_k$ is positive semi-definite.

2. In the sequel we write $\varphi\geqq 0$ for positive forms, and $\varphi_1\geqq\varphi_2$ if $\varphi_1-\varphi_2$ is a positive form.

Let $N=\left[\dfrac{n!}{p!(n-p)!}\right]^2$ and φ a (p,p)-form. If $\Lambda=\{L_s^{n-p}\}$ is a system of N complex linear subspaces, we consider the N linear equations $\varphi\wedge\tau(L_s^{n-p})=C_{\varphi,s}(z)\,\beta_n$ in order to calculate the N coefficients of φ as linear combinations $\varphi_{I,J}(z)=\sum_s\lambda_{I,J}^s C_{\varphi,s}(z)$ of the $C_{s,\varphi}$ with coefficients $\lambda_{I,J}^s\in\mathbb{C}$ depending on Λ but not on φ. More precisely, given (1,0) forms

$$\alpha_{k,s}=\sum_{j=1}^n a_{k,s}^j dz_j,\qquad 1\leqq k\leqq n-p;\ 1\leqq s\leqq N$$

and

$$\omega_s=i\alpha_{1,s}\wedge\bar\alpha_{1,s}\wedge\ldots\wedge i\alpha_{n-p,s}\wedge\bar\alpha_{n-p,s}=\sum A_s^{I,J}dz_I\wedge d\bar z_J,$$

the $A_s^{I,J}$ are distinct monomials in the $a_{k,s}^j$, $\bar a_{k,s}^j$. Let $\Delta=\Delta(\omega_s)=\det[A_s^{I,J}]$. We have $\Delta=\lambda P$, $\lambda\neq 0$, and P is a real valued polynomial in the space $\mathbb{R}^{2N\cdot n}$ of the x_q, real and imaginary parts of the $a_{k,s}^j$, $1\leqq q\leqq 2Nn$, and hence either $\Delta\equiv 0$ or $W=\{x\in\mathbb{R}^{2Nn}\colon\Delta=0\}$ is an algebraic variety A of real dimension $2Nn-1$. But $\Delta\not\equiv 0$, since it is a sum of distinct monomials each with non-zero coefficient. Hence, in every open set of $\mathbb{R}^{2N\cdot n}$, we can find points such that $\Delta\neq 0$. Now the sets e_s defined in $\mathbb{R}^{2Nn}$ by $\omega_s\equiv 0$ are real analytic subvarieties of A and $\Omega=\mathbb{R}^{2Nn}-\bigcup_s e_s$ is a dense open set in $\mathbb{R}^{2Nn}$. If we suppose $\omega_s\not\equiv 0$ for $1\leqq s\leqq N$, the equations $\alpha_{k,s}=0$ for $1\leqq k\leqq n-p$ define a subspace L_s^{n-p}, and $\omega_s=b_s\tau(L_s^{n-p})$ for $b_s>0$, $1\leqq s\leqq N$. Then $\Delta=\Delta(\omega_s)=B\Delta[\tau(L_s^{n-p})]$ for $B=b_1\ldots b_N>0$ and in Ω, $\Omega\subset\mathbb{R}^{2N}$, the conditions $\Delta\neq 0$ and $\Delta_1=\Delta[\tau(L_s^{n-p})]\neq 0$ are equivalent; we shall say that the system $\Lambda=\{L_s^{n-p}\}$ is regular. We conclude that in each open set of $\mathbb{R}^{2Nn}$ we can find points such that $\Delta_1\neq 0$. If $G_{n-p}(\mathbb{C}^n)$ is the $p(n-p)$ dimensional complex Grassmannian manifold of $(n-p)$ dimensional linear subspaces of $\mathbb{C}^n$ (cf. [H]) then in each open set of $G_{n-p}(\mathbb{C}^n)$, we can find a regular system $\Lambda=\{L_s^{n-p}\}$ which allows us to calculate $\varphi_{I,J}(z)$ as a linear combination of $C_{\varphi,s}(z)$. If φ is a positive form, the $C_{\varphi,s}(z)$ are positive functions (later we use the same algebraic process on positive currents, and $C_{\varphi,s}\cdot\beta_n$ will be a positive measure).

Proposition 2.6. *Let M be a complex submanifold of $\Omega\subset\mathbb{C}^n$ of dimension p (cf. Definition 2,33) and $\varphi\in\Phi_p^+(\Omega)$ with compact support in Ω. Then*

$$\int_M \varphi=[M](\varphi)\geqq 0. \tag{2,4}$$

If $\varphi=d\psi$ for a form ψ with $\mathscr{C}^1$ coefficients, then $\int_M d\psi=[M](d\psi)=0$.

Proof. Let $\{U_k\}$ be a locally finite covering of M by relatively compact local coordinate patches, and let $\{\alpha_k\}$ be a partition of unity subordinate to $\{U_k\}$. Then

$$\int_M \varphi=\sum_k \int_M \alpha_k\varphi=\sum_k [M](\alpha_k\varphi)$$

(where the sum is finite since φ has compact support).

For each U_k, there exists a holomorphic homeomorphism F_k of U_k onto V_k, an open neighborhood of the origin in $\mathbb{C}^p$, and

$$\int_M \alpha_k\varphi=\int_{V_k} F_k^\star(\alpha_k\varphi)=\int_{V_k}\alpha'_k\varphi'_k.$$

Since $\alpha'_k=\alpha_k\circ F^{-1}>0$ and $\varphi'_k=\varphi\circ F^{-1}\in\Phi_p^+(V_k)$, we have $\int_{V_k}\alpha'_k\varphi'_k\geqq 0$, from which (2,4) follows. If $\varphi=d\psi$ then

$$[M](d\psi)=\sum_k\int_{V_k}\alpha'_k d\psi'_k=\sum_k\int_{V_k} d(\alpha'_k\psi'_k)-\sum_k\int_{V_k} d\alpha'_k\wedge\psi'_k.$$

It follows from Stokes' Theorem that each summand in the first sum is zero, since $\operatorname{supp}\alpha'_k$ is compact in V_k. On the other hand $\sum_k\int_{V_k} d\alpha'_k\wedge\psi'_k=\int_M\left(\sum d\alpha_k\right)\wedge\psi=0$, since $\sum\alpha_k\equiv 1$. □

The area of a manifold is a positive measure σ defined for $f\in\mathscr{C}_0^\infty(\Omega)$ by

$$\sigma(f)=[M](f\beta_p) \tag{2,5}$$

where $\beta=\frac{i}{2}\sum_{k=1}^n dz_k\wedge d\bar{z}_k$ and $\beta_p=(p!)^{-1}\beta^p$.

This leads us to:

Proposition 2.7. *If M is a complex submanifold of $\Omega\subset\mathbb{C}^n$, then the area of M defined by* (2,5) *is the sum of its projections on the coordinate spaces*

$$\mathbb{C}^p(I)=\mathbb{C}^p(dz_{i_1},\dots,dz_{i_p}),\qquad I=(i_1<i_2<\dots<i_p).$$

Proof. We have $\beta_p=(p!)^{-1}\beta^p=\left(\frac{i}{2}\right)^p(-1)^{\frac{p(p-1)}{2}}\sum_I dz_I\wedge d\bar{z}_I=\sum_I\beta_I$ so that from (2,5) we obtain $\sigma(f)=\sum_I[M](f\beta_I)=\sum_I\sigma_I(f)$, where σ_I, given by the integration of β_I on M, is the projection with multiplicity of M on $\mathbb{C}^p(I)\subset\mathbb{C}^n$, where $\mathbb{C}^p(I)$ is defined by equations $z_j=0$ for $j\notin I$. □

Remark. The positive measure $[M]\wedge\tau(L^p)$ is the projection of the area of M on the subspace L^p.

We recall that $\mathscr{C}^k_{0,(p,q)}(\Omega)$ is the space of differential forms φ of degree (p,q) with coefficients in $\mathscr{C}^k_0(\Omega)$. We let $\mathscr{E}^\infty_0(\Omega)=\bigcup_{p,q}\mathscr{C}^\infty_{0,(p,q)}(\Omega)$. Let Ω_i be an exhaustion of Ω by compacts sets, $\Omega_i\Subset\Omega_{i+1}\Subset\Omega$ and $\Omega=\bigcup_{i=1}^\infty\Omega_i$. We equip $\mathscr{C}^\infty_{0,(p,q)}(\Omega_i)$ with the topology of uniform convergence of the coefficients $\varphi_{I,J}$ and all of their derivatives. With this topology, $\mathscr{C}^\infty_{0,(p,q)}(\Omega_i)$ is a Fréchet space. Finally, we equip $\mathscr{C}^\infty_{0,(p,q)}(\Omega)$ with the strict inductive limit topology, $\mathscr{C}^\infty_{0,(p,q)}(\Omega)=\lim\limits_{\to}\mathscr{C}^\infty_{0,(p,q)}(\Omega_i)$. The dual space of $\mathscr{C}^\infty_{0,(p,q)}(\Omega)$ is the set of linear functionals $t(\varphi)$ for which $\lim\limits_{m\to\infty}t(\varphi_m)=0$ for every sequence $\{\varphi_m\}$ which tends to zero in $\mathscr{C}^\infty_{0,(p,q)}(\Omega)$; a sequence $\{\varphi_m\}$ tends to zero in $\mathscr{C}^\infty_{0,(p,q)}(\Omega)$ if and only if

i) there exists Ω_i such that, $\operatorname{supp}\varphi_m\subset\Omega_i$ for every m,

ii) for every multi-index α, $\lim\limits_{m\to\infty}[\sup\limits_{z\in\Omega}\sup\limits_{I,J}|D^\alpha\varphi_{m,I,J}(z)|]=0$, where D^α is a differential operator with respect to the underlying real coordinates in $\mathbb{C}^n$. For further details, we refer the reader to [E, F].

Definition 2.8. The elements of the dual space of $\mathscr{C}^\infty_{0,(p,q)}(\Omega)$ are the currents of type (p',q') with $p'=n-q$ and $q'=n-q$.

If in addition, we have $\lim\limits_{m\to\infty}t(\varphi_m)=0$ whenever $\{\varphi_m\}$ satisfies (i) and (ii'): $\lim\limits_{m\to\infty}[\sup\limits_{z}\sup\limits_{I,J}|\varphi_{m,I,J}(z)|]=0$, we say that t is continuous of order zero; t then extends as a linear functional to the space $\mathscr{C}^0_{0,(p,q)}(\Omega)$ of differential forms of order (p,q) with continuous coefficients having compact support. Currents of type (n,n) apply to functions in $\mathscr{C}^\infty_0(\Omega)$ and are thus distributions; if they are continuous of order zero, they are measures. Currents of type $(0,0)$ apply to forms of maximum type and play the role of density-distributions. We will call them generalized functions; the product of such a function with β_n, the volume element, is a distribution.

Definition 2.9. A current t will be said to be *positive of degree* $(n-p)$ if

i) t is zero on the subspaces $\mathscr{C}^\infty_{0,(r,s)}(\Omega)$ if $(r,s)\neq(p,p)$ (i.e. $t\in[\mathscr{C}^\infty_{0,(p,p)}(\Omega)]'$),

ii) for every system $\alpha_1,\ldots,\alpha_p$ of complex linear forms in dz_j with constant coefficients and every $\varphi\in\mathscr{C}^\infty_0(\Omega)$ with $\varphi\geqq 0$,

$$T(t,\alpha)(\varphi)=t[\varphi i\alpha_1\wedge\bar\alpha_1\wedge\ldots\wedge i\alpha_p\wedge\bar\alpha_p]\geqq 0. \tag{2,6}$$

In particular, a form $\psi\in\Phi^+_{n-p}(\Omega)$ defines a positive current via integration $\varphi\in\mathscr{C}^\infty_{0,(p,p)}(\Omega)\to\int_\Omega\psi\wedge\varphi$. We will let $T^+_{n-p}(\Omega)$ be the space of positive currents of degree $n-p$ in Ω. As in Proposition 2.4, we have:

Proposition 2.10. *A current t defined in Ω belongs to $T^+_{n-p}(\Omega)$ if and only if for every linear subspace L^p of dimension p, $t \wedge \tau(L^p)$ is a positive distribution (hence a positive measure).*

We see that an element $t \in T^+_{n-p}(\Omega)$ can be identified with a measure in Ω (depending on L^p). A current $t \in T^+_{n-p}(\Omega)$ is represented by a differential form homogeneous of type $(n-p, n-p)$. We can express it in a canonical form

$$t = k'_p \sum_{I,J} t_{I,J} dz_I \wedge d\bar{z}_J \qquad I = (i_1 < \ldots < i_{n-p}),\ J = (j_1 < \ldots < j_{n-p}) \tag{2,7}$$

where $k'_p = 2^{-(n-p)}$ if $n-p$ is even and $k'_p = i2^{-(n-p)}$ if $(n-p)$ is odd. The distributions $t_{I,I}\beta_n$ are positive measures.

Proposition 2.11. *A current $t \in T^+_{n-p}(\Omega)$ is the limit on every compact subset of Ω of a sequence $\{t_m\}$ of positive currents represented by forms $t_m \in \Phi^+_{n-p}(\Omega)$.*

Proof. Let $\rho \in \mathscr{C}^\infty_0(B(0,1))$ such that $\rho \geqq 0$, $\int \rho(z) d\tau(z) = 1$, and set $\rho_\varepsilon(z) = \rho\left(\frac{z}{\varepsilon}\right)\varepsilon^{-2n}$, $t_\varepsilon = t \star \rho_\varepsilon$ given by $t_\varepsilon = k'_p \sum_{I,J} (t_{I,J})_\varepsilon dz_I \wedge d\bar{z}_J$, where $(t_{I,J})_\varepsilon = (t_{I,J}\beta_n) \star \rho_\varepsilon$. Then $(t_{I,J})_\varepsilon \in \mathscr{C}^\infty(\Omega_\varepsilon)$ and for $\varphi \in \mathscr{C}^\infty_{0,(p,q)}(\Omega)$

$$t(\varphi) = \lim_{\varepsilon \to 0} t[\rho_\varepsilon \star \varphi] = \lim_{\varepsilon \to 0} t_\varepsilon(\varphi). \tag{2,8}$$

□

§2. Exterior Product

Theorem 2.12. i) *if $t \in T^+_p(\Omega)$ and $\varphi \in \Phi^+_1(\Omega)$, then $t \wedge \varphi \in T^+_{p+1}(\Omega)$;*

ii) *if $t \in T^+_1(\Omega)$ and $\varphi \in \Phi^+_p(\Omega)$, then $t \wedge \varphi \in T^+_{p+1}(\Omega)$;*

iii) *in particular, if $t_1 \in T^+_p(\Omega)$ and $t_j \in \Phi^+_1(\Omega)$, $j = 2, \ldots, q$ then $t_1 \wedge t_2 \wedge \ldots \wedge t_q \in T^+_{p+q}(\Omega)$.*

Proof. Let us first suppose that $t \in \Phi^+_p(\Omega)$ and $\varphi \in \Phi^+_1(\Omega)$. Then for $z_0 \in \Omega$ fixed, $\varphi(z_0) = i \sum_{j=1}^{n} C_j(z_0)\alpha'_j \wedge \bar{\alpha}'_j$ with $C_j(z_0) \geqq 0$, where the α'_j are (1,0) forms with constant coefficients. Thus, for any system $\alpha_1, \ldots, \alpha_{p-1}$,

$$t \wedge \varphi \wedge i\alpha_1 \wedge \bar{\alpha}_1 \wedge \ldots \wedge i\alpha_{n-p-1} \wedge \bar{\alpha}_{n-p-1} = \psi(z)\beta_n$$

with $\psi(z_0) \geqq 0$. Since this is true for every point in Ω, $t \wedge \varphi \in \Phi^+_{p+1}(\Omega)$. To prove the general case, we choose a sequence $t_m \in \Phi^+_p(\Omega)$ (resp. $\Phi^+_1(\Omega)$) such that $t_m \to t$, by Proposition 2.11. □

Definition 2.13. If $\varphi \in \mathscr{C}^\infty_{0,(p,q)}(\Omega)$, we define the *norm* of φ by $\|\varphi\| = \sup_{z \in \Omega} \sup_{I,J} |\varphi_{I,J}(z)|$ and the *modulus* of φ by $|\varphi|(z) = \sup_{I,J} |\varphi_{I,J}(z)|$.

For a measure μ defined in a domain Ω, we define $|\mu|_\Omega = \sup|\mu(f)|$, for $f \in \mathscr{C}_0(\Omega)$ and $|f|_\Omega = \sup\limits_{z\in\Omega} |f(z)| \leqq 1$.

Definition 2.14. Let t be a current defined in a domain Ω of C^n and continuous of order zero. We shall say that the positive measure μ *dominates* the current t if there exists a constant C_μ such that for every $\varphi \in \mathscr{C}^\infty_{0,(p,q)}(\Omega)$

$$|t(\varphi)| \leqq C_\mu \mu(|\varphi|).$$

Definition 2.15. Let $t \in T^+_{n-p}(\Omega)$ and $\sigma_t = t \wedge \beta_p$; then σ_t is called the trace of the current t.

Theorem 2.16. *Let $t \in T^+_{n-p}(\Omega)$. Then the coefficients $t_{I,J}$ of the current t in the representation* (2,7) *are associated with complex measures $T_{I,J}$ and there exists a constant $C(n,p)$ depending only of the dimensions n, p, such that*

$$|T_{I,J}|_{\Omega'} = |t_{I,J}\beta_n|_{\Omega'} \leqq C(n,p)|\sigma_t|_{\Omega'}$$

in any domain $\Omega' \subset \Omega$. If $\Lambda = \{L^p_s\}$, $s = 1, \ldots, N$ is a regular system, let $\mu_s = t \wedge \tau(L^p_s)$ be the system of the positive measures associated with Λ. Then there exist constants C_1, C_2 depending only on Λ, n, p, such that for $\varphi \in \mathscr{C}_{0,(p,q)}(\Omega)$

$$|t(\varphi)| \leqq C_1 \sum_{s=1}^{N} \mu_s(|\varphi|) \tag{2,9}$$

$$\sigma_t(|\varphi|) \leqq C_2 \sum_{s=1}^{N} \mu_s(|\varphi|) \leqq C_2 N \sigma_t(|\varphi|). \tag{2,10}$$

Thus $\sum \mu_s$ dominates t and σ_t dominates t.

Proof. Let Λ be a regular system. Then we can solve for $T_{I,J}$ as a linear combination of the positive measures $\mu_s = t \wedge \tau(L^p_s)$, $s = 1, \ldots, N$, that is $T_{I,J} = \sum_{s=1}^{N} C^s_{I,J}\mu_s$, where the $C^s_{I,J}$ are complex constants depending on Λ. This proves the first part of (2,10). Since β is invariant with respect to rotations and

$$\beta_p = \sum_I (i/2)^p \, dz_{i_1} \wedge \ldots \wedge dz_{i_p} \wedge d\bar{z}_{i_1} \wedge \ldots \wedge d\bar{z}_{i_p},$$

we see that $\sigma_t = t \wedge \beta_p \geqq t \wedge \tau(L^p_s) = \mu_s$ for all s, which terminates the proof of (2,10); (2,9) is an immediate consequence of (2,10). □

Corollary 2.17. *If $t \in T^+_{n-p}(\Omega)$, then* $\operatorname{supp} t = \operatorname{supp} \sigma_t$.

Proposition 2.18. *Let $t \in T^+_{n-p}(\Omega)$. Then for every system $\alpha_1, \ldots, \alpha_p$ of* (1,0) *forms with continuous coefficients whose support is compact, condition* (ii) *of Definition* 2.9 *holds.*

Proof. If $t \in \Phi_{n-p}^{+}(\Omega)$, then this is true, and according to Proposition 2.11 and Theorem 2.16, it remains true for $t \in T_{n-p}^{+}(\Omega)$ by passing to the limit. □

Remarks. From the definition and from the fact that positive currents are continuous of order zero, we see that

i) $T_p^{+}(\Omega)$ and $\Phi_p^{+}(\Omega)$ are cones over the set of positive continuous functions (i.e. $t_1, t_2 \in T_p^{+}(\Omega)$ and $\alpha_1, \alpha_2 \in \mathscr{C}^0(\Omega)$, $\alpha_1 \geqq 0$, $\alpha_2 \geqq 0$, then $\alpha_1 t_1 + \alpha_2 t_2 \in T_p^{+}(\Omega)$);

ii) Since the degree of a positive current is even, $\varphi \wedge \psi = \psi \wedge \varphi$ if $\varphi \in \Phi_p^{+}(\Omega)$ and $\psi \in \Phi_q^{+}(\Omega)$, but in general $\varphi \wedge \psi \notin \Phi_{p+q}^{+}(\Omega)$.

iii) A current $t = i \sum t_{p,q} dz_p \wedge d\bar{z}_q$ is positive if and only if the distribution $\theta(\lambda) = (\sum t_{p,q} \lambda_p \bar{\lambda}_q) \beta_n = \sum T_{p,q} \lambda_p \bar{\lambda}_q$ is a positive measure for $\lambda \in \mathbb{C}^n$; $T_{p,q} = t_{p,q} \beta_n$, is a complex measure associated with $t_{p,q}$; $T_{p,p} \geqq 0$ and $T_{p,q} = \bar{T}_{p,q}$ are complex conjugate measures. For forms, it is an obvious consequence of Proposition 2.5 and of remark 1 after Proposition 2.5. To prove the properties for a current t, apply Proposition 2.11 and the corresponding property of the forms.

iv) If ω is a homogeneous form of type $(p, 0)$, then $\varphi = k'_p \omega \wedge \bar{\omega}$ is a positive form (it is a consequence of the definition of positive forms). Then for $t \in T_1^{+}(\Omega)$ and $\varphi \in \mathscr{C}_{0,(p,p)}(\Omega)$, we have $t \wedge \varphi \in T_{p+1}^{+}(\Omega)$ if $\varphi = k'_p \sum_j c_j \omega_j \wedge \bar{\omega}_j$, $c_j \geqq 0$, $\omega_j \in \mathscr{E}_{0,(p,0)}(\Omega)$; $k'_p = 2^{-(n-p)}$ if $(n-p)$ is even and $k'_p = i 2^{-(n-p)}$ if $(n-p)$ is odd.

§3. Positive Closed Currents

If t is a current defined on a domain Ω, we define the operators d, ∂ and $\bar{\partial}$ on t by duality; if t is an element of $[\mathscr{C}_{0,(p,q)}^{\infty}(\Omega)]'$, then $dt(\varphi) = (-1)^{p+q} t(d\varphi)$, $\partial t(\varphi) = (-1)^{p+q} t(\partial \varphi)$, and $\bar{\partial} t(\varphi) = (-1)^{p+q} t(\bar{\partial}\varphi)$. If the current t is defined by integration of forms, that is $t(\varphi) = \int \psi \wedge \varphi$ for a form ψ, then $dt(\varphi) = (-1)^{p+q} t(d\varphi) = (-1)^{p+q} \int \psi \wedge d\varphi = \int d\psi \wedge \varphi$. Thus dt is represented by $d\psi$ and the definition is consistent.

Definition 2.19. We say that the current t is *closed* if $dt = 0$, that is $t(d\varphi) = 0$ for $\varphi \in \mathscr{E}_0^{\infty}(\Omega)$.

We will let $\tilde{T}_{n-p}^{+}(\Omega)$ represent the set of positive closed currents of degree $(n-p)$ in Ω.

Proposition 2.20. *It $t \in \tilde{T}_{n-p}^{+}(\Omega)$, then $\partial t = \bar{\partial} t = 0$.*

Proof. Since t is homogeneous of type $(n-p, n-p)$, and $\partial t(\varphi) = t(\partial \varphi)$, it is sufficient to prove $\partial t(\varphi) = 0$ for $\varphi \in \mathscr{C}_{0,(p-1,p)}^{\infty}(\Omega)$. In this case $0 = dt(\varphi) = t(d\varphi)$

$=t(\partial\varphi)+t(\bar{\partial}\varphi)$ and since $\bar{\partial}\varphi\in\mathscr{C}^\infty_{0,(p-1,p+1)}$, $t(\bar{\partial}\varphi)=0$ by i) of Definition 2.9. The proof for $\bar{\partial}t=0$ is similar.

Remark. If $\rho_\varepsilon\in\mathscr{C}^\infty_0(B(0,\varepsilon))$, supp $\varphi\in\Omega_\varepsilon$ and t is closed, then

$$d(t\star\rho_\varepsilon)(\varphi)=(t\star\rho_\varepsilon)(d\varphi)=t((d\varphi)\star\rho_\varepsilon)=t(d(\varphi\star\rho_\varepsilon))=0.$$

Thus, it follows from Proposition 2.11 that $t\in\tilde{T}^+_{n-p}(\Omega)$ is the limit on every compact subset of Ω of a sequence $\{t_m\}$ of elements in $\tilde{T}^+_{n-p}(\Omega)$ with $\mathscr{C}^\infty$ coefficients.

We set $\alpha_a=\frac{i}{2\pi}\partial\bar{\partial}\log\|z-a\|^2$ and $\gamma=\frac{i}{2}\partial\|z\|^2\wedge\bar{\partial}\|z\|^2$, and we will write α instead of α_0 for simplicity. The form α is associated with the positive semi-definite Hermitian form

$$ds^2=\|z\|^{-4}\left[\|z\|^2\sum_{k=1}^{n}dz_k d\bar{z}_k-\left(\sum_{k=1}^{n}\bar{z}_k dz_k\right)\left(\sum_{k=1}^{n}z_k d\bar{z}_k\right)\right]$$

which is the metric on the projective space $\mathbb{P}(\mathbb{C}^n)$ (cf. [*H*]). Moreover since $\log\|z-a\|^2\in\mathrm{PSH}(\mathbb{C}^n)$, the exterior differential form $i\partial\bar{\partial}\log\|z-a\|^2$ is positive in $\mathbb{C}^n-\{a\}$. Thus $\alpha_a\in\tilde{\Phi}^+_1(\mathbb{C}^n-\{a\})$, and it follows from Theorem 2.12 that $\alpha_a^p\in\tilde{\Phi}^+_p(\mathbb{C}^n-\{a\})$, since $d\alpha_a^p=d\alpha_a\wedge\alpha_a^{p-1}+\alpha_a\wedge d\alpha_a^{p-1}=0$ follows by induction. We obtain the simple expression:

$$\alpha^p=\pi^{-p}[\|z\|^{-2p}\beta^p-p\|z\|^{-2p-2}\beta^{p-1}\wedge\gamma], \tag{2,11}$$

since $\gamma\wedge\gamma=0$.

Proposition 2.21. *We have* $\alpha_a^n=0$ *in* $\mathbb{C}^n-\{a\}$.

Proof. Let $\omega_i=\{z\in\mathbb{C}^n\colon z_i\neq 0\}$ and let $\xi_k=\frac{z_k}{z_i}$, $k\neq i$. Then

$$\alpha=\frac{i}{2\pi}\left[\partial\bar{\partial}\log|z_i|^2+\partial_\xi\bar{\partial}_\xi\log\left(1+\sum_{k=1}^{n-1}\xi_k\bar{\xi}_k\right)\right].$$

Since $\partial\bar{\partial}\log|z_i|=0$ for $z_i\neq 0$, α^n is a form of type (n,n) in the space $\mathbb{C}^{n-1}$ and hence $\alpha^n=0$. Since $\mathbb{C}^n-\{0\}=\bigcup_{i=1}^{n}\omega_i$, $\alpha^n=0$ in $\mathbb{C}^n-\{0\}$. Since ∂ and $\bar{\partial}$ are translation invariant, $\alpha_a^n=0$ in $\mathbb{C}^n-\{a\}$. □

Let $t\in\tilde{T}^+_{n-p}(\Omega)$ and suppose $0\in\Omega$. We set

$$v_t=t\wedge\alpha^p=t\wedge(i/2\pi\partial\bar{\partial}\log\|z\|^2)^p. \tag{2,12}$$

It follows from Theorem 2.13 that v_t is a positive (n,n) current, hence a measure, in $\Omega-\{0\}$.

Theorem 2.22. *Let* $t \in \tilde{T}^{+}_{n-p}(\Omega)$, $0 \in \Omega$ *and* $B(0,R) \Subset \Omega$. *Let* $r_1 < r < r_2 < R$, $\sigma_t(r) = \int\limits_{\|z\| \leqq r} t \wedge \beta_p$ *and* $v_t(r_1, r_2) = \int\limits_{r_1 < \|z\| \leqq r_2} t \wedge \alpha^p$. *Then*

$$v_t(r_1, r_2) = \tau_{2p}^{-1} \left[\frac{\sigma_t(r_2)}{r_2^{2p}} - \frac{\sigma_t(r_1)}{r_1^{2p}} \right]. \tag{2,13}$$

Proof. We first assume that t has $\mathscr{C}^\infty$ coefficients. Then, since $dt = 0$, by Poincaré's Lemma, there exists θ such that $d\theta = t$. Thus, we obtain by Stokes' Theorem

$$\begin{aligned} v_t(r_1, r_2) &= \int\limits_{r_1 < \|z\| \leqq r_2} t \wedge \alpha^p = \int\limits_{r_1 < \|z\| \leqq r_2} d\theta \wedge \alpha^p = \int\limits_{r_1 < \|z\| \leqq r_2} d(\theta \wedge \alpha^p) \\ &= \int\limits_{\|z\| = r_2} \theta \wedge \alpha^p - \int\limits_{\|z\| = r_1} \theta \wedge \alpha^p. \end{aligned}$$

On a surface $\|z\| = \text{constant}$, $d\|z\|^2 = \partial \|z\|^2 + \bar{\partial} \|z\|^2 = 0$, so $\gamma = \frac{i}{2} \partial \|z\|^2 \wedge \bar{\partial} \|z\|^2 = 0$ and (2,11) shows that $\alpha^p = \pi^{-p} \|z\|^{-2p} \beta^p$. Hence for $j = 1, 2$, by Stokes' Theorem

$$\begin{aligned} \int\limits_{\|z\| = r_j} \theta \wedge \alpha^p &= \tau_{2p}^{-1} r_j^{-2p} \int\limits_{\|z\| = r_j} \theta \wedge \beta_p = \tau_{2p}^{-1} r_j^{-2p} \int\limits_{B(0, r_j)} t \wedge \beta_p \\ &= \tau_{2p}^{-1} r_j^{-2p} \sigma(r_j), \end{aligned}$$

which gives (2,13) for this case. For the general case, we consider $t_\varepsilon = t \star \rho_\varepsilon$ constructed in Proposition 2.11. Since $\sigma_t(r)$ is the measure carried by the compact ball $\overline{B(0,r)}$, we have $\sigma_t(r) \leqq \sigma_{t_\varepsilon}(r + \varepsilon) \leqq \sigma_t(r + 2\varepsilon)$, and hence $\lim\limits_{\varepsilon \to 0} \sigma_{t_\varepsilon}(r + \varepsilon) = \sigma_t(r)$. We apply (2,13) to t_ε and the balls $\overline{B(0, r_1 + \varepsilon)}$, $\overline{B(0, r_2 + \varepsilon)}$ and let ε go to zero. □

Theorem 2.23. *Let* $t \in \tilde{T}^{+}_{n-p}(\Omega)$ *and suppose* $a \in \Omega$. *Let* $\sigma_t(a, r) = \int\limits_{\|z - a\| \leqq r} t \wedge \beta_p$ *for* $r < d_\Omega(a)$ *and let* v_t^a *be the positive measure defined by* (2,12) *relative to the point* a (i.e. $v_t^a = t \wedge \alpha_a^p$). *Then*

i) $r^{-2p} \sigma_t(a, r)$ *is an increasing function of* r *for* $r < d_\Omega(a)$ *and*

$$\lim_{r \to 0} \tau_{2p}^{-1} r^{-2p} \sigma_t(a, r) = v_t(a) \text{ exists and is positive;} \tag{2,14}$$

ii) v_t^a *can be extended to all of* Ω, *that is it remains bounded in every neighborhood of* a;

iii) *if we extend* v_t^a *as the point mass* $v_t(a) \delta(a)$, *where* $v_t(a)$ *is defined by* (2,14), *then*

$$\int\limits_{\|z - a\| \leqq r} v_t^a = \tau_{2p}^{-1} r^{-2p} \sigma_t(a, r).$$

Proof. Since $v_t(r_1, r_2) \geqq 0$, it follows from (2,13) that $\tau_{2p}^{-1} r^{-2p} \sigma_t(a, r)$ is increasing and that the limit in (2,14) exists. The same formula shows that

$$\lim_{\varepsilon\to 0} \nu_t(\varepsilon, r) = \tau_{2p}^{-1} r^{-2p} \sigma_t(a, r) - \nu_t(a) = \int_{0<\|z-a\|\leq r} \nu_t^a$$

or equivalently

$$\nu_t(a) + \int_{0<\|z-a\|\leq r} \nu_t^a = \tau_{2p}^{-1} r^{-2p} \sigma_t(a, r) = \nu_t^a(r). \qquad \square$$

The existence of the number $\nu_t(a)$, called the *Lelong number* of the current t, is an essential property of positive closed currents.

Definition 2.24. Let $t \in T_{n-p}^+(\mathbb{C}^n)$. We call the function

(2,15) $$\nu_t(r) = \tau_{2p}^{-1} r^{-2p} \sigma_t(0, r)$$

the indicator of growth function of t; it is the mass of the measure ν_t carried by the ball $\overline{B(0, r)}$.

§4. Positive Closed Currents of Degree 1

Proposition 2.25. *Let $V \in \mathrm{PSH}(\Omega)$ for Ω a domain in $\mathbb{C}^n$. Then*

$$t = i\partial\bar{\partial} V = i \sum_{j,k} \frac{\partial^2 V}{\partial z_j \partial \bar{z}_k} dz_j \wedge d\bar{z}_k$$

taken as a distribution defines an element of $\tilde{T}_1^+(\Omega)$.

Proof. First since V belongs to $L^1_{\mathrm{loc}}(\Omega)$ the derivatives $\dfrac{\partial^2 V}{\partial z_p \partial \bar{z}_q}$ are defined as distributions and $t = i\partial\bar{\partial} V$ is defined as a current. It is closed since for $\varphi \in \mathscr{E}_0(\Omega)$ we have

$$dt(\varphi) = t(d\varphi) = i\int \partial\bar{\partial} V \wedge d\varphi = i\int d\bar{\partial} V \wedge d\varphi = i\int -\bar{\partial} V \wedge dd\varphi = 0.$$

To prove the positivity of t, let us first consider $V \in \mathscr{C}^2 \cap \mathrm{PSH}(\Omega)$; since $t = i\partial\bar{\partial} V$ is a form, its restriction to a complex line $z = z_0 + wu$, $u \in \mathbb{C}$, for given $w \in \mathbb{C}^n$ and $z_0 \in \Omega$ is $i\left[\sum \dfrac{\partial^2 V}{\partial z_p \partial \bar{z}_q} w_p \bar{w}_q\right] du \wedge d\bar{u} = h(V, w)\, i\, du \wedge d\bar{u}$. The Hermitian form $h(V, w)$ is positive semi-definite, because $V_w(u) = V(z_0 + wu)$ is subharmonic of class $\mathscr{C}^2$ and $\Delta_u V_w = 4h(V, w) \geq 0$ for all $w \in \mathbb{C}^n$. Then the exterior form $i\partial\bar{\partial} V$ is positive (see Remark (iii) after Proposition 2.18). For the general case we proceed by regularization using the convoluter ρ_ε (see Proposition 2.11): $t_\varepsilon = i\partial\bar{\partial} V_\varepsilon$, for $V_\varepsilon = V \star \rho_\varepsilon$ is positive by the preceeding argument and $V_\varepsilon \in \mathrm{PSH}(\Omega_\varepsilon)$ where $\Omega_\varepsilon \Subset \Omega$ is the open set $\{z \in \Omega : d_\Omega(z) > \varepsilon\}$. Then on each $\Omega' \Subset \Omega$, $t = \lim_{\varepsilon \to 0} t_\varepsilon$ is the weak limit of the positive currents t_ε; it is a positive current and $t \in \tilde{T}_1^+(\Omega)$. $\square$

We shall now show that the converse of this is true, at least locally, that is, for every $t \in \tilde{T}_1^+(\Omega)$ and every $z \in \Omega$, there exists a neighborhood U_z of every $z \in \Omega$ such that $t = i\partial\bar{\partial}V$ for $V \in \mathrm{PSH}(U_z)$. For the proof, we shall use an integral operator which gives solutions of the $\bar{\partial}$-equation with special regularity properties.

Let $k(z) \in \mathscr{C}^2(\mathbb{C}^n)$ be a strictly convex function, that is

$$\sum_{i,j=1}^{2n} \frac{\partial^2 k(z)}{\partial x_i \partial x_j} t_i t_j \geqq C(z)\|t\|^2$$

for every vector $t \in \mathbb{R}^{2n}$ and some function $C(z) > 0$, where $x = (x_1, \ldots, x_{2n})$ are the underlying real coordinates. Let $\Omega = \{z \in \mathbb{C}^n : k(z) < 0\}$; we orient the boundary of Ω, $bd\Omega$, so that Stokes' Theorem is valid. Let $\chi(t)$ be a non-negative $\mathscr{C}^\infty$ function of the real variable t such that $\chi(t) \equiv 0$ for $t \leqq 1/2$ and $\chi(t) \equiv 1$ for $t \geqq 3/4$ and set $\varphi(z, \xi) = \chi\left(\frac{k(\xi)}{k(z)}\right)$, $g_i(z, \xi) = (\bar{\xi}_i - \bar{z}_i)\varphi(z, \xi) + (1 - \varphi(z, \xi))\frac{\partial k}{\partial \xi_i}(\xi)$.

For ξ fixed, the set $K_\xi = \{z : k(z) < k(\xi)\}$ is strictly convex and contains $\operatorname{supp}\varphi$; let us consider the tangent plane $\mathbb{R}e \sum_{i=1}^n (\xi_i^0 - z_i)\frac{\partial k}{\partial \xi_i}(\xi^0) = 0$ for $\xi^0 \in bdK_\xi$. Then $\mathbb{R}e \sum_{i=1}^n (\xi_i^0 - z_i)\frac{\partial k}{\partial \xi_i}(\xi^0) > 0$ for $z \in K_\xi$.

Thus $\mathbb{R}e \sum_{i=1}^n (\xi_i - z_i) g_i(z, \xi) > 0$ for $z \in \operatorname{supp}\varphi$. We set

$$K(z, \xi) = C_n \frac{\sum_{j=1}^n (-1)^{j+1} g_j \bigwedge_{\substack{i=1 \\ i \neq j}}^n \bar{\partial}_\xi g_i \bigwedge_{i=1}^n d\xi_i}{\left[\sum_{i=1}^n (\xi_i - z_i) g_i\right]^n}$$

where $C_n = \frac{(n-1)!(-1)^{\frac{n(n-1)}{2}}}{(2\pi i)^n}$. An easy calculation shows that $\bar{\partial}_\xi K(z, \xi) = 0$ whenever the denominator is different from zero, that is for $\xi \neq z$. We note that $K(z, \xi) = -\frac{(n-2)!\, i\partial_\xi}{2\pi^n}\|z - \xi\|^{2-2n} \wedge \beta_{n-1}$ for $k(\xi) > \frac{1}{2}k(z)$, and in particular, this holds if ξ is in a small enough neighborhood of z.

Theorem 2.26 (Cauchy-Fantappié Formula). *Let $k(z) \in \mathscr{C}^2(\mathbb{C}^n)$ be a strictly convex function and $\Omega = \{z \in \mathbb{C}^n : k(z) < 0\}$. Let $h(z)$ be a function in $\mathscr{C}^1$ in a neighborhood of $\bar{\Omega}$. Then*

$$h(z) = -\int_{bd\Omega} h(\xi) K(z, \xi) + \int_\Omega \bar{\partial} h(\xi) \wedge K(z, \xi).$$

Proof. Let z be fixed. Applying Stokes' Theorem, we obtain

$$\begin{aligned}\int_{\Omega} \bar{\partial} h(\xi) \wedge K(z, \xi) &= \lim_{\varepsilon \to 0} \int_{\Omega - B(z,\varepsilon)} \bar{\partial} h(\xi) \wedge K(z, \xi) \\ &= \lim_{\varepsilon \to 0} \int_{\Omega - B(z,\varepsilon)} d(h(\xi) K(z, \xi)) \\ &= \int_{bd\Omega} h(\xi) K(z, \xi) + \lim_{\varepsilon \to 0} \int_{bd B(z,\xi)} h(\xi) K(z, \xi) \\ &= \int_{bd\Omega} h(\xi) K(z, \xi) + h(z)\end{aligned}$$

since

$$\lim_{\varepsilon \to 0} \int_{bd B(z,\varepsilon)} \frac{-(n-2)!\, i\partial_{\xi}}{2\pi^n} \|z-\xi\|^{2-2n} \wedge \beta_{n-1} = 1. \qquad \square$$

Corollary 2.27. *Let β be a $(0,1)$ form in a neighborhood of $\overline{B(0,1)}$ with $\mathscr{C}^\infty$ coefficients such that $\bar{\partial}\beta = 0$ and let*

$$\alpha(z) = \int_{B(0,1)} \beta(\xi) \wedge K(z, \xi).$$

Then $\bar{\partial}\alpha = \beta$ in $B(0,1)$.

Proof. Choose $k(z) = \|z\|^2 - 1$. Let $\tilde{\alpha}$ be any $\mathscr{C}^\infty$ function such that $\bar{\partial}\tilde{\alpha}(z) = \beta(z)$ in a neighborhood of $\overline{B(0,1)}$ (cf. Appendix III). Then

$$\tilde{\alpha}(z) = -\int_{bdB} \tilde{\alpha}(\xi) K(z, \xi) + \int_{B} \beta(z) \wedge K(z, \xi).$$

But $K(z, \xi)$ is holomorphic in $z \in B$ for $\xi \in bdB$. Thus $\bar{\partial}\alpha(z) = \beta(z)$. $\square$

Theorem 2.28. *Let θ be a positive closed current of degree 1 defined in a neighborhood of $\overline{B(0,1)}$. Then there exists a plurisubharmonic function V defined in $B(0,1)$ such that $i\partial\bar{\partial} V = \theta$ as a current.*

Proof. Let $\theta = i \sum_{k,j=1}^{n} \theta_{kj} dz_k \wedge d\bar{z}_j$ and set $\theta^{\varepsilon}_{kj} = \theta_{kj} \star \rho_\varepsilon$, which, for ε small enough, is $\mathscr{C}^\infty$ in a neighborhood of $\overline{B(0,1)}$. Let $\theta^\varepsilon = \theta \star \rho_\varepsilon$. Since θ is a positive closed current, $d\theta^\varepsilon = 0$, and we find v_1^ε and v_2^ε, a $(0,1)$ and $(1,0)$ form respectively, such that $d(v_2^\varepsilon - v_1^\varepsilon) = \theta^\varepsilon$. We can calculate v_2^ε and v_1^ε explicitly in terms of the coefficients of θ^ε:

$$v_2^\varepsilon = \sum_{j,k} \left[\int_0^1 t\,\theta^\varepsilon_{jk}(tz)\,dt\right] z_j d\bar{z}_k; \quad v_1^\varepsilon = \sum_{j,k} \left[\int_0^1 t\,\theta^\varepsilon_{jk}(tz)\right] \bar{z}_k dz_j.$$

It follows from degree considerations that $\bar{\partial} v_2^\varepsilon = \partial v_1^\varepsilon = 0$, and $v_2^\varepsilon = \bar{v}_1^\varepsilon$. Let

$$V_\varepsilon = [\textstyle\int v_2^\varepsilon(\xi) \wedge K(z, \xi) + \int v_1^\varepsilon(\xi) \wedge \overline{K(z, \xi)}] = 2\,\mathbb{R}e \int v_2^\varepsilon(\xi) \wedge K(z, \xi).$$

Then $i\partial\bar{\partial} V_\varepsilon = i[\partial v_2^\varepsilon - \bar{\partial} v_1^\varepsilon] = i\,d(v_2^\varepsilon - v_1^\varepsilon) = \theta^\varepsilon$.

Let $\sigma=\sum_{i=1}^{n}\theta_{ii}\,d\tau$, $\sigma^{\varepsilon}=\sum_{i=1}^{n}\theta_{ii}^{\varepsilon}\,d\tau$. Choose δ so small that θ is defined in a neighborhood of $\overline{B(0,1+\delta)}$ and set $\tilde{V}(z)=-\int_{B(0,1)}\frac{(n-2)!}{2\pi^n}\|z-\xi\|^{2-2n}d\sigma(\xi)$. Then $\tilde{V}(z)$ and $\tilde{V}_{\varepsilon}(z)=\tilde{V}\star\rho_{\varepsilon}$ are subharmonic functions such that $\tilde{V}_{\varepsilon}(z)$ decreases to $\tilde{V}(z)$ as ε goes to zero.

By Stokes' Theorem, we see that, setting $C_n'=\frac{(n-2)!}{2\pi^n}$,

$$\begin{aligned}\int_{bdB(0,1)} & -iC_n'\|z-\xi\|^{2-2n}v_2^{\varepsilon}\wedge\beta_{n-1}\\ &=iC_n'\int_{B(0,1)}\partial v_2^{\varepsilon}\wedge(-\|z-\xi\|^{2-2n})\wedge\beta_{n-1}\\ &\quad+iC_n'\int_{B(0,1)}v_2^{\varepsilon}\wedge\partial\|z-\xi\|^{2-2n}\wedge\beta_{n-1}\\ &=C_n'\int_{B(0,1)}-\|z-\xi\|^{2-2n}d\sigma^{\varepsilon}+iC_n'\int_{B(0,1)}v_2^{\varepsilon}\wedge\partial\|z-\xi\|^{2-2n}\wedge\beta_{n-1}.\end{aligned}$$

Hence

$$\begin{aligned}(2,16)\quad V_{\varepsilon}(z)=2\mathbb{R}e\int_{B(0,1)} & v_2^{\varepsilon}(\xi)\wedge[K(z,\xi)\\ &+iC_n'\partial\|z-\xi\|^{2-2n}\wedge\beta_{n-1}]+2C_n'\int_{B(0,1)}-\|z-\xi\|^{2-2n}d\sigma^{\varepsilon}\\ &+2\mathbb{R}e\int_{bdB(0,1)}iC_n'\|z-\xi\|^{2-2n}v_2^{\varepsilon}\wedge\beta_{n-1}.\end{aligned}$$

Since $\int_{B(0,1)}d\sigma^{\varepsilon}$ is bounded independently of ε, the coefficients of v_2^{ε} have bounded L^1 norms on $bdB(0,1)$ independently of ε, so we can find a sequence $\varepsilon_m\downarrow 0$ such that each coefficient of $v_2^{\varepsilon_m}$ converges weakly to a measure. Since $K(z,\xi)=-C_n'\partial\|z-\xi\|^{2-2n}\wedge\beta_{n-1}$ in a neighborhood of z and since $C_n'\int_{B(0,1)}-\|z-\xi\|^{2-2n}d\sigma^{\varepsilon}\geqq\tilde{V}_{\varepsilon}(z)-C$ for some constant $C<+\infty$, we can find a subsequence ε_m' of ε_m such that $V_{\varepsilon_m'}\to V$ pointwise. Furthermore, we have

$$(2,17)\qquad V_{\varepsilon}(z)\geqq C_K+\tilde{V}_{\varepsilon}(z)\geqq C_K+\tilde{V}(z)\qquad\text{for } z\in K\Subset B(0,1).$$

Thus, if $\varphi\in\mathscr{C}^{\infty}_{0,(n-1,n-1)}(\Omega)$, it follows from (2,17) and the Lebesgue Dominated Convergence Theorem that

$$\begin{aligned}i\partial\bar{\partial}V(\varphi)&=\int_{\Omega}Vi\partial\bar{\partial}\varphi=\lim_{\varepsilon\to0}\int_{\Omega}V_{\varepsilon}i\partial\bar{\partial}\varphi=\lim_{\varepsilon\to0}\int_{\Omega}i\partial\bar{\partial}V_{\varepsilon}\wedge\varphi\\ &=\lim_{\varepsilon\to0}\int_{\Omega}\theta^{\varepsilon}\wedge\varphi=\int_{\Omega}\theta\wedge\varphi.\end{aligned}$$

□

Remark. We sketch here a second proof of Theorem 2.28 which is of historical interest and provides motivation for much of the technics to be developped in Chapter 3.

1) Suppose that θ' is a (1,1) form whose coefficients $\theta'_{p,q}$ are polynomials in the $2n$ real variables x, y;

(a) $$\theta'_{p,q} = \sum_{\lambda,\mu} P_{p,q,\lambda}(z) Q_{p,q,\mu}(\bar{z})$$

where the $P_{p,q,\lambda}$ and $Q_{p,q,\mu}$ are homogeneous polynomials of degree λ and μ respectively. If there exists a solution V of the equations

(b) $$\frac{\partial^2 V}{\partial z_p \partial \bar{z}_q} = \theta'_{p,q}$$

then we set $W(z, \bar{z}, t, t') = V(zt, \bar{z}t')$ for $t, t' \in \mathbb{C}$. It then follows from (b) that $\frac{\partial^2 W}{\partial t \partial t'}(z, \bar{z}, t, t') = \sum_{p,q} \theta'_{p,q}(tz, t'z) z_p \bar{z}_q$, which is equivalent to

$$W(z, \bar{z}, 1, 1) = V(z, \bar{z}) = \sum_{\lambda,\mu} (\lambda+1)^{-1}(\mu+1)^{-1} P_{p,q,\lambda}(z) Q_{p,q,\mu}(\bar{z}).$$

It follows from the compatibility conditions

(c) $$\left\{\frac{\partial \theta'_{p,q}}{\partial z_s} = \frac{\partial \theta'_{s,q}}{\partial z_p} \text{ and } \frac{\partial \theta'_{p,q}}{\partial \bar{z}_k} = \frac{\partial \theta'_{p,k}}{\partial \bar{z}_q}\right\}$$

and power series arguments that V is a solution of (b). This technic is due to H. Poincaré. A simple adaptation of the above argument shows that the conclusion still holds in the case of $\theta'_{p,q}$ defined and real analytic in a ball $B(0, R)$. The solution V will then also be given in a neighborhood of the origin.

2) To treat the general case where $\theta = i \sum \theta_{p,q} dz_p \wedge d\bar{z}_q$ is a positive closed current, we set $\sigma = \frac{(n-2)!}{\pi^{n-1}} \sum_{m=1}^{n} \theta_{m,m}$, the trace of θ, and we form the potential

$$U(z) = -\int_{\|a\| \leqq R+\delta} \|z-a\|^{-2n+2} d\sigma(a)$$

where θ is defined in the ball of radius $R+2\delta$. Then $\Delta U = \sum_{m=1}^{n} \theta_{m,m}$. Set $\theta'_{p,q} = \theta_{p,q} - \frac{\partial^2 U}{\partial z_p \partial \bar{z}_q}$. Then $\theta'_{p,q}$ is a distribution represented by a harmonic function in $B(0, R)$ since

$$\begin{aligned}\frac{\partial^2 \theta'_{p,q}}{\partial z_m \partial \bar{z}_m} &= \frac{\partial^2}{\partial z_m \partial \bar{z}_m} \theta_{p,q} - \frac{\partial^2}{\partial z_m \partial \bar{z}_m} \frac{\partial^2 U}{\partial z_p \partial \bar{z}_q} \\ &= \frac{\partial^2}{\partial z_p \partial \bar{z}_q} \theta_{m,m} - \frac{\partial^2}{\partial z_p \partial \bar{z}_q} \frac{\partial^2 U}{\partial z_m \partial \bar{z}_m}\end{aligned}$$

by (c). Thus $\sum_{m=1}^{n} \frac{\partial^2 \theta'_{p,q}}{\partial z_m \partial \bar{z}_m} = 0$. By (1), there exists R', $0 < R' < R$ such that we can find V' solution of $\frac{\partial^2 V'}{\partial z_p \partial \bar{z}_q} = \theta'_{p,q}$ in $B(0, R')$. We then set $V = U + V'$ in $B(0, R')$.

Proposition 2.29. *Let h be a pluriharmonic function defined in a neighborhood of the ball $B(0,1)$. Then there exists a function f holomorphic in $B(0,1)$ such that $h = \mathbb{R}e\, f$.*

Proof. If h is pluriharmonic, then $\partial\bar{\partial}h = 0$ and so $d(\partial h - \bar{\partial}h) = 0$. Thus, by Poincaré's Lemma, there exists a function g such that $dg = \partial h - \bar{\partial}h$. Since h is real valued, $\bar{\partial}h = \overline{\partial h}$. Hence $i dg = i\partial h - \overline{i\partial h}$, which implies that $i\bar{\partial}g = -i\bar{\partial}h = \overline{(i\partial h)} = \overline{(i\partial g)}$, so that ig is also real valued. If $f = h + g$, then $\bar{\partial}f = \bar{\partial}h + \bar{\partial}g = \bar{\partial}h - \bar{\partial}h = 0$, so f is holomorphic and $h = \mathbb{R}e\, f$. □

Corollary 2.30. *Let θ be a positive closed current of degree 1 in $\mathbb{C}^n$. Then there exists $V \in \mathrm{PSH}(\mathbb{C}^n)$ such that $i\partial\bar{\partial}V = \theta$.*

Proof. By Theorem 2.29, we can find V_m in $B(0,m)$ such that $i\partial\bar{\partial}V_m = \theta$. Then $V_{m+1} - V_m$ is pluriharmonic in $B(0,m)$, hence the real part of a function $h_m(z) \in \mathscr{H}(B(0,m))$. Since the entire functions are dense in $\mathscr{H}(B(0,m))$, we can find $h'_m(z) \in \mathscr{H}(\mathbb{C}^n)$ such that $|h_m(z) - h'_m(z)| < 2^{-m}$ on $B(0,m-1)$. Set $V_0 \equiv 0$; we set $V_m \equiv 0$ on $\complement B(0,m)$, and $V(z) = \sum_{k=0}^{\infty} [(V_{k+1}(z) - V_k(z)) - \mathbb{R}e\, h'_k(z)]$. Then $V \in \mathrm{PSH}(\mathbb{C}^n)$, since in $B(0,m)$,

$$V(z) = V_m(z) + \mathbb{R}e \sum_{k=m+1}^{\infty} [h_k(z) - h'_k(z)] - \sum_{k=1}^{m} \mathbb{R}e\, h'_m(z). \qquad \square$$

If V is a plurisubharmonic function with $t = i\partial\bar{\partial}V$, then $\sigma_t = t \wedge \beta_{n-1} = \frac{1}{2\pi}\Delta V$. This permits an easy calculation of $v_t(a)$:

Proposition 2.31. *Let $V \in \mathrm{PSH}(\Omega)$. Then if $t = i\partial\bar{\partial}V$*

i) $\sigma_t = \frac{1}{2\pi}\Delta V$;

ii) *$v_t(a)$ is the density (in real dimension $(2n-2)$) of the measure $1/2\pi\Delta V$ computed on balls centered at a;*

iii) $v_t(a) = \lim_{r\to 0} \frac{\partial}{\partial \log r} \lambda(a,r,V) = \lim_{r\to 0} \frac{\lambda(a,r,V)}{\log r}$.

Proof. From Gauss' Theorem, we have

$$\int_{B(0,r)} \Delta V \beta_n = \omega_{2n} r^{2n-1} \frac{\partial}{\partial r} \lambda(a,r,V)$$

$$= 2\pi(\tau_{2n-2} r^{2n-2}) \frac{\partial}{\partial \log r} \lambda(a,r,V).$$

The result now follows from the fact that $\lambda(a,r,V)$ is an increasing convex function of $\log r$ (Proposition I.17). □

Proposition 2.32. *Let* $F_s(z_1, \ldots, z_n) \in \mathscr{H}(\Omega)$, $s=1, \ldots, m$ *and let*

$$V = \tfrac{1}{2} \log \left(\sum_{s=1}^{m} |F_s(s)|^2 \right) \in \mathrm{PSH}(\Omega), \qquad t = i\partial\bar{\partial} V.$$

Then

(2,18) $\nu_t(a) = \min_s \nu_s$, *where* ν_s = *multiplicity of the zero of* F_s *at* a. *In particular, if* $V = \log|F(z)|$, ν_t *is the multiplicity of the zero at* a, *that is the degree* $q(a)$ *of the first homogeneous polynomial in the Taylor series expansion of* F *at* a *which is not identically zero (i.e.* $F(z+a) = \sum_{m \geq q(a)} P_m(z)$*).*

Proof. We suppose that $a=0$. Then $F_s(uz_1, \ldots, uz_n) = u^\nu F_s'(u, z_1, \ldots, z_n)$ where F_s' is holomorphic in the $(n+1)$ variables (u, z). Thus

$$V(u, z) = \nu \log|u| + \tfrac{1}{2} \log \left(\sum_{s=1}^{m} |F_s'|^2 \right) = \nu \log|u| + W(u, z),$$

where W is a plurisubharmonic function of (u, z). For $u=0$,

$$\psi(z) = \sum_{s=1}^{m} |F_s'(0, z)|^2 \not\equiv 0.$$

Thus, when r goes to zero $\lambda(0, r, V) = \nu \log r + \frac{1}{2} A + \varepsilon(r)$ where $A = \lambda(0, 1, \log\psi) > -\infty$ and $\varepsilon(r)$ goes to zero with r, from which (2,18) follows. □

§ 5. Analytic Varieties and Currents of Integration

Since we are interested in studying the properties of the common zero set of several holomorphic functions, we recall here some of the complex analytic structure of such sets.

Definition 2.33. Let $\Omega \subset \mathbb{C}^n$ be a domain. A closed subset $M \subset \Omega$ is said to be a *complex submanifold of dimension* p if for every $z \in M$, there exists a neighborhood $U_z \subset \Omega$ of z and holomorphic functions $f_i \in \mathscr{H}(U_z)$, $i=1, \ldots, n$ such that

$$M \cap U_z = \{z' : f_1(z') = \ldots = f_{n-p}(z') = 0\} \quad \text{and} \quad \operatorname{rank} \left[\frac{\partial f_i}{\partial z_j}(z) \right]_{i,j=1}^{n} = n-p.$$

Definition 2.34. Let $\Omega \subset \mathbb{C}^n$ be a domain. A subset $Y \subset \Omega$ is an *analytic variety* in Ω if for every $z \in \Omega$, there exists a neighborhood U_z and functions $f_i \in \mathscr{H}(U_z)$, $i=1, \ldots, t_z$ such that $Y \cap U_z = \{z' \in U_z : f_1(z') = \ldots = f_{t_z}(z') = 0\}$.

If $U_z \cap U_{\tilde{z}} \neq \emptyset$, the set $Y \cap U_z \cap U_{\tilde{z}}$ is defined by two different systems, $f_j(z') = 0$ in U_z and $g_j(z') = 0$ in $U_{\tilde{z}}$, and the number of the equations can be different.

Definition 2.35. Let Ω be a domain in $\mathbb{C}^n$ and $Y \subset \Omega$ an analytic variety. Then for $z \in Y$, the *complex dimension of* Y *at* z, $\dim_z Y$, is the minimal number of linear equations $\sum_{i=1}^{n} a_{is}(z'_i - z_i) = 0$ which, when added to the $f_j = 0$, $j = 1, \ldots, t_z$, give z as an isolated solution of the system of equations $\left\{ f_j(z') = 0, \ \sum_{i=1}^{n} a_{is}(z'_i - z_i) = 0 \right\}$. We say that Y is of *pure dimension* p if $\dim_z Y = p$ for all $z \in Y$.

Remark. If Y is of pure dimension 0 in Ω, then Y is just a discrete set of points.

Definition 2.36. If Ω is a domain in $\mathbb{C}^n$ and Y is an analytic variety in Ω, then Y is said to be a *complete intersection* if Y is of pure dimension p and $Y = \{z \in \Omega \colon f_1(z) = \ldots = f_{n-p} = 0, \ f_i \in \mathscr{H}(\Omega)\}$.

Definition 2.37. An analytic variety $Y \subset \Omega$ is said to be *irreducible* in Ω if for every pair of analytic varieties Y_1 and Y_2 such that $Y = Y_1 \cup Y_2$, either $Y = Y_1$ of $Y = Y_2$.

The complex structure of analytic varieties can now be resumed as follows (cf. [A, B]): suppose that Y is an analytic variety in $\Omega \subset \mathbb{C}^n$; then Y is a union (finite in each $\Omega' \Subset \Omega$) of irreducible analytic varieties Y_k such that

i) there exists a proper subvariety $Y'_k \subset Y_k$ called the set of the *singular points* of Y_k such that $\tilde{Y}_k = Y_k - Y'_k$ is a connected analytic complex submanifold of $(\Omega - Y'_k)$ of dimension p_k; $z \in \tilde{Y}_k$ is called a regular point of Y_k;

ii) for Ω' relatively compact in Ω, $Y_k \cap \Omega' = \emptyset$ for $k \geqq k_{\Omega'}$.

iii) $\dim Y = \sup \dim Y_k$, and the closed set $Y' = \bigcup_k Y'_k \bigcup_{\delta \neq j} (Y_\delta \cap Y_j)$ is an analytic subvariety of dimension at most $(\dim Y - 1)$; $Y - Y'$ is a union of disjoint complex manifolds in $(\Omega - Y')$.

We note that if $Y = \{z \in \Omega \colon f_i(z) = 0, \ i = 1, \ldots, p, \ f_i \in \mathscr{H}(\Omega)\}$ is defined by p holomorphic functions, then $\dim_z \tilde{Y}_k \geqq (n-p)$ for all z. In particular, for a holomorphic function $f \not\equiv 0$, $f \in \mathscr{H}(\Omega)$, if $Y_f = \{z \in \Omega \colon f(z) = 0\}$, then the dimension of the regular points of Y_f is exactly $(n-1)$.

Definition 2.38. Let $\Omega \subset \mathbb{C}^n$ be a domain. Then a Weierstrass pseudo-polynomial $P(u; z) \in \mathscr{H}(\mathbb{C} \times \Omega)$ is a function $P(u; z) = u^k + \sum_{i=0}^{k-1} a_i(z) u^i$, $a_i \in \mathscr{H}(\Omega)$, $a_i(0) = 0$.

Now, we recall a classical result.

Proposition 2.39 (Weierstrass Preparation Theorem, cf. [A, B]). *Let f be holomorphic in a neighborhood Ω of 0 in C^n and assume that $z_n^{-p} f(0, z_n)$ is holo-*

morphic and does not vanish at 0. *Then we can write f in one and only one way in the form $f=hP_p$, where h and P are holomorphic in a neighborhood of* 0, $h(0)\neq 0$, *and P_p is a Weierstrass polynomial, that is:* $P(z)=z_n^p+\sum_0^{p-1} a_j(z')z_n^j$, *where the a_j are holomorphic functions in a neighborhood of* 0 *in* $\mathbb{C}^{n-1}$ *and vanish when* $z'=(z_1,\dots,z_{n-1})=0$.

The following proposition is a first step in the proof of the existence of a precise notion of "area" for analytic varieties: first we have to prove the boundedness of the area of $\tilde{Y}_k$ in a neighborhood of the singular points $z\in Y_k'$. We use the property that Y_k can be locally imbedded in a complete intersection.

Proposition 2.40. *Let $\Omega\subset\mathbb{C}^n$ be a domain and $Y\subset\Omega$ an analytic variety such that $0\in Y$ and*

$$Y=\{z\in\Omega\colon f_j(z)=0,\ 1\leqq j\leqq t,\ f_j\in\mathscr{H}(\Omega)\}$$

$\dim_0 Y=p$. *Then*

1) *for every system of axes (not necessarily orthogonal) such that* 0 *is an isolated point of* $\mathbb{C}^{n-p}(z_{p+1},\dots,z_n)\cap Y$ *there exists a domain*

$$D=\{z\in\Omega\colon |z_i|<\delta,\ i=1,\dots,p,\ |z_j|<\delta',\ j=p+1,\dots,n\}$$

and an analytic subvariety $\hat{Y}$ of D with $(Y\cap D)\subset(\hat{Y}\cap D)$.

$\hat{Y}$ is a complete intersection in D given by $(n-p)$ pseudo-polynomials defined in $\mathbb{C}^{n-p}\times\{(z_i,\dots,z_p)\colon |z_i|<\delta\}$,

$$\hat{Y}=\{z\colon P_{p+1}(z_{p+1};z_1,\dots,z_p)=\dots=P_n(z_n;z_1,\dots,z_p)=0\}.$$

2) *if in addition, Y is of pure dimension p in Ω, then the projection $\pi\colon\mathbb{C}^n\to\mathbb{C}^p$ has the following property: Setting* $W=\{(z_1,\dots,z_p)\in\pi(D);\ \prod_{j=p+1}^{n} R_j(z)=0\}$ *where R_j is the discriminant of the pseudo-polynomial P_j, then for every $z\in Y\cap D$ such that $\pi(z)\notin W$, there exists a neighborhood U_z of z such that $z_{p+1},\dots,z_n$ are holomorphic functions of $(z_1,\dots,z_p)$ on $U_z\cap Y$ and $(z_1,\dots,z_p)$ are local coordinates on $U_z\cap Y$.*

Proof. We suppose that $p\geqq 1$, since if $p=0$, $Y=\{0\}$ is defined by $z_j=0$, $j=1,\dots,n$. For p fixed, we use induction on n. For $p=n$, the statement of the theorem is obvious, since in that case $Y=\Omega$. Thus, we assume the statement proved for $(n-1)$ and prove it for n under the assumption $n\geqq p+1$. Since $\dim_0 Y=p$, we can find a subspace L^{n-p} such that $Y\cap L^{n-p}$ has 0 as an isolated point. By a linear change of coordinates, we can suppose that $L^{n-p}=\mathbb{C}^{n-p}(z_{p+1},\dots,z_n)$. Thus the complex line containing $(0,\dots,0,z_n)$ is in L^{n-p}. Now we consider the equation $f_1=\dots=f_t=0$ which defines Y. There exists f_j, which we suppose to be f_1, such that $\varphi(z_n)=f_1(0,\dots,0,z_n)\not\equiv 0$. Suppose that $B(0,r)\Subset\Omega$.

Set $z'=(z_1, \dots, z_{n-1})$ and $D_1=\{(z', z_n): \|z'\|^2<r^2-r_n^2, |z_n|<r_n\}$, where we choose $r_n<\frac{1}{2}r$ such that $\varphi(z_n)$ has only $z_n=0$ as zero in the disc $|z_n|<r_n$ and $|\varphi(z_n)|\geqq a>0$ for $|z_n|=r_n$. Let k be the multiplicity of $z_n=0$ as a zero of $\varphi(z_n)$. By Proposition 2.39, we can find a domain $\Delta_1'\subset D_1$ such that $f_1(z', z_n) =Q(z_n; z')\tilde{f}_1(z)$ for $z\in D_1'=\Delta_1'\times\{z_n: |z_n|<r_n\}$, and $\tilde{f}_1\in\mathscr{H}(D_1')$ has no zeros in D_1'; $Q(z_n; z')$ is a pseudo-polynomial in z_n of degree k. Thus

$$Y\cap D_1'=\{z\in D_1': Q(z_n; z')=0, f_j(z)=0, 2\leqq j\leqq t\}.$$

Let $Q(z_n; z')=\prod_\alpha Q_\alpha^{m_\alpha}(z_n; z')$ be the factorization of Q into irreducible pseudo-polynomials in z_n, and set $\tilde{Q}=\prod_\alpha Q_\alpha(z_n; z')$. Then $\deg\tilde{Q}=\tilde{k}\leqq k$ and what is more $\tilde{R}(z')=\text{discriminant } \tilde{Q}\not\equiv 0$ for $z'\in\Delta_1'$;

$$Y\cap D_1'=\{z\in D_1': \tilde{Q}(z_n; z')=0, f_j(z)=0, 2\leqq j\leqq t\}.$$

For $z'\in\Delta_1'$ and $\tilde{R}(z')\neq 0$, the k roots $\psi_\nu(z')$ of $\tilde{Q}(z_n, z')=0$ are distinct and holomorphic in z'. Let W_1 be the analytic variety defined for $z'\in\Delta_1'$ by $\tilde{R}(z') =0$. If f_j is one of the functions defining Y, $2\leqq j\leqq t$, we define the function $\tilde{f}_j(z')=\prod_{\nu=1}^{\tilde{k}} f_j(z', \psi_\nu(z'))$. Then $\tilde{f}_j(z')$ is holomorphic in z' for $z'\in\Delta_1'\setminus W_1$, and since the $\psi_\nu(z')$ remain in a compact subset of Ω, $\tilde{f}_j(z')$ is uniformly bounded in a neighborhood of every point W_1 and thus can be continued as an analytic function to Δ_1' (cf. Corollary I.23).

We claim that $\pi_n(Y)=Y_{n-1}=\{z'\in\Delta_1': \tilde{f}_j(z')=0, 2\leqq j\leqq t\}$. Indeed, if $(z', z_n)\in Y$, then $\tilde{Q}(z_n; z')=0$ and so $f_j(z', z_n)=0$, $2\leqq j\leqq t$, so $\tilde{f}_j(z')=0$, $2\leqq j\leqq t$. On the other hand, if $z_0'\in Y_{n-1}$ and $z_0'\in\Delta_1'\setminus W_1$, we can find $\tilde{z}_n\in\mathbb{C}$ solution of the equation $\tilde{Q}(z_n; z_0')=0$ with $|\tilde{z}_n|<r_n$ and a function $\psi_\nu(z')$ holomorphic in a neighborhood of z_0' such that $z_n=\psi_\nu(z')$, $|\psi_\nu(z')|<r_n$, $\tilde{z}_n=\psi_\nu(z_0')$ and $\tilde{f}_j(z', \psi_\nu(z'))=0$ for all j. Thus $f_j(z_0', \tilde{z}_n)=0$, $1\leqq j\leqq t$ and $(z_0'; \tilde{z}_n)\in Y$. If $z'\in W_1\cap\Delta_1'$, we can find a sequence $z_q=(z'^{(q)}, z_n^{(q)})\in Y$, $z'^{(q)}\to z_0'$, $|z_n^{(q)}|<r_n$, $\pi(z_q)\notin W_1$. By compactness, we can find a subsequence which converges to a point $\tilde{z}=(z_0', \tilde{z}_n)$, $|\tilde{z}_n|\leqq r_n$. Since Y is closed, $\tilde{z}\in Y$ and $\pi(\tilde{z})=z_0'$. In particular, $\mathbb{C}^{n-p-1}(z_{p+1}, \dots, z_{n-1})\cap Y_{n-1}=\{0\}$.

We now apply the induction hypothesis: we can find

$$P_{p+1}(z_{p+1}; z_1, \dots, z_p), \dots, P_{n-1}(z_{n-1}; z, \dots, z_p)$$

and D_2 such that

$$Y_{n-1}\cap D_2=\{z\in D_2: P_{p+j}=0, 1\leqq j\leqq n-p-1\}.$$

Set

$$\tilde{P}_n(z_n; z')=\prod_{s_1, \dots, s_{n-1-p}} \tilde{Q}(z_n; z_1, \dots, z_p, \xi_{p+1}^{(s_1)}, \dots, \xi_{n-1}^{(s_{n-1-p})})$$

where the $\xi_{p+j}^{(s_k)}(z_1, \dots, z_p)$ are distinct roots of $P_{p+j}(z_{p+j}; z_1, \dots, z_p)=0$ for $(z_1, \dots, z_p)\notin W_2$ and

$$W_2=\left\{(z_1, \dots, z_p): \prod_{j=p+1}^{n-1} R_j(z_1, \dots, z_p)=0\right\}.$$

By the same argument as above, we see that $\tilde{P}_n$ continues as a holomorphic function to D_2. Since $\tilde{P}_n$ is a pseudopolynomial in z_n, we can replace it by $\hat{P}_n$ which has the same zeros as $\tilde{P}_n$ but no multiple factors. Then $Y \subset \hat{Y} = \{z \in D_2: P_{p+1} = \ldots = P_{n-1} = \tilde{P}_n = 0\}$ and $\pi(Y \cap D) = \pi(\hat{Y} \cap D) = \pi(D_2)$, which establishes (1).

We now show (2). Let $z_0 = (z_1^0, \ldots, z_n^0) \in Y$ such that $z_0 \notin W$. Then each of the equations $P_{p+j}(z_{p+j}; z_1, \ldots, z_p) = 0$ has in a connected neighborhood V of $(z_1^0, \ldots, z_p^0)$ in $\mathbb{C}^p$ a unique root $z_{p+j} = \xi_{p+j}(z_1, \ldots, z_p)$ which is holomorphic with value $\xi_{p+j}(z_1^0, \ldots, z_p^0) = z_{p+j}^0$. Thus, there exists a neighborhood $U = \{z: |z_k - z_k^0| < r_k\} \subset D$ such that $\pi(U) \cap W = \emptyset$ and such that $\hat{Y} \cap U$ is a connected analytic submanifold of U defined by the mapping of V given by the equations $z_{p+j} = \xi_{p+j}(z_1, \ldots, z_p)$, $(z_1, \ldots, z_p) \in V \subset \pi(U)$. Then $\hat{Y}$ is an analytic subvariety which is locally irreducible; Y and $\hat{Y}$ have the same dimension p, and $Y \subset \hat{Y}$. Thus $Y \cap U = \hat{Y} \cap U$ and $(z_1, \ldots, z_p)$ are local coordinates. □

Proposition 2.41. *Under the hypotheses of Proposition* 2.40, *the set* $Y' \subset Y$ *of singular points of* Y *is contained in a proper analytic variety of dimension at most* $(p-1)$ *at each of its points.*

Proof. By Proposition 2.40, we have $\pi(Y') \subset W = \left\{(z_1, \ldots, z_p): \prod_{j=p+1}^{n} R_j = 0\right\}$ and $\prod_{j=p+1}^{n} R_j \not\equiv 0$ in $\pi(D)$ since the P_{p+j} have no multiple factors. Thus $\dim_{z'} W < p$ for all $z' \in W$. Since the fiber $\pi^{-1}(y)$ is discrete, we see that $\dim_z \pi^{-1}(W) < p$ for $z \in \pi^{-1}(W)$. □

Let $G_{n-p}(\mathbb{C}^n)$ be the Grassmannian of $(n-p)$ dimensional subspaces of $\mathbb{C}^n$. Then $G_{n-p}(\mathbb{C}^n)$ is a compact analytic manifold of complex dimension $p(n-p)$ (cf. [H]).

Proposition 2.42. *Let* X *be an analytic variety of pure dimension* p *defined in* $B(0,r)$ *such that* $0 \in X$. *Then there exists* $\tilde{r} < r$ *such that the set* $\eta = \{L \in G_{n-p}(\mathbb{C}^n): L \cap X \cap B(0, \tilde{r})$ *is not a discrete set*$\}$ *is an analytic variety in* $G_{n-p}(\mathbb{C}^n)$.

Proof. Let $F_t(z_1, \ldots, z_n) = 0$, $t = 1, \ldots, T$, be a system of equations which define X in $B(0,r)$. Let $L_0 \in G_{n-p}(\mathbb{C}^n)$, which for simplicity we take to be $L_0 = \{z: z_1 = \ldots = z_p = 0\}$. Finally, let U_ε be the neighborhood of L_0 given by

$$U_\varepsilon = \left\{z_k = \sum_{j=p+1}^{n} C_k^j z_j,\ k = 1, \ldots, p,\ |C_k^j| < \varepsilon\right\}. \tag{2,19}$$

We consider the equations

$$F_t\left(\sum_{j=p+1}^{n} C_1^j z_j, \ldots, \sum_{j=p+1}^{n} C_p^j z_j, z_{p+1}, \ldots, z_n\right) = 0, \qquad t = 1, \ldots, T \tag{2,20}$$

as defining an analytic variety χ in $G_{n-p}(\mathbb{C}^n)\times\mathbb{C}^{n-p}$ in a neighborhood of $L_0\times\{0\}$. More precisely, we consider on $G_{n-p}(\mathbb{C}^n)$ the open set determined by (2.19) for $\varepsilon<1$ and on $\mathbb{C}^{n-p}$ the polydisc $\Delta=\{(z_{p+1},\ldots,z_n)\colon |z_j|<r/n,\ j=p+1,\ldots,n\}$, so that the system (2,20) defines χ in $U_\varepsilon\times\Delta$. We now restrict ourselves to $\omega=U_{\varepsilon'}\times\Delta'$, where $\varepsilon'<\varepsilon$ and $\Delta'\Subset\Delta$. The irreducible branches of χ which meet ω are finite in number. Let χ_s be one of them of complex dimension d_s.

Let $p\colon(L,z)\to L$ be the projection onto $G_{n-p}(\mathbb{C}^n)$, and let $d_s'(\xi)$ be the dimension of the fiber $p^{-1}(p(\xi))$ for $\xi\in\chi_s$. Then $d_s'(\xi)=0$ at $(L,0)$ if and only if $\xi\notin\eta$. On the other hand, the rank $r(\xi)$ at the point ξ satisfies $r(\xi)=d_s(\xi)-d_s'(\xi)$, and $d_s(\xi)$ is constant on χ_s. The condition $r(\xi)\leqq d_s-1$ defines an analytic subvariety $\chi_s'\subset\chi_s$, and $\eta\cap U_{\varepsilon'}=\bigcup_s\chi_s'\cap(U_{\varepsilon'}\times\{0\})$, which is an analytic variety. Thus, η is an analytic variety in $G_{n-p}(\mathbb{C}^n)$, so it remains to show that $\eta\neq G_{n-p}(\mathbb{C}^n)$. Since $\dim X=p$, there exists $\tilde r$ and L such that $X\cap B(0,\tilde r)\cap L=(0)$. □

The following proposition is a second step in the proof of the existence of a precise notion of "area" for an analytic variety Y: we prove that the area of $\tilde Y$ is bounded in a neighborhood of a singular point $z\in Y'$.

Theorem 2.43. *Let Ω be a domain in $\mathbb{C}^n$ and Y an analytic variety of pure dimension p in Ω. Suppose that K is a compact subset of Ω. Then there exists $C(K,Y)>0$ such that the area $\sigma_{\tilde Y}$ of the manifold $\tilde Y\subset Y$ in K satisfies:*

$$\text{(2,21)}\qquad \int_{B(z,r)}\sigma_{\tilde Y}\leqq C(K,Y)r^{2p}\qquad \textit{for } B(z,r)\subset K.$$

Proof. It is enough to prove (2,21) for a compact neighborhood D of a point $z\in Y\cap\Omega$. For simplicity, we assume $z=0$. Let L_0^{n-p} be an $(n-p)$-dimensional subspace such that $L_0^{n-p}\cap Y$ contains 0 as an isolated point. By Proposition 2.42, we can find a neighborhood ω of L_0 in $G_{n-p}(\mathbb{C}^n)$ such that for $L^{n-p}\in\omega$, $L^{n-p}\cap Y$ contains 0 as an isolated point. For $N=\dfrac{n!}{p!(n-p)!}$, we choose a system $\Lambda=\{L_1^{n-p},\ldots,L_N^{n-p}\}\subset\omega^N\subset G_{n-p}^N(\mathbb{C}^n)$ such that $\Lambda'=\{L_1^p,\ldots,L_N^p\}$ forms a regular system, where L_i^p is the subspace orthogonal to L_i^{n-p}.

Let us first consider the form $\tau(L_1^p)$ and $[Y](\tau(L_1^p))=\int_{\tilde Y}\tau(L_1^p)$. By Proposition 2.40, we can find a neighborhood D_1 of the origin and a system of axes for which we can choose $\mathbb{C}^{n-p}(z_{p+1},\ldots,z_n)=L_1^{n-p}$ and $\mathbb{C}^p(z_1,\ldots,z_p)=L_1^p$ such that $(Y\cap D_1)\subset(\hat Y\cap D)$ where

$$\hat Y\cap D=\{z\colon P_{j+p}(z_{j+p};z_1,\ldots,z_p)=\ldots=P_n(z_n;z_1,\ldots,z_p)=0\}$$

and $\deg P_{j+p}=\nu_{j+p}$. Thus, for $z'=(z_1,\ldots,z_p)$ fixed, there are at most $\gamma=\prod_{j=1}^{n-p}\gamma_{j+p}$ points in $\pi^{-1}(z')\cap\hat Y$ in D_1; the same is then true for any ball

$B(z,r)\subset D_1$. It follows from Corollary I.12 and the fact that $\pi^{-1}(W)\cap\tilde{Y}$ is a subvariety of $\tilde{Y}$ that

$$\int_{\tilde{Y}\cap B(z,r)-\pi^{-1}(W)} \tau(L_1^p) = \int_{\tilde{Y}\cap B(z,r)} \tau(L_1^p).$$

Thus

$$\int_{\tilde{Y}\cap B(z,r)} \tau(L_1^p) \leqq \gamma \int_{B(z',r)} \frac{i}{2} dz_1 \wedge d\bar{z}_1 \wedge \ldots \wedge \frac{i}{2} dz_p \wedge d\bar{z}_p$$

where $z'=\pi(z)$ is the projection on $\mathbb{C}^p(z_1,\ldots,z_p)$. Thus

$$\int_{\tilde{Y}\cap B(z,r)} \tau(L_1^p) \leqq \gamma\tau_{2p} r^{2p} = C_1 r^{2p} \qquad \text{for } B(z,r)\subset D_1.$$

We proceed in the same way for $L_2^p, \ldots, L_N^p$ and obtain similar bounds $\int_{\tilde{Y}\cap B(z,r)} \tau(L_s^p) \leqq C_s r^{2p}$ for every ball $B(z,r)\subset D_s$, where C_s is independent of $B(z,r)$.

Let Δ be a connected open neighborhood of 0 in $\bigcap_{s=1}^{N} D_s$. Then for $B(z,r)\subset\Delta$ we obtain, $\int_{\tilde{Y}\cap B(z,r)} \tau(L_s^p) \leqq (\sup_{1\leqq s\leqq N} C_s) r^{2p}$.

By Theorem 2.16, there exists a constant C'' depending only on the system $\Lambda=\{L_s^p\}$ such that for every positive current t, $\sigma_t \leqq C''\left(\sum_{s=1}^{N} t\wedge\tau(L_s^p)\right)$. Applying this to the current of integration $[\tilde{Y}]$, we obtain

$$\sigma_{\tilde{Y}}[B(z,r)] \leqq C(\Delta, Y) r^{2p} \qquad \text{for } B(z,r)\subset\Delta. \tag{2,22}$$

Since K is compact, we can cover it by a finite number of domains Δ_i for which (2,22) is true for a constant $C(\Delta_i, Y)$. If we let $C=\sup C(\Delta_i, Y)$, we obtain (2.21). □

Let us remark that the definition of a current t is local and the same is truc for the current dt, the closure of t defined by $dt(\varphi)=t(d\varphi)$: *if* $\{U_j\}$ is a locally finite covering of a domain $\Omega\subset C^n$ by subdomains and $\rho_j\in\mathscr{C}_0^\infty(U_j)$ such that $\sum\rho_j(x)=1$ on Ω, we write for φ with coefficients in $\mathscr{C}_0^\infty(\Omega)$: $t(\varphi)=\sum t(\rho_j\varphi)=\sum t(\varphi_j)$ for $\varphi_j=\rho_j\varphi$. We prove now a generalization of Stokes' Theorem and use assumptions on the mass of a closed positive current t defined in a domain $\Omega\subset R^m$ to obtain a continuation of t as a closed current to R^m. The problem is local, therefore we suppose Ω relatively compact.

Theorem 2.44. *Let t be a current continuous of order zero defined in a bounded domain $\Omega\subset\mathbb{R}^m$.*

i) *in order that t extends across the boundary of Ω to a current t' continuous of order zero, it is necessary and sufficient that t be bounded in Ω, that is that the measure coefficients of t be of finite total mass in Ω. In this case, the simple extension $\tilde{t}$ of t, which has no mass on $\complement\Omega$, is obtained by*

$$\tilde{t}(\varphi) = \lim_{q\to\infty} t[\alpha_q\varphi] \tag{2,23}$$

where $\alpha_q(x)$ *is a family of functions in* $\mathscr{C}_0^\infty(\Omega)$ *such that* $0 \leqq \alpha_q(x) \leqq 1$, $\alpha_{q+1}(x) \geqq \alpha_q(x)$ *and* $\lim_{m\to\infty} \alpha_q(x) = \chi_\Omega$, *the characteristic function of* Ω;

ii) *if* t *is closed, the simple extension* $\tilde{t}$ *of* t, *defined by* (2,23) *is closed if and only if*

$$\lim_{q\to\infty} t \wedge d\alpha_q = 0$$

for one sequence $\alpha_q(x)$ *with the properties stated in* i).

Proof. If t extends to a current t' defined in $\Omega' \supset \bar{\Omega}$ and continuous of order zero in Ω', obviously the mass $\|t'\|_G$ of t' in $G \Subset \Omega'$ is an upper bound of the mass of t in $G \cap \Omega$, and for $G = \Omega$, the mass $\|t\|_\Omega$ must be bounded. Conversely, if t is of finite total mass in Ω, (2,23) by convergence defines a current $\tilde{t}$ in $\mathbb{R}^m$ of bounded mass and so $\tilde{t}$ is seen to be continuous of order zero, and the definition of $\tilde{t}$ does not depend on the particular sequence $\alpha_q(x)$ with the above properties. Furthermore, for a form φ with coefficients in $\mathscr{C}_0^\infty(\mathbb{R}^m)$

$$\tilde{t}(d\varphi) = \lim_{q\to\infty} t(\alpha_q d\varphi) = \lim_{q\to\infty} [t(d(\alpha_q \varphi)) - t(d\alpha_q \wedge \varphi)].$$

The first term vanishes since t is closed and $\operatorname{supp}(\alpha_q\varphi)$ is compact in Ω. Then $\tilde{t}(d\varphi) = -\lim_{q\to\infty} t(d\alpha_q \wedge \varphi)$ for each sequence $\{\alpha_q\}$ with the given properties, and ii) is proved. □

Corollary 2.45. *Let* Ω *be a domain in* $\mathbb{R}^m = \mathbb{R}^p \times \mathbb{R}^{m-p}$, $0 \leqq p < m$, $y = (x_1, \ldots, x_p)$, $y' = (x_{p+1}, \ldots, x_m)$ *and* $\Omega_1 = [x \in \Omega \colon \|y'\| \neq 0]$. *Then if* t *is a closed current continuous of order zero in the open set* Ω_1 *(it is defined on the forms* φ *with coefficients in* $\mathscr{C}_0^\infty(\Omega_1)$*), a sufficient condition for the simple extension* $\tilde{t}$ *of* t *to be a closed current in* Ω *is that for each domain* $G \Subset \Omega$

$$\lim_{r\to 0} r^{-1} \|t\|_G^r = 0 \tag{2,24}$$

where $\|t\|_G^r$ *is the mass of* t *in* $G \cap \Omega_1 \cap [\|y'\| < r]$.

Proof. Let $l_r(x) \in \mathscr{C}_0^\infty(\Omega)$ be a family of functions, $0 \in l_r(x) \leqq 1$, defined for $r > 0$, $l_\rho(x) \leqq l_r(x)$ if $\rho \leqq r$, $l_r(x) = 1$ if $d_\Omega(x) \geqq r$, and $\lim_{r\to\infty} l_r(x) = \chi_\Omega(x)$ the characteristic function of Ω. For a given form φ with coefficients in $\mathscr{C}^\infty(\Omega)$ we consider a domain G such that $\operatorname{supp}\varphi \subset G \Subset \Omega$ and we have $l_r(x) = 1$ for $x \in G$ and $r < r_0$. Let $g(t)$ be a $\mathscr{C}^\infty$ decreasing function of t defined for $t \geqq 0$ with values $g(t) = 1$ if $0 \leqq t \leqq 1/2$, and $g(t) = 0$ if $t \geqq 1$. Then we consider for $x = (y, y')$, the family $\alpha_r(x) = l_r(x)\left[1 - g\left(\frac{\|y'\|}{r}\right)\right]$ and $\tilde{t} = \lim_{r\to 0} t\alpha_r$. By Theorem 2.44,

$$\tilde{t}(d\varphi) = \lim_{r\to 0} [-t(d\alpha_r \wedge \varphi)].$$

For $r < r_0$, we have $l_r(x) \equiv 1$ for $x \in G$, and $|dg| \leqq \frac{m}{r} \sup_t |g'(t)| = \frac{C}{r}$. Then $|\tilde{t}(d\varphi)| \leqq \frac{C}{r} \|t\|_G^r$ for $r < r_0$ and $\tilde{t}(d\varphi) = 0$ as a consequence of (2,24). □

Theorem 2.46. *Let Ω be a domain in $\mathbb{R}^m = \mathbb{R}^p \times \mathbb{R}^{m-p}$, $0 \leqq p < m$, and $\Omega_1 = (\mathbb{R}^m \setminus \mathbb{R}^p) \cap \Omega$. A sufficient condition for a closed current t continuous of order zero in Ω_1 to have a closed simple extension $\tilde{t}$ is that for every domain $G \Subset \Omega$, there exists a constant C_G and $\gamma > p+1$ such for every ball $B = B(x, r) \subset G$, the following bound holds:*

$$\|t\|_B \leqq C_G r^\gamma. \tag{2,25}$$

Proof. Given a domain $G \Subset \Omega$ and $G_r = \Omega \cap [\|y'\| \leqq r]$, $x = (y, y')$, there exists $r_0 > 0$ such that the image G'_r of G_r by the projection $\mathbb{R}^m \to \mathbb{R}^p$, $x = (y, y') \to (y, 0)$ is compact in $G \cap \mathbb{R}^p_1$. We cover G_r by cubes A_i of side $2r$ and axes parallel to the coordinate axes of $\mathbb{R}^m$, such that $\mathbb{R}^p \cap A_i$ is a face of A_i. There exists C_1 such that the total number of the cubes A_i with $A_i \cap G_r \neq \emptyset$ is at most $C_1 2^{-p}$. Then by (2,25) we have

$$\|t\|^r_G \leqq C_1 C_G r^{\gamma - p}$$

and the result now follows from Corollary 2.45 since $\gamma > p+1$. □

We now arrive at the principal result of this section:

Theorem 2.47. *Let Y be an analytic variety of pure complex dimension $p \geqq 1$ defined in the domain $\Omega \subset \mathbb{C}^n$. Let $t = [\tilde{Y}]$ be the current of integration on the regular points of Y, defined in $\Omega - Y'$. Then the simple extension of t to Ω exists and is a positive closed current $\tilde{t} \in \tilde{T}^+_{n-p}(\Omega)$. We will denote $\tilde{t}$ by $[Y]$.*

Proof. It follows from Proposition 2.6 that $[\tilde{Y}]$ is an element of $\tilde{T}^+_{n-p}(\Omega - Y')$. Furthermore, the set Y' is an analytic variety of dimension at most $(p-1)$. Let $\tilde{Y}_1$ be the regular points of Y' and Y'_1 the singular points of Y'. Then in $(\Omega - Y')$, Y'_1 is a union of complex submanifolds Y_s. Let $Y''_s = \bigcup_{s \neq s'} (Y_s \cap Y_{s'})$, which is an analytic set of dimension $(p-2)$ at most and set $\Omega'' = (\Omega - Y'_1 - \bigcup_s Y''_s)$. Let $\{U_i\}$ be a locally finite covering of Y_1 in Ω'' such that for each U_i, there exists a mapping $\gamma_i \colon U_i \to \mathbb{C}^n$ with $\gamma_i(U_i \cap Y_i)$ a neighborhood V_i of 0 in $\mathbb{C}^{p_i}$, $p_i \leqq (p-1)$. Let β_i be a partition of unity subordinate to U_i. Let $\varphi \in \mathscr{C}^\infty_{0,(p,q-1)}(\Omega'') \cup \mathscr{C}^\infty_{0,(p-1,q)}(\Omega'')$. Then $d\tilde{t}(\varphi) = \tilde{t}(d\varphi) = \sum \tilde{t}(d(\beta_i \varphi))$, since $\sum \beta_i \equiv 1$. It follows from Corollary 2.45 and Theorem 2.43 that $\tilde{t}$ extends to a closed current in U_i, since $p_i \leqq (p-1)$ and thus $2p_i + 1 < 2p$. Thus $t(d(\beta_i \varphi)) = 0$ for all i and $\tilde{t}$ is closed in Ω''. Since $\hat{Y} = (Y'_1 \cup Y''_s)$ is an analytic set of dimension at most $(p-2)$, we repeat the above reasoning to extend $\tilde{t}$ to $\Omega'' - \hat{Y}'$ where $\hat{Y}'$ is an analytic set of dimension at most $(p-3)$. By iterating this process, we arrive after at most p steps at an extension $\tilde{t}$ of t as a positive closed current to all of Ω. □

Remark 1. In the statement of Theorem 2.47, we did not consider the case $p = 0$. In this case $t = \sum \delta(a_i)$, where $Y = \{a_i\}$ and $\delta(a_i)$ is the Dirac measure at a_i.

Remark 2. The essential point of the proof of Theorem 2.47 is Theorem 2.43 which corresponds to a property of the area of analytic sets. This also leads to the following:

Proposition 2.48. *The area of an analytic variety $Y \subset \Omega$ of pure dimension p exists in the real dimension $2p$ and is given by*

$$\sigma = \int_Y \beta_p = \int [Y] \wedge \beta_p = [Y] \beta_p,$$

is finite on every compact subset of Ω, and has the property $\sigma = \sum_I \sigma_I$, where σ_I is the projection of σ on the subspace $\mathbb{C}^p(z_I)$.

Thus we see that for $t = [Y]$, σ_t, the trace of t, is just the area of Y.

Proposition 2.49. *If $t \in \tilde{T}^+_{n-p}(\Omega)$, $v_t(a)$ is upper semi-continuous.*

Proof. Since $\sigma_t(a, r)$ is the mass of σ_t carried by the closed ball of radius r and center a, $\sigma_t(a, r)$ is upper-semicontinuous for r fixed, and since $v_t(a) = \inf_{r \to 0} \tau_{2p}^{-1} r^{-2p} \sigma_t(a, r)$, it is also upper semi-continuous. □

Theorem 2.50. *Let $t' = F_\star t$ be the image of a positive closed current $t \in \tilde{T}^+_{n-p}$ by an application $z' = F(z)$ which is biholomorphic from a neighborhood of z_0 onto a neighborhood of $z'_0 = F(z_0)$. Then the Lelong numbers $v_{t'}(z'_0)$ of t' at z'_0 and $v_t(z_0)$ of t at z_0 are equal.*

Proof. We shall divide the proof into several steps.

i) For simplicity, we assume that $z_0 = 0$, $z'_0 = 0$ and that F is biholomorphic between the two open neighborhoods Ω and Ω' of z_0 and z'_0 respectively in $\mathbb{C}^n$, $n \geqq 2$. The current $t' = F_\star t$ on a form φ with coefficients in $\mathscr{C}_0^\infty(\Omega)$ is given by

$$t'(\varphi) = F_\star t(\varphi) = t(F^\star \varphi) \tag{2,26}$$

where $F^\star \varphi$ is obtained by replacing in φ the variables z' as functions of z. We set $\psi(z) = F^\star(\|z'\|^2)$, that is, for $F = (F_k)$

$$\psi(z) = \sum_{i=1}^n F_i(z) \cdot \bar{F}_i(z). \tag{2,27}$$

We then obtain from the definition of $v_{t'}(0)$ in Ω' that

$$v_{t'}(0) = \lim_{r \to 0} \frac{1}{(\pi r^2)^p} \int_{\psi(z) < r^2} t \wedge \left(\frac{i}{2} \partial \bar{\partial} \psi\right)^p.$$

ii) This definition leads us to study the definition of the Lelong number with respect to a function $\psi(z)$ having properties similar to chose of $\varphi(z) = \|z\|^2$.

Let $L(\Omega)\subset \mathrm{PSH}(\Omega)$ be those functions $V(z)$ such that

a) $V(z)\geqq 0$

b) $V(z)\in\mathscr{C}^2(\Omega)\cap \mathrm{PSH}(\Omega)$

c) $V(z)\leqq r$ is compact in Ω for $0<r<R(\varphi)$

d) $\log V(z)\in \mathrm{PSH}(\Omega)$.

Clearly if V_1 and V_2 are in $L(\Omega)$, then $V_1 V_2\in L(\Omega)$ and $V_1^l\in L(\Omega)$ for $l>0$. But we also have $V_1+V_2\in L(\Omega)$. To see this, we note that if $V\in L(\Omega)$, then $\log V+\mathrm{Re}\sum_{k=1}^{n} A_k z_k\in \mathrm{PSH}(\Omega)$ for every $A=(A_1,\dots,A_n)\in\mathbb{C}^n$. Thus $V\left|\exp\sum_{k=1}^{n} A_k z_k\right|\in \mathrm{PSH}(\Omega)$ for every $A\in\mathbb{C}^n$. A simple calculation shows that this condition is necessary and sufficient to have (d) (cf. Lemma 3.46). But if it is verified for V_1 and V_2, it is verified for V_1+V_2. In particular, if we let

(2,28) $$\tilde{\psi}_\varepsilon(z)=\psi^k(z)+\varepsilon\varphi^l(z) \quad \text{for} \quad \varphi=\sum_{k=1}^{n} z_k \bar{z}_k$$

and ψ defined by (2.27), $\tilde{\psi}_\varepsilon(z)\in L(\Omega)$ if it is $\mathscr{C}^2$, that is if $k\geqq 2$, $l\geqq 2$.

Given a function $h\in L(\Omega)$, we define the number $v_{h,t}(0)$, the Lelong number of t with respect to h, by

(2,29) $$v_{h,t}(0)=\lim_{r\to 0}(\pi r^2)^{-p}\int_{h(z)<r^2} t\wedge\left(\frac{i}{2}\partial\bar{\partial}h\right)^p.$$

We then have $v_{\varphi,t}(0)=v_t(0)$. The existence of the limit in (2,29) follows, as in Theorem 2.23, from the inequality

(2,30) $$\begin{aligned}0\leqq v_{h,t}(r_1,r_2)&=\int_{r_1^2<h(z)<r_2^2} t\wedge\alpha_h^p\\ &\leqq\pi^{-p}\left[r_2^{-2p}\int_{h(z)<r_2^2} t\wedge\left(\frac{i}{2}\partial\bar{\partial}h\right)^p-r_1^{-2p}\int_{h(z)<r_1^2} t\wedge\left(\frac{i}{2}\partial\bar{\partial}h\right)^p\right]\end{aligned}$$

where $\alpha_h=\dfrac{i}{2\pi}\partial\bar{\partial}\log h$ is a positive closed current with continuous coefficients. Thus $v_{h,t}(r_1,r_2)$ is positive. If we let $h(z)=\psi(z)$ defined by (2,27), we obtain $v_{h,t}(0)=v_{t'}(0)$.

iii) We have the relationship

(2,31) $$v_{h^l,t}(0)=l^p v_{h,t}(0).$$

We first assume that $l\geqq 2$ so that $h^l\in L(\Omega)$. Then

$$\partial\bar{\partial}h^l=lh^{l-1}\partial\bar{\partial}h+l(l-1)h^{l-2}\partial h\wedge\bar{\partial}h.$$

On an h-sphere defined by $h(z)=r^2$, we have

$$\partial\bar{\partial}h^l=lh^{l-1}\partial\bar{\partial}h=lr^{2(l-1)}\partial\bar{\partial}h.$$

Thus

$$v_{h^l,t}(0)=\lim_{r\to 0}\pi^{-p}r^{-2p}\int_{h^l(z)<r^2} t\wedge\left(\frac{i}{2}\partial\bar{\partial}h^l\right)^p.$$

As in Theorem 2.22, if we set $t=d\theta$, by Stokes' Theorem, we have

$$\begin{aligned}(\pi r^2)^{-p} &\int_{h^l(z)<r^2} t\wedge\left(\frac{i}{2}\partial\bar{\partial}h^l\right)^p\\ &=(\pi r^2)^{-p}\int_{h^l(z)=r^2}\theta\wedge\left(\frac{i}{2}\partial\bar{\partial}h^l\right)^p\\ &=(\pi r^2)^{-p}l^p r^{2p(l-1)}\int_{h(z)=r^2}\theta\wedge\left(\frac{i}{2}\partial\bar{\partial}h\right)^p\\ &=l^p\left[(\pi r^2)^{-p}\int_{h(z)<r^2} t\wedge\left(\frac{i}{2}\partial\bar{\partial}h\right)^p\right]\end{aligned}$$

which proves (2,31).

In order to prove that $v_{\psi,t}(0)=v_{\varphi,t}(0)$ (i.e. $v_t(0)=v_{t'}(0)$), we shall prove that $v_{\psi,t}(0)\geqq v_{\varphi,t}(0)$. The inverse inequality then follows from applying the same reasoning to F^{-1}, from which we deduce the equality. From (2,31), we see that it is enough to show that for $k>l\geqq 2$, we have $v_{\psi^k,t}(0)\geqq v_{\varphi^l,t}(0)$. Let $\tilde{\psi}_\varepsilon(z)$ be the function defined by (2,28). For every $\varepsilon>0$, there exist $r_\varepsilon>0$, $C_1(\varepsilon)$ and $C_2(\varepsilon)$ positive constants such that for $\|z\|<r_\varepsilon$, $C_1(\varepsilon)<\dfrac{\tilde{\psi}_\varepsilon(z)}{\varphi^l(z)}<C_2(\varepsilon)$. Let $r>2$ be fixed, $0<r<R-\frac{1}{2}\inf(R(\varphi),R(\psi))$ and $0<\varepsilon<1$. Consider the integral

$$I_\varepsilon(r)=(\pi r^2)^{-p}\int_{\tilde{\psi}_\varepsilon(z)<r^2} t\wedge\left(\frac{i}{2}\partial\bar{\partial}\tilde{\psi}_\varepsilon\right)^p.$$

We show that $I_\varepsilon(r)\geqq v_{\varphi^l,t}(0)$. For $0<\rho<r$

$$\begin{aligned}I_\varepsilon(r)&\geqq\lim_{\rho\to 0}\pi^{-p}\rho^{-2p}\int_{\tilde{\psi}_\varepsilon(z)<\rho^2} t\wedge\left(\frac{i}{2}\partial\bar{\partial}\tilde{\psi}_\varepsilon\right)^p\\ &\geqq\lim_{\rho\to 0}\pi^{-p}\rho^{-2p}\int_{\varphi^l(z)<\rho^2/\varepsilon} t\wedge\left(\frac{i}{2}\partial\bar{\partial}\tilde{\psi}_\varepsilon\right)^p.\end{aligned}$$

Set $\gamma_\varepsilon=\varepsilon\beta_l+\gamma'$ where $\beta_l=\frac{i}{2}\partial\bar{\partial}\varphi^l$ and $\gamma'=\frac{i}{2}\partial\bar{\partial}\psi^k$. Since γ_ε is a positive form, so is γ_ε^p. Hence

$$I_\varepsilon(r)\geqq\lim_{\rho\to 0}\pi^{-p}\rho^{-2p}\int_{\varphi^l(z)<\rho^2/\varepsilon}\varepsilon^p t\wedge\left(\frac{i}{2}\partial\bar{\partial}\varphi^l\right)^p\geqq v_{\varphi^l,t}(0)=l^p v_{\varphi,t}(0).$$

Furthermore, by the Lebesgue Dominated Convergence Theorem, we have

$$\lim_{\varepsilon\to 0}I_\varepsilon(r)=I_0(r)=(\pi r^2)^{-p}\int_{\psi^k(z)<r^2} t\wedge\left(\frac{i}{2}\partial\bar{\partial}\psi^k\right)^p$$

so that $I_0(r)\geqq l^p v_{\varphi,t}(0)$. From this it follows that

$$\lim_{r\to 0}I_0(r)=v_{\psi^k,t}(0)=k^p v_{\psi,t}(0)\geqq l^p v_{\varphi,t}(0)\qquad\text{for all } k>l\geqq 2.$$

Thus in fact $v_{t'}(0)=v_{\psi,t}(0)=v_{\varphi,t}(0)=v_t(0)$. □

Corollary 2.51. *If X is an analytic variety of pure dimension p, for every point $z_0 \in X$ which is regular, $\nu_{[X]}(z_0)$, the Lelong number with respect to the current of integration on X, is equal to 1.*

Proof. We can find neighborhoods U of z_0 and V of $0 \in \mathbb{C}^n$ and a biholomorphic map $F(U) \to V$ such that $F(z_0)=0$ and $F(U \cap X) = V \cap \mathbb{C}^p(z_1, \ldots, z_p) = Y$. By Theorem 2.50,

$$\nu_{[X]}(z_0) = \nu_{[Y]}(0) = \lim_{r \to 0} \tau_{2p}^{-1} r^{-2p} \sigma_{[Y]}(r) = 1. \qquad \square$$

Remark 3. The area σ of an analytic variety in Ω is the trace of $t=[Y]$. By (2,13), $\sigma_t(r)$, the area of Y in the ball $B(a,r) \subset \Omega$ has the property that the quotient $(\tau_{2p} r^{2p})^{-1} \sigma_t(r)$ is an increasing function of r. Then Proposition 2.50 gives a lower bound for $\sigma_t(r)$

$$\sigma_t(r) \geqq \tau_{2p} r^{2p} \nu_t(a) \geqq \tau_{2p} r^{2p}.$$

Historical Notes

The first attempt to study the "area" of an analytic set X goes back to Poincaré [2], who showed that if f is holomorphic, then $\log|f|$ is locally the sum of an R^{2n}-harmonic function and a potential $-C_n \int \|a-z\|^{2-2n} d\sigma(a)$, where σ is the "area" of the divisor $f=0$. In 1938, Kneser [1], in an attempt to generalize the Weierstrass product to $\mathbb{C}^n$, used the projective area and constructed a (locally convergent) representation of the holomorphic function $\log f$, where f defines a given divisor X, and in 1952 Stoll [2] by this method succeeded in giving a bound for a solution of the Cousin problem in $\mathbb{C}^n$ for X of finite order. In 1950 Rütishauser [1] showed that for X an analytic manifold in $\mathbb{C}^2$, $\sigma_x(r) \geqq \pi r^2$. In 1950, Lelong [8] gave a convergent representation for $\log|P|$, P a polynomial in $\mathbb{C}^n$, by a potential. Using the technics of distributions [F] and currents [E], in 1954 Lelong gave a modern formulation for integration over a divisor X by $[X](\varphi) = \int \frac{i}{\pi} \partial\bar{\partial} \log|f| \wedge \varphi$ and gave the current associated to a Cousin data of zeros.

The general problem of proving the existence and closure of the current of integration $[X]$ for X an analytic variety was different for co-dimension $X>1$, since X cannot be supposed to be a complete intersection. The positive currents and closed positive currents were introduced by Lelong [10] in 1957, and the closure was obtained as a consequence of bounds for measures. Now the positive closed currents are a classical notion in complex analysis, as will be illustrated in the following chapters.

Chapter 3. The Relationship Between the Growth of an Entire Function and the Growth of its Zero Set

The problem of constructing a holomorphic function of one complex variable with a given zero set was solved by Weierstrass in the middle of the nineteenth century. He showed that if Ω is a domain in the complex plane, if $\{a_\nu\}$ is a sequence of points without limit point in Ω, and if $\{m_\nu\}$ is a sequence of positive integers, then there exists a function $f(z)\in\mathscr{H}(\Omega)$ which has a zero of order exactly m_ν at every point a_ν. The equivalent problem for several complex variables is Cousin's Second Problem, which we state as follows: if Ω is a domain in $\mathbb{C}^n$, then for every zero set defined locally in Ω, does there exist a global holomorphic function which defines the same zero set? More specifically, if U_i is an open covering of Ω and $f_i\in\mathscr{H}(U_i)$ are such that $f_i f_j^{-1}\in\mathscr{H}(U_i\cap U_j)$ and $f_j f_i^{-1}\in\mathscr{H}(U_i\cap U_j)$ for all pairs i,j, does there exist $f\in\mathscr{H}(\Omega)$ such that $ff_i^{-1}\in\mathscr{H}(U_i)$ and $f^{-1}f_i\in\mathscr{H}(U_i)$ for all i? The answer is in general negative, even when Ω is a domain of holomorphy, and depends upon the topological as well as the complex analytic properties of Ω (cf. [A, B]); however, when Ω is a simply connected domain of holomorphy (as in the case of $\mathbb{C}^n$), the answer is always affirmative.

We shall be interested in studying a quantitative version of Cousin's Second Problem. The Cousin data $X=(U_i, f_i)$ defines a divisor in $\mathbb{C}^n$ composed of an analytic variety $Y(X)$ of dimension $(n-1)$, which is just the zero set of f_i in U_i, and a set of non-negative integers m_k, the multiplicity of $Y_k(X)$ in the Cousin data, where $Y_k(X)$ is an irreducible branch of $Y(X)$; m_k is the order of f_i on $\tilde{Y}_k(X)$, the regular points of $Y_k(X)$ (see below). It follows from the compatibility conditions $f_j f_i^{-1}$ holomorphic in $U_i\cap U_j$ and the connectivity of $\tilde{Y}_k(X)$ that the m_k are well defined. This permits us to define an "area with multiplicity" for the Cousin data X in the ball $B(0,r)$ by $\sigma(r) = \sum_k m_k$ area $(Y_k(X)\cap B(0,r))$. The problem then is to find an entire function f such that ff_i^{-1} and $f_i f^{-1}$ are holomorphic in U_i and $\log M_f(r)$ has minimal asymptotic growth. For entire functions of finite order, this is equivalent to constructing a solution of minimal order of growth. The solution f for Cousin data of finite order and the properties of f generalize to several complex variables the well known results of E. Borel, J. Hadamard, and E. Lindelöf for one complex variable (cf. [D]).

§1. Positive Closed Currents of Degree 1 Associated with a Positive Divisor

Let $X=(f_i, U_i)$ be a set of Cousin data in Ω, $Y(X)$ the analytic variety given by $Y(X)\cap U_i=\{z\in U_i\colon f_i(z)=0\}$, and $\tilde{Y}(X)$ the $(n-1)$ dimensional complex manifold of regular points of $Y(X)$.

Theorem 3.1. *A Cousin data $X=(f_i, U_i)$ in a domain Ω defines in a canonical way a positive closed current θ_X, which is called the current associated with the Cousin data. The value of θ_X is given by $\theta_X=\frac{i}{\pi}\partial\bar{\partial}\log|f_j|$ in U_j. Moreover if $Z=(f_i', U_i')$ is a Cousin data and if Z is equivalent to X, then $\theta_Z=\theta_X$.*

Proof. Let $t_j=\frac{i}{\pi}\partial\bar{\partial}\log|f_j|$ in U_j and define θ_X in Ω by $\theta_X=t_j$ in U_j. Since $f_j f_k^{-1}$ and $f_k f_j^{-1}$ are holomorphic in $U_j\cap U_k$, $\log|f_j|-\log|f_k|$ is pluriharmonic in $U_k\cap U_j$. Thus $t_j-t_k=\frac{i}{\pi}\partial\bar{\partial}[\log|f_j|-\log|f_k|]=0$, so θ_X is well defined and is positive and closed, since each t_j is positive and closed. By the same method, in a covering V_s finer then $\{U_i\cap U_j'\}$, we prove $\partial\bar{\partial}[\log|f_j'|-\log|f_i|]=0$ and $\theta_Z=\theta_X$. □

In the classical case where $n=1$, if $f(z)$ is holomorphic in Ω and a is a zero of f, then there exists a neighborhood U_a of a such that $f(z)=(z-a)^q g(z)$, where $g(z)\neq 0$ in U_a. Then

$$\pi^{-1}i\partial\bar{\partial}\log|f(z)|=\frac{q}{\pi}i\partial\bar{\partial}\log|z-a|=\frac{q}{2\pi}\Delta\log|z-a|\cdot\beta=q\delta(a)$$

where $\delta(a)$ is the Dirac measure of the point a. Thus for $n=1$, the current associated with a Cousin data $\{a_k, m_k\}$ is $\theta_X=\sum_k m_k\delta(a_k)$. For $n>1$, we show:

Theorem 3.2. *Let $X=(f_i, U_i)$ be a Cousin data and θ_X the associated positive closed current. Then*

$$\theta_X=\sum_k m_k[Y_k(X)], \tag{3,1}$$

where the $Y_k(X)$ are the irreducible branches of $Y(X)$, $[Y_k(X)]$ is the positive closed current of integration over the connected submanifold $\tilde{Y}_k(X)$ of regular points of $Y_k(X)$, and $m_k=v_X(z)$ is a positive integer, the multiplicity of $Y_k(X)$ in the Cousin data X. For every form $\varphi\in\mathscr{C}^\infty_{0,(n-1,n-1)}(\Omega)$, we have

$$\theta_X(\varphi)=\sum m_k\int_{\tilde{Y}_k(X)}\varphi \tag{3,2}$$

where the sum is taken over those $Y_k(X)$ which intersect the support of φ.

First we remark that as a consequence of Proposition 2.31 and 2.32, at each point $x \in \tilde{Y}_k(X)$, the Lelong number $\nu(x, [Y])$ has the value 1. By using an analytic isomorphism, we can suppose $x=0$ and $Y_k(X)$ defined by $z_n=0$. Then

$$\int_{z_n=0} \varphi = \int \frac{1}{2\pi} \Delta \log |z_n| d\tau_n \wedge \varphi; \qquad d\tau_n = \frac{i}{2} dz_n \wedge d\bar{z}_n.$$

For $\varphi \in \mathscr{C}_{0,(n-1,n-1)}$

$$[Y_k](\varphi) = \int \frac{i}{\pi} \partial\bar{\partial} \log |z_n| \wedge \varphi,$$

and σ, the trace measure $\sigma(r)$ in the ball $\|z\| \leqq r$ of the characteristic function χ, is just $[Y_k](\beta_{n-1}\chi) = \tau_{2n-2} r^{2n-2}$. Thus $\nu(0, [Y_k]) = 1$.

Now we consider the Lelong number $\nu_X(z)$ for the current θ_X associated with the Cousin data $X = (f_i, U_i)$ and prove:

Proposition 3.3. *The number $\nu_X(z)$ for the current θ_X associated with the Cousin data $X = (t_i, U_i)$ is a positive integer m_k which is constant for $z \in \tilde{Y}_k(X)$.*

Proof. Since $\tilde{Y}_k(X)$ is connected, it is enough to show that $\nu_X(z)$ is locally constant on $\tilde{Y}_k(X)$. Let $z_0 \in \tilde{Y}_k(X)$. There exists a holomorphic map $w = H(z)$ which is a homeomorphism of a neighborhood U_{z_0} of z_0 onto a neighborhood V of the origin such that $H(z_0)=0$ and $w_n(z)=0$ if and only if $z \in \tilde{Y}_k(X) \cap U_{z_0}$; if f_j defines $\tilde{Y}_k(X)$ in U_{z_0}, then $f_j'(w) = f_j \circ H^{-1}(w)$ is zero in V if and only if $w_n = 0$. By the Weierstrass Preparation Theorem (Proposition 2.39) we can find a neighborhood V' of the origin with $V' \subset V$ and $f_j'(w) = \left[w_n^q + \sum_{j=1}^{q-1} a_j(w') w_n^j \right] g_j'(w)$, where $g_j' \neq 0$ in V'. Since $f_j'(w) = 0$ if and only if $w_n = 0$, it follows that $f_j'(w) \equiv w_n^q g_j'(w)$. Thus the current θ_X has the value $\theta_X = q \frac{i}{\pi} \partial\bar{\partial} \log |w_n|$ and has q for Lelong number on the set $\{f_j'(w)=0\}$ in a neighborhood of the origin. Since it is an invariant (with respect to the complex analytic isomorphisms), we have $\nu_X(z) = q$ in a neighborhood of z_0 on $\tilde{Y}_k(X)$.

Proof of Theorem 3.2. For $\varphi \in \mathscr{C}^\infty_{0(n-1,n-1)}$ with support in U_{z_0}, we obtain

$$\theta_X(\varphi) = \frac{i}{\pi} \int \partial\bar{\partial} \log |f_j| \wedge \varphi = \frac{i}{\pi} \int \partial\bar{\partial} \log |f_j'| \wedge (\varphi \circ H^{-1})$$
$$= q \int_{w_n=0} \varphi \circ H^{-1} = q \int_{\tilde{Y}_k(X)} \varphi,$$

and q is a positive integer which is associated by Proposition 3.3 with the Y_k. Then (3,2) is proved for φ with support in U_{z_0}. To end the proof, we proceed by a partition of unity subordinate to the open covering U_z of $Y(X) - \bigcup_k Y_k(X)'$, where $Y_k(X)'$ is the set of singular points of $Y(X)$ on $Y_k(X)$.

It follows from Theorems 2.46 and 2.47 that if $Y'=\bigcup_k Y_k(X)'$, $\theta_X=\theta_X|_{\Omega-Y'}$ $=\sum_k m_k[\tilde{Y}_k(X)]$ by Proposition 3.3. □

Definition 3.4. For $X=(f_i, U_i)$ a Cousin data, the positive integers m_k which appear in the expression (3,1) for θ_X are called the *multiplicities* of the $Y_k(X)$ in X, and the current θ_X associated with the Cousin data is the *current of integration with multiplicities.*

Definition 3.5. The measure $\sigma_X=\theta_X\wedge\beta_{n-1}$, trace of the current θ_X, will be called the *area of* $Y(X)$ *with multiplicities.* This definition is justified by the fact that $\int_{Y_k(X)}\beta_{n-1}$ is just the $(2n-2)$ dimensional area of the complex manifold $\tilde{Y}_k(X)$.

Remark 1. The majoration of $\|\theta_X\|$ by $C_n\sigma_X$ can be interpreted as the majoration of the current of integration over the analytic variety $Y(X)$ by the *area* of the analytic variety $Y(X)$.

Remark 2. In the same way, ν_X can be interpreted as the *projective area,* and $\nu_X(r)$ is the measure relative to the metric of $\mathbb{P}(\mathbb{C}^n)$ of the cone of the complex lines through the origin which intersect $X\cap B(0,r)$; $\nu_X(0)$ is the degree of the cone of the directions of the complex tangents to X at 0. Thus, $\nu_X(0)$ can be interpreted as a mean value relative to the Haar measure $d\omega_{2n}$ on the unit sphere. More precisely, we state:

Proposition 3.6. *Suppose that X is defined in a neighborhood of zero by $f(z)=0$. Then $\nu_X(r)=\sigma_X(r)[\tau_{2n-2}r^{2n-2}]^{-1}$ has the properties:*

i) $\nu_X(r)=\dfrac{\partial}{\partial\log r}\lambda(0,r,\log|f|)$,

ii) $\nu_X(r)=\omega_{2n}^{-1}\int_{\|\alpha\|=1}n(\alpha,r)\,d\omega_{2n}(\alpha)$,

where $n(\alpha,r)$ is the number of zeros of $f(u\alpha)$ in the complex line $z=\alpha u$ of modulus at most r (with multiplicities).

Proof. Since $\sigma_X(r)=i/\pi\partial\bar\partial\log|f|\wedge\beta_{n-1}=(2\pi)^{-1}\Delta(\log|f|)\beta_n$, by Gauss' Theorem,

$$\sigma_X(r)=(2\pi)^{-1}\int_{\|z\|\le r}\Delta\log|f|\beta_n=\frac{r^{2n-1}}{2\pi}\frac{\partial}{\partial r}\lambda(0,r,\log|f|)\omega_{2n}$$

$$=[\tau_{2n-2}r^{2n-2}]\frac{\partial}{\partial\log r}\lambda(0,r,\log|f|).$$

On the other hand, on the complex line $z=\alpha u$, for $|u|=r$,

$$n(\alpha,r)=\frac{\partial}{\partial\log r}(2\pi)^{-1}\int_0^{2\pi}\log|f(\alpha re^{i\theta})|\,d\theta,$$

and hence averaging over all α such that $\|\alpha\|=1$, we have

$$\omega_{2n}^{-1} \int n(\alpha, r)\, d\omega_{2n}(\alpha) = \frac{\partial}{\partial \log r} \lambda(0, r, \log|f|). \qquad \square$$

§2. Indicators of Growth of Cousin Data in $\mathbb{C}^n$

The Euclidean area $\sigma_X(r)$ and projective area $v_X(r)$ give indicators of growth for the Cousin data X. They are related by the formula (2,15)

$$v_X(r) = (\tau_{2n-2} r^{2n-2})^{-1} \sigma_X(r).$$

It is rather the projective indicator that we shall use.

If X is defined globally by a polynomial $P(z)$ of degree m, it follows from Proposition 3.6 that $v_X(t) = \frac{\partial}{\partial \log t} \lambda(0, t, \log|P|)$ and hence $\lim\limits_{t\to\infty} v_X(t) = m$ $=$ degree P. As in the study of entire functions, we shall be interested in the scale of finite order.

Definition 3.7. The indicator $v_X(r)$ will be said to be of *finite order* ρ if $\limsup\limits_{r\to\infty} \frac{\log v_X(r)}{\log r} = \rho < +\infty$.

Definition 3.8. If $\rho(r)$ is a proximate order, *the type* λ of $\frac{v_X(r)}{r^{\rho(r)}}$ with respect to $\rho(r)$ is $\lambda = \limsup\limits_{r\to\infty} \frac{v_X(r)}{r^{\rho(r)}}$ and $v_X(r)$ is said to be of minimal, normal or maximal type according to whether $\lambda = 0$, $0 < \lambda < +\infty$, or $\lambda = +\infty$.

Proposition 3.9. *For $a>0$, $s>0$ and $n \geqq 1$, the following conditions are equivalent*

i) $\int_a^\infty t^{-s} dv_X(t) < +\infty$,

ii) $\int_a^\infty t^{-s-1} v_X(t)\, dt < +\infty$,

iii) $\int_a^\infty t^{-s+2-2n} d\sigma_X(t) < +\infty$,

iv) $\int_a^\infty t^{-s+1-2n} d\sigma_X(t) < +\infty$,

and any one of these conditions implies that $\lim\limits_{r\to\infty} v_X(r) r^{-s} = 0$.

Proof. Integrating by parts, we obtain

$$\int_a^r t^{-s} dv_X(t) = [t^{-s} v_X(t)]_a^r + s \int_a^r v_X(t) t^{-s-1} dt. \tag{3,3}$$

Since the non-constant terms on the right hand side are positive and $v_X(t)$ is positive, i) implies ii) and the existence of $\lim_{t\to\infty} t^{-s} v_X(t) = C$. Then $C=0$ follows from ii). Conversely, ii) implies that $\lim_{r\to\infty} \int_r^{2r} t^{-s-1} v_X(t)\,dt = 0$ or, since $v_X(t)$ is increasing, $\lim_{r\to\infty} v_X(r) \int_r^{2r} t^{-s-1}\,dt = \lim_{r\to\infty} C_s \frac{v_X(r)}{r^s} = 0$, so ii) implies i) by (3,3). The equivalence of ii) and iv) as well as i) and iii) follows from (2,15). □

Definition 3.10. The number $\tau = \inf\{s\}$ such that i) holds will be called the *convergence exponent* of the Cousin data X.

Definition 3.11. The smallest integer q for which $\int_1^\infty t^{-q-1}\,dv_X(t) < +\infty$ is called the *genus* of the Cousin data. We have, from Proposition 3.9, that $\int_1^\infty t^{-q-2} v_X(t)\,dt < +\infty$.

Proposition 3.12. *The order ρ of $v_X(t)$ is equal to the convergence exponent of $v_X(r)$. If ρ is not an integer, the genus q of $v_X(t)$ is the largest integer less than ρ; if ρ is an integer, we have $q-1 \leqq \rho \leqq q$. If $\rho = q-1$, then X is of minimal type with respect to the order ρ.*

Proof. The proof follows immediately from the definitions and Proposition 3.9. □

§ 3. Canonical Potentials in $\mathbb{R}^m$

For $x \in \mathbb{R}^m$, we let

$$h_p(a,x) = \|a-x\|^{-p} \qquad 1 \leqq p \leqq m-2$$
$$h_0(a,x) = -\log \|a-x\| \qquad \text{for } p=0.$$

For $p=m-2$, $-h_{m-2}(a,x)$ is the Newtonian kernel in $\mathbb{R}^m$ and

$$\Delta_x h_{m-2}(a,x) = 2\pi\tau_{m-2}\delta(a).$$

For q a non-negative integer, we define:

$$e_p(a,x,q) = -h_p(a,x) + h_p(a,0) + \ldots + \frac{1}{q!}\frac{\partial^q h_q}{\partial t^q}(a,tx)\Big|_{t=0},$$

which we call the canonical kernel of genus q and dimension p in $\mathbb{R}^m$. Then

$$\frac{1}{q!}\left.\frac{\partial^q h_p(a,tx)}{\partial t^q}\right|_{t=0} = \frac{P_q(a,x,p)}{\|a\|^p},$$

where $P_q(a,x,p)$ is a homogeneous polynomial of x in $\mathbb{R}^m$ of degree q. For $p=m-2$, the functions $P_q(a,x,m-2)$ are harmonic polynomials in $\mathbb{R}^m$, and for $p=m-2s$, $2\leqq 2s\leqq m$, they verify $\Delta_x^s P_q(a,x,p)=0$, where Δ^s is the Laplacian iterated s times. We then have, for $a\neq 0$, $0\leqq p\leqq m-2$

$$e_p(a,x,q)=-h_p(a,x)+\|a\|^{-p}[1+P_1(a,x,p)+\ldots+P_q(a,x,p)] \tag{3,4}$$

$$=-\|a\|^{-p}\sum_{j=q+1}^{\infty} P_j(a,x,p) \tag{3,5}$$

where the latter expression converges uniformly on every compact subset of the ball $\|x\|<\|a\|$.

Let $\|x\|=t\|a\|$ for $t>0$ and let θ be the angle between the vectors $(0,a)$ and $(0,x)$ in $\mathbb{R}^m$. Then

$$\|a-x\|^2=\|a\|^2[1-2t\cos\theta+t^2]=\|a\|^2(1-te^{i\theta})(1-te^{-i\theta}),$$

and hence (3,5) is majorized term by term by the series

$$(1-t)^{-p}=\sum_{s=0}^{\infty} b_{p,s}t^s \quad \text{with } b_{p,s}=(s!)^{-1}p(p+1)\ldots(p+s-1). \tag{3,6}$$

The case $p=0$ corresponds to the classical case studied by Weierstrass for $\mathbb{C}=\mathbb{R}^2$ of the potential related to the kernel $-\log\|a-x\|$. The series is then dominated by $-\log(1-u)=\sum_{s=1}^{\infty} u^s/s$, and the estimates obtained in the classical case are valid for the kernel $e_0(a,x,q)$ for all $\mathbb{R}^m$:

Proposition 3.13. *For $p=0$ and every $m\geqq 2$, the kernel $e_0(a,x,q)$ satisfies the following estimates for all $\mathbb{R}^m$ $\left(\text{where } a\neq 0 \text{ and } u=\dfrac{\|x\|}{\|a\|}>0\right)$:*

i) *if $q\geqq 1$, $|e_0(a,x,q)|\leqq u^{q+1}$ for $u\leqq\dfrac{q}{q+1}$. For $q=0$, $|e_0(a,x,q)|\leqq eu$ for $u<e^{-1}$.*

ii) *if $q\geqq 1$, $e_0(a,x,q)\leqq eu^q(2+\log q)$ for $u\geqq\dfrac{q}{q+1}$, and if $q=0$ $e_0(a,x,0)\leqq\log(1+u)$ for all $u>0$.*

Proof. The first part of i) stems for the fact that

$$\sum_{s=q+1}^{\infty}\frac{u^s}{s}\leqq\frac{u^{q+1}}{1-u}\cdot\frac{1}{q+1}\leqq u^{q+1} \quad \text{if } u<\frac{q}{q+1}.$$

The second part stems from $u^{-1}|\log(1-u)|\leqq e$ for $u<e^{-1}$. For ii) we write $\log|1-u|\leqq\log(1+u)$; then for $q\geqq 1$

$$e_0(a,x,q)\leqq u^q\left[\frac{1}{q}+\frac{1}{q-1}u^{-1}+\ldots+\tfrac{1}{2}u^{-q+2}+2u^{-q+1}\right].$$

Therefore

$$e_0(a,x,q)\leqq u^q\left(1+\frac{1}{q}\right)^q\left(2+\sum_2^q s^{-1}\right)\leqq eu^q(2+\log q). \qquad \square$$

Proposition 3.14. *For* $p \geqq 1$, *we let* $\tau(p,q)=\left(\frac{p+q}{p+q+1}\right)^p$. *Then for* $a \neq 0$ *and* $u=\frac{\|x\|}{\|a\|}$,

i) (3,7) $\quad |e_p(a,x,q)| \leqq C_1(p,q)\,\|a\|^{-p} u^{q+1} \quad$ if $u \leqq \tau(p,q)$;

ii) (3,8) $\quad e_p(a,x,q) \leqq C_2(p,q)\,\|a\|^{-p} u^q \quad$ if $u \geqq \tau(p,q)$

where

$$C_1(p,q)=[(p-1)!]^{-1}(p+q)^{p-1}(p+q+1)$$

and

$$\begin{aligned} C_2(p,q) &= [(p-1)!]^{-1}(q+1)(q+1)\ldots(q+p-1)\exp\left(\frac{pq}{p+q}\right) \\ &< e^p(p+q-1)^p[(p-1)!]^{-1} \end{aligned}$$

for $p \geqq 1$ *and* $q>0$. *For* $p \geqq 1$ *and* $q=0$,

$$C_1(p,0)=[(p-1)!]^{-1}(p+1)^p \quad \textit{and} \quad C_2(p,0)=1.$$

Proof. From (3,5), (3,6) and $0<u<\tau=\left(\frac{p+q}{p+q+1}\right)^p<1$, we have

$$|e_p(a,x,q)| \leqq \|a\|^{-p}\sum_{q+1}^{\infty} b_{p,s}u^s$$

a) For $p=1$, (3,6) gives $b_{p,s}=1$ and

$$\begin{aligned} |e_p(a,x,q)| &\leqq \|a\|^{-p}u^{q+1}(1+\tau+\ldots)=\|a\|^{-p}u^{q+1}(1-\tau)^{-1} \\ &= \|a\|^{-p}u^{q+1}(p+q+1) \end{aligned}$$

and (i) is proved with the value $C(1,q)$.

b) For $p \geqq 2$, we write

$$b_{p,s}=\frac{(p+s-1)!}{(p-1)!s!}=\frac{(s+1)\ldots(s+p-1)}{(p-1)!} \leqq \frac{(p+s-1)^{p-1}}{(p-1)!}$$

and

$$|e_p(a,x,q)| \leqq \|a\|^{-p}\frac{u^{q+1}}{(p-1)!}\sum_{m=0}^{\infty}(p+q+m)^{p-1}\tau^m$$

We remark that for $m \geqq 1$ and $p \geqq 1$, we have

$$1<\left(\frac{p+q+m}{p+q+m-1}\right)^{p-1} \leqq \left(\frac{p+q+1}{p+q}\right)^{p-1}=\sigma$$

from which it follows that

$$\begin{aligned} |e_p(a,x,q)| &\leqq \frac{\|a\|^{-p}}{(p-1)!}(p+q)^{p-1}(1+\sigma\tau+\ldots) \\ &\leqq \frac{\|a\|^{-p}}{(p-1)!}(p+q)^{p-1}(1-\sigma\tau)^{-1}. \end{aligned}$$

Thus $\sigma\tau = \dfrac{p+q}{p+q+1}$ and $(1-\sigma\tau)^{-1} = p+q+1$, which proves (3,7) with the value $C_1(p,q)$.

In order to calculate $C_2(p,q)$, we use the first equality of (3,4). Since $-h_p(a,x)$ is negative, we obtain

$$\begin{aligned} e_p(a,x,q) &\leqq \|a\|^{-p}[1+b_{p,1}u+\ldots+b_{p,q}u^q] \\ &\leqq \|a\|^{-p}\tau^{-q}u^q[1+b_{p,1}\tau+\ldots+b_{p,q}\tau^q]. \end{aligned}$$

This gives immediately $C_2(p,0)=1$. In the general case, since $\tau<1$ and there are $(q+1)$ terms in the brackets and since $b_{p,s} \leqq b_{p,q}$, we have:

$$\begin{aligned} e_p(a,x,q) &\leqq u^q(q+1)b_{p,q}\left(\frac{p+q+1}{p+q}\right)^{pq} \\ &\leqq u^q(q+1)\frac{(p+q-1)!}{(p-1)!\,q!}\exp\left(\frac{pq}{p+q}\right). \end{aligned}$$

□

Proposition 3.15. *Let $a, x \in \mathbb{R}^n$ with $a \neq 0$ and $m \geqq 2$ and let p, q be positive integers. Then*

(3,9)
$$e_p(a,x,q) \leqq C(p,q)\frac{\|x\|^{q+1}}{\|a\|^{p+q}(\|x\|+\|a\|)}$$

a) *for* $p \geqq 1$, $C(p,q) = \dfrac{e^p}{(p-1)!}(p+q+1)^p$;

b) *for* $p=0$ *and* $q \geqq 1$, $C(0,q) = 3e(2+\log q)$;

c) *for* $p=q=0$, $C(0,0)=1$.

Proof. We choose $C(p,q) = \sup[(1+\tau)C_1(p,q),\ (1+\tau)^{-1}C_2(p,q)]$ and use Proposition 3.14 to obtain the estimate. If we replace τ by $\left(\dfrac{p+q}{p+q+1}\right)^p$, we obtain the value for $p \geqq 1$. The case $p=0$ follows easily from Proposition 3.13. □

Remark. The bounds for the kernel $e_p(a,x,q)$ do not depend on m in $\mathbb{R}^m$. In the sequel, we use these in $\mathbb{C}^n = \mathbb{R}^{2n}$.

§4. The Canonical Representation of Entire Functions of Finite Order

If $X=(f_j, U_j)$ is a Cousin data in $\mathbb{C}^n$ and $\nu_X(r)$ is of finite order ρ, we are interested in finding an entire function $F(z)$ whose zero set is exactly X such that $M_F(r) = \sup_{\|z\| \leqq r} \log|F(z)|$ is minimal. In fact, we shall treat this problem as a special case of a larger problem. For $V=\log|F|$, we have

$$\frac{i}{\pi}\partial\bar{\partial}V=\frac{i}{\pi}\partial\partial\log|F|=\theta_X \tag{3,10}$$

where θ_X is the current associated with the Cousin data X. Then we have to determine a plurisubharmonic function V which is a solution of $\frac{i}{\pi}\partial\bar{\partial}V=\theta$, where θ is a given (1,1) positive closed current of degree 1. From (3,10) we deduce, with $\theta=\theta_X$:

$$\frac{i}{\pi}\partial\bar{\partial}V\wedge\beta_{n-1}=\theta\wedge\beta_{n-1}=\sigma_\theta; \tag{3,11}$$

σ_θ is a positive distribution, so it is a positive measure, the area of the Cousin data X with multiplicity (see Theorem 3.2). We first construct V as a potential in $\mathbb{R}^{2n}=\mathbb{C}^n$. Then we prove that for θ of finite order, the solution V of (3,11) is actually a solution of (3,10).

As in the classical case for $n=1$ and the Hadamard Theorem, we shall show that the kernel $e_{2n-2}(a,z,q)$ of genus q guarantees the convergence of the potential $I_q(z)=k_{2n-2}^{-1}\int e_{2n-2}(a,z,q)d\sigma_X(a)$. An essential step in the proof will be to show that $I_q(z)$ is in fact plurisubharmonic and gives the solution of (3,10). Moreover, for X a Cousin data, we shall see that $I_q(z)=\log|F(z)|$ for $F(z)$ an entire function.

Proposition 3.16. *If θ is of genus q $\left(\text{that is } \int_a^\infty t^{-s-1}dv_\theta(t)<\infty \text{ for } s\geq q\right)$ and $0\notin\operatorname{supp}\theta$, then*

$$I_q(z)=k_{2n-2}^{-1}\int_{r_0}^{\infty} e_{2n-2}(a,z,q)d\sigma_\theta(a), \tag{3,12}$$

with $k_{2n-2}=\dfrac{2\pi^{n-1}}{(n-2)!}$, converges uniformly on every compact subset of $\mathbb{C}^n$ and gives a solution of the equation (3,11). If we write ΔV for the distribution $\frac{i}{\pi}\partial\bar{\partial}V\wedge\beta_{n-1}=\frac{1}{2\pi}\Delta V\cdot\beta_n$, we have

$$\Delta I_q=2\pi\sigma_\theta. \tag{3,13}$$

Remark. We call $I_q(z)$ the canonical potential of genus q for θ_X.

Proof. Suppose $\|z\|\leq R$ and $R'>R\tau^{-1}$ for τ chosen as in Proposition 3.14. Then for (3,12), we obtain

$$\begin{aligned}&k_{2n-2}^{-1}\int_{\|a\|>R'}|e_{2n-2}(a,z,q)|d\sigma_\theta(a)\\&\leq k_{2n-2}^{-1}C_1(2n-2,q)\|z\|^{q+1}\int_{\|a\|>R'}\|a\|^{2-2n-q-1}d\sigma_\theta(a),\end{aligned}$$

and the right hand side converges by Proposition 3.9, since $\nu_\theta(t)$ is of genus q. This establishes the uniform convergence. On the other hand, $e_{2n-2}(a,z,q)$ differs from $h_{2n-2}(a,z)$ by a finite sum of harmonic polynomials, from which it follows that $\Delta e_{2n-2}(a,z,q)=k_{2n-2}2\pi\delta(a)$, and (3,13) holds. □

Theorem 3.17. *The canonical potential $I_q(z)$ defined by* (3,12) *with respect to a positive closed* (1,1) *current θ of genus q and such that $B(0,r_0)\cap\operatorname{supp}\theta=\emptyset$ satisfies the inequality*

(3,14) $$I_q(z)\leqq A(n,q)r^q\left[\int_{r_0}^{r} t^{-q-1}\nu_\theta(t)\,dt+r\int_{r}^{\infty} t^{-q-2}\nu_\theta(t)\,dt\right]$$

for $\|z\|=r$. We can choose $A(n,q)=(2n-2)^{-1}C(2n-2,q)(q+2n-1)$.

Proof. From (3,9), we obtain that

$$\sup_{\|z\|=r} I_q(z)=M(r)\leqq k_{2n-2}^{-1}C(2n-2,q)r^{q+1}\int_{r_0}^{\infty}\frac{d\sigma_\theta(t)}{(t+r)t^{q+2n-2}}.$$

We integrate by parts in order to express the right hand side in terms of $\sigma_\theta(t)$ and hence $\nu_\theta(t)$:

$$\int_{r_0}^{\infty}\frac{d\sigma_\theta(t)}{(t+r)t^{q+2n-2}}=\left[\frac{\sigma_\theta(t)}{(t+r)t^{q+2n-2}}\right]_{r_0}^{\infty}+\int_{r_0}^{\infty}\frac{(at+br)}{(t+r)^2t^{q+2n-1}}\sigma_\theta(t)\,dt$$

with $a=q+2n-1$, $b=q+2n-2$. The first term is zero, since $\sigma_\theta(r_0)=0$ and $\lim_{t\to\infty} t^{-q-2n-1}\sigma_\theta(t)=\lim_{t\to\infty} t^{-q-1}\nu_\theta(t)=0$, and $\nu_\theta(t)$ is of genus q.

Thus

$$\int_{r_0}^{\infty}\frac{d\sigma_\theta(t)}{(t+r)t^{q+2n-2}}\leqq(2n+q-1)\int_{r_0}^{\infty}\frac{\sigma_\theta(t)\,dt}{(t+r)t^{q+2n-1}}$$
$$=(2n+q-1)\tau_{2n-2}\int_{r_0}^{\infty}\frac{\nu_\theta(t)\,dt}{(t+r)t^{q+1}}.$$

It then follows that

$$M(r)\leqq k_{2n-2}^{-1}\tau_{2n-2}(2n+q-1)C(2n-2,q)r^{q+1}\int_{r_0}^{\infty}\frac{\nu_\theta(t)\,dt}{(t+r)t^{q+1}}$$
$$\leqq A(n,q)r^{q+1}\int_{r_0}^{\infty}\frac{\nu_\theta(t)\,dt}{(t+r)t^{q+1}},$$

and

$$\int_{r_0}^{\infty}\frac{\nu_\theta(t)\,dt}{(t+r)t^{q+1}}\leqq r^{-1}\int_{r_0}^{r}\frac{\nu_\theta(t)\,dt}{t^{q+1}}+\int_{r}^{\infty}\frac{\nu_\theta(t)\,dt}{t^{q+2}}.$$ □

Remark. A better estimate can be obtained by distinguishing between the intervals $r_0<t<\tau^{-1}r$ and $\tau^{-1}r<t<+\infty$. In this way, we have

$$I_q(z) \leqq (2n+q-1)(2n-2)^{-1}\left[C_1(2n-2,q)r^{q+1}\int_{r\tau^{-1}}^{\infty} v_\theta(t)t^{-q-2}dt\right.$$
$$\left.+C_2(2n-2,q)\,r^q\int_{r_0}^{r\tau^{-1}} v_\theta(t)\,t^{-q-1}dt\right]$$
$$+\tau^q\frac{v_\theta(r\tau^{-1})}{(2n-2)}[C_2(2n-2,q)-C_1(2n-2,q)\,\tau].$$

The canonical potential is $\mathbb{R}^{2n}$ subharmonic. Thus, using (3,12) and Gauss' Theorem, we have

$$2\pi\sigma_\theta(r)=\int_{\|z\|\leqq r}\Delta I_q\cdot\beta_n=\int\frac{\partial I_q}{\partial r}(r\alpha)r^{2n-2}d\omega_{2n}(\alpha)$$
$$=\omega_{2n}r^{2n-2}\frac{\partial}{\partial\log r}\lambda(0,r,I_q)$$

where, since $\lambda(0,r,I_q)$ is a convex function of $-r^{2-2n}$, it has a derivative except perhaps for a countable set of values of r. Thus

$$v_\theta(r)=\frac{\partial\lambda(0,r,I_q)}{\partial\log r}. \tag{3,15}$$

Theorem 3.18. *The canonical potential $I_q(z)$ with respect to the positive closed (1,1) current θ whose support does not contain the origin has the following properties:*

i) *$M(r)$ and $v_\theta(r)$ are of the same order ρ,*

ii) *if ρ is not an integer and if $v_\theta(r)$ is of minimal, normal, or maximal type with respect to r^ρ, then so is $M(r)$ and the integrals $\int_{r_0}^{\infty} M(t)t^{-\rho-1}dt$ and $\int_{r_0}^{\infty} v_\theta(t)t^{-\rho-1}dt$ converge or diverge together;*

iii) *if ρ is an integer, $M(r)$ and $v_\theta(r)$ are not necessarily of the same type, but if $\int v_\theta(t)t^{-\rho-1}dt<+\infty$, that is, if the genus of θ is $q=\rho-1$, then $M(r)$ is of minimal type with respect to r^ρ.*

Proof. i) From (3,15), we see that $\lambda(0,r,I_q)$ is a convex increasing function of $\log r$. Thus, since $\lambda(0,r,I_q)\geqq I_q(0)=0$, we obtain

$$v_\theta(r)\leqq\lambda(0,er,I_q)-\lambda(0,r,I_q)\leqq\lambda(0,er,I_q)\leqq M(er), \tag{3,16}$$

and hence $\rho'=\text{order } I_q\geqq\rho=\text{order } v_\theta(r)$. In the other direction, we use (3,14). If $v_\theta(t)\leqq C(\varepsilon)t^{\rho+\varepsilon}$ for $\varepsilon>0$, then (3,14) gives $M(r)\leqq C'(\varepsilon)A(n,q)r^{\rho+\varepsilon}$, so that $\rho'\leqq\rho$.

ii) If ρ is not an integer, the genus q of θ satisfies $q<\rho<q+1$. If γ is the type of $v_\theta(t)$ then $v_\theta(t)\leqq(\gamma+\varepsilon)t^\rho$ for $t\geqq R$, and we obtain from (3,14), letting

$$\alpha = \int_{r_0}^{R} t^{-q-1} v_\theta(t)\,dt,$$

(3,17) $$M(r) \leqq A(n,q)[\alpha + (\gamma+\varepsilon)]\left(\frac{r^\rho}{\rho-q} + \frac{r^\rho}{q+1-\rho}\right),$$

which shows that the type γ' of $M(r)$ is at most $C_1\gamma$. On the other hand, from (3,16), we see that $\gamma \leqq e\gamma'$. In the same way, (3,16) shows that the convergence of $\int_{r_0}^{\infty} M(t) t^{-\rho-1} dt$ implies that of $\int_{r_0}^{\infty} v_\theta(t) t^{-\rho-1} dt$. Conversely, using (3,14)

$$\int_{r_0}^{R} M(r) r^{-\rho-1} dr \leqq A(n,q)\left[\int_{r_0}^{R} r^{q-\rho-1} dr \int_{r_0}^{r} v_\theta(t)^{-q-1} dt \right.$$
$$\left. + \int_{r_0}^{R} r^{q-\rho} dr \int_{r}^{\infty} v_\theta(t) t^{-q-2} dt\right].$$

By changing the order of integration and observing that $q-\rho<0<q+1-\rho$, we obtain

$$\int_0^R r^{q+\rho-1} dr \int_0^r t^{-q-1} v_\theta(t)\,dt = \int_0^R t^{-q-1} v_\theta(t)\,dt \int_t^R r^{q-\rho-1} dr$$
$$\leqq (\rho-q)^{-1} \int_0^R t^{-\rho-1} v_\theta(t)\,dt < (\rho-q)^{-1} \int_0^\infty t^{-\rho-1} v_\theta(t)\,dt$$

and

$$\int_R^\infty r^{q-\rho} dr \int_r^\infty t^{-q-2} v_\theta(t)\,dt = \int_R^\infty t^{-q-2} v_\theta(t)\,dt \int_R^t r^{q-\rho} dr$$
$$\leqq (q-\rho+1)^{-1} \int_R^\infty t^{-\rho-1} v_\theta(t)\,dt < +\infty,$$

which proves (ii).

iii) If ρ is an integer and $\int v_\theta(t)\, t^{-\rho-1} dt < +\infty$, that is $q=\rho-1$, then by Proposition 3.9, $\lim_{t\to\infty} v_\theta(t)\, t^{-\rho} = 0$. Let $R>r_0$ be such that for $\varepsilon>0$, $v_\theta(t) < \varepsilon\, t^\rho$ for $t>R$ and $\int_R^\infty v_\theta(t) t^{-\rho-1} dt < \varepsilon$. Then for $r>R$, we obtain from (3,17) that

$$r^{-\rho} M(r) \leqq A(n,q)\left[\frac{1}{r}\int_{r_0}^{R} v_\theta(t) t^{-\rho} dt + \frac{\varepsilon}{r}\int_R^r dt + \int_R^\infty v_\theta(t) t^{-\rho-1} dt\right]$$
$$\leqq A(n,q)\left[C r^{-1} + \frac{\varepsilon(r-R)}{r} + \varepsilon\right],$$

which shows that $M(r)$ is of minimal type of order ρ. □

We now develop an analogue of Theorem 3.18 for proximate orders.

Theorem 3.19. *Let θ be a closed positive* (1,1) *current such that $0 \notin \operatorname{supp}\theta$ and such that its indicator $v_\theta(t)$ is of finite order ρ which is not an integer and* *normal type with respect to the proximate order* $\rho(r)$. *Then* $I_q(z)$, *its canonical potential, is also of normal type with respect to the proximate order* $\rho(r)$.

We shall need the following Lemma:

Lemma 3.20. *If* $\rho(r)$ *is a proximate order, then for* $\lambda<\rho+1$ *and* $r>R>r_0$,

$$l(r)=\int_R^r t^{\rho(t)-\lambda}dt=(\rho+1-\lambda)^{-1}r^{\rho(r)+1-\lambda}+o(r^{\rho(r)+1-\lambda})$$

and for $\lambda>\rho+1$,

$$l'_\lambda(r)=\int_r^\infty t^{\rho(t)-\lambda}dt=(\lambda-\rho-1)^{-1}r^{\rho(r)+1-\lambda}+o(r^{\rho(r)+1-\lambda}).$$

Proof. After an integration by parts, we obtain

$$\int_R^r t^{\rho-\lambda}t^{\rho(t)-\rho}dt=(\rho+1-\lambda)^{-1}[t^{\rho(t)+1-\lambda}]_R^r$$

$$-(\rho+1-\lambda)^{-1}\int_R^r [t^{\rho(t)-\lambda}(\rho(t)-\rho)$$

$$+t^{\rho(t)+1-\lambda}\rho'(t)\log t]dt=I_1+I_2.$$

It follows from Definition 1.15 that given $\varepsilon>0$, there exists T_ε such that $|I_2|\leqq\varepsilon\int_R^r t^{\rho(t)-\lambda}dt$ for $r>T_\varepsilon$ (we recall that $\rho-\lambda>-1$ implies that $\lim_{r\to\infty}\int_R^r t^{\rho(t)-\lambda}dt=+\infty$). Hence

$$\left|\int_R^r t^{\rho(t)-\lambda}dt-(\rho+1-\lambda)^{-1}r^{\rho(r)+1-\lambda}\right|\leqq\varepsilon(1+\varepsilon)^{-1}[r^{\rho(r)+1-\lambda}+C]$$

where $C=(\rho+1-\lambda)^{-1}R^{\rho(R)+1-\lambda}$.

For $\lambda>\rho+1$, we obtain

$$\int_r^\infty t^{\rho-\lambda}t^{\rho(t)-\rho}dt=(\lambda-\rho-1)^{-1}r^{\rho(r)+1-\lambda}$$

$$+(\lambda-\rho-1)\int_r^\infty t^{\rho(t)-\lambda}[(\rho(t)-\rho)$$

$$+\rho'(t)\cdot t\log t]dt=\tilde I_1+\tilde I_2.$$

It follows from Definition 1.15 that given $\varepsilon>0$, there exists T'_ε such that $|\tilde I_2|\leqq\varepsilon\int_r^\infty t^{\rho(t)-\lambda}dt$ and thus

$$\left|\int_r^\infty t^{\rho(t)-\lambda}dt-(\lambda-\rho-1)^{-1}r^{\rho(r)+1-\lambda}\right|\leqq\varepsilon(1-\varepsilon)^{-1}r^{\rho(r)+1-\lambda}. \qquad \square$$

Proof of Theorem 3.19. From (3,16), we have $v_\theta(t) \leqq M(et)$ and hence

$$v_\theta(t)t^{-\rho(t)} \leqq M(et)t^{-\rho(t)} \leqq [M(et)(et)^{-\rho(et)}][et]^{\rho(et)-\rho(t)} \cdot e^{\rho(t)}.$$

It follows from Definition 1.15 that $\lim_{t\to\infty} e^{\rho(t)} = e^\rho$ and from Theorem 1.18 that $\lim_{t\to\infty} [et]^{\rho(et)-\rho(t)} = 1$. Thus $\limsup_{t\to\infty} v_\theta(t)t^{-\rho(t)} \leqq e^\rho \limsup_{t\to\infty} M(t)t^{-\rho(t)}$. In the other direction, with $q < \rho < q+1$ and $v_\theta(t) < (C+\varepsilon)t^{\rho(t)}$ for $r > R > r_0$, we obtain via (3,17)

$$M(r) \leqq A(n,q)r^q \left\{ \int_{r_0}^{R} v_\theta(t)^{-q-1} dt + (C+\varepsilon)[l_{q+1}(r) + l'_{q+2}(r)r] \right\}$$

$$\leqq A(n,q)(C+\varepsilon)\left[\frac{1}{\rho-q} + \frac{1}{q+1-\rho}\right] r^{\rho(r)} + o(r^{\rho(r)})$$

by Lemma 3.20. □

Remark. Starting with bounds for the growth of $M(r) = \sup_{\|z\|=r} I_q(z)$, it is easy to obtain a control of the mean values $\lambda(0,r,I_q)$ on $\|z\| = r$ and of

$$A(0,r,I_q) = (\tau_{2n} r^{2n})^{-1} \int_{\|z\| \leqq r} |I_q(z)| \, d\tau_{2n}.$$

We set $I_q^+ = \sup(I_q, 0)$, $I_q^- = \sup(-I_q, 0)$. From the subharmonicity of I_q we obtain $0 \leqq \lambda(0,r,I_q) = \lambda(0,r,I_q^+) - \lambda(0,r,I_q^-)$ from which it follows that

$$0 = \lambda(0,r,I_q) \leqq \lambda(0,r,I_q^+) \leqq M(r)$$

and hence

$$\lambda(0,r,|I_q|) \leqq 2M(r) \quad \text{and} \quad A(0,r,|I_q|) \leqq 2M(r). \tag{3,18}$$

§5. Solution of the $\partial\bar{\partial}$ Equation

We have already seen that the canonical potential $I_q(z)$ associated with a positive closed (1,1) current θ of genus q and such that $0 \notin \operatorname{supp}\theta$ solves the equation $\frac{1}{2\pi}\Delta I_q = \sigma_\theta$. In this paragraph, we shall show that it solves in fact the more restrictive condition of equation (3,10).

Let $\theta = \frac{i}{\pi}\sum_{p,q} \theta_{p,q} dz_p \wedge d\bar{z}_q$, where the $\theta_{p,q}$ are complex measures. Then $\theta' = \frac{i}{\pi}\partial\bar{\partial} I_q - \theta = \frac{i}{\pi}\sum_{p,q} \theta'_{p,q} dz_p \wedge d\bar{z}_q$ has the following properties:

i) its trace $\sum \theta'_{pp}$ is the zero measure;
ii) θ' is ∂ and $\bar{\partial}$ closed.

Proposition 3.21. *If a current θ' of type (1,1) is closed and has zero trace, then it can be represented by a differential form with harmonic coefficients.*

Proof. Let us first suppose that the coefficients $\theta'_{p,q}$ are twice continuously differentiable. Then, since $d\theta' = \partial\theta' + \bar{\partial}\theta' = 0$, we obtain

$$\frac{\partial \theta'_{p,q}}{\partial z_m} = \frac{\partial \theta'_{m,q}}{\partial z_p}; \quad \frac{\partial \theta'_{p,q}}{\partial \bar{z}_m} = \frac{\partial \theta'_{p,m}}{\partial \bar{z}_q} \quad \text{for } m \neq p, q.$$

Thus

$$4\Delta\theta'_{p,q} = \sum_m \frac{\partial^2 \theta'_{p,q}}{\partial z_m \partial \bar{z}_m} = \frac{\partial^2}{\partial z_p \partial \bar{z}_q} \sum \theta'_{m,m} = 0.$$

To treat the general case, we take $\alpha \in \mathscr{C}_0^\infty(B(0,1))$ such that $\int \alpha(z)\, d\tau_{2n} = 1$ and $\alpha \geqq 0$ and set $\alpha_\varepsilon(z) = \alpha\left(\frac{z}{\varepsilon}\right)\varepsilon^{-2n}$, where α depends only on $\|z\|$. Then $\theta'_{p,q} \star \alpha_\varepsilon = \int [\theta'_{p,q} d\tau_{2n}(u)][\alpha_\varepsilon(z-u)]$ is a $\mathscr{C}^\infty$ function, and the current $\sum_{p,q} (\theta'_{p,q} \star \alpha_\varepsilon) dz_p \wedge d\bar{z}_q$ satisfies the hypotheses of the Proposition. Hence the coefficients $\theta'_{p,q} \star \alpha_\varepsilon$ are harmonic functions. Using the mean value property for harmonic functions we obtain $[\theta'_{p,q} \star \alpha_{\varepsilon'}] \star \alpha_\varepsilon = \theta'_{p,q} \star \alpha_{\varepsilon'}$, so $[\theta'_{p,q} \star \alpha_{\varepsilon'} - \theta'_{p,q}] \star \alpha_\varepsilon = 0$ for every $\varepsilon, \varepsilon'$. Hence when $\varepsilon \to 0$, we obtain $\theta'_{p,q} = \theta'_{p,q} \star \alpha_{\varepsilon'}$, which shows that as a current $\theta'_{p,q}$ is equivalent to a form with harmonic coefficients. □

Lemma 3.22. *Let $h(x)$ be a harmonic function for $\|x\| < R$ in $\mathbb{R}^p$. Then $f(X)$, its complexification, is holomorphic for $\|X\| < R/\sqrt{2}$ in the space $\mathbb{C}^p$ and, setting $m_c(r') = \sup_{\|X\| \leqq r'} |f(X)|$ and $m(r) = \sup_{\|x\| \leqq r} |h(x)|$ for $r'\sqrt{2} < r < R$, we have*

$$m_c(r') \leqq \left(1 + \frac{r'^2}{r^2}\right)\left(\frac{1}{2} - \frac{r'^2}{r^2}\right)^{-p/2} m(r).$$

Proof. The Poisson Integral Representation of $h(x)$ for $\|x\| \leqq r$ defines $f(X)$ by $f(X) = \omega_p^{-1} \int f(r\alpha) r^{p-2} \frac{r^2 - \sum X_k^2}{[\sum (r\alpha_k - X_k)^2]^{p/2}} d\omega_p(\alpha)$; it is the unique holomorphic function in $\mathbb{C}^p$ which takes on the values $h(x)$ on $\mathbb{R}^p$. A simple calculation shows that for $\|x\| = r'$ and $r'\sqrt{2} < r$

$$(3{,}19) \quad \left|\sum_{k=1}^p (r\alpha_k - X_k)^2\right| \geqq \left|\operatorname{Re} \sum_{k=1}^p (r\alpha_k - X_k)^2\right| \geqq (r - r'\cos\varphi)^2 - r'^2 \sin^2\varphi,$$

where $\|x\| = \|X\| \cos\varphi = r'\cos\varphi$ defines φ. Thus, for $\|X\| < r/\sqrt{2}$, (3,19) is never zero and so $f(X)$ is holomorphic for $\|X\| < r/\sqrt{2}$ in $\mathbb{C}^p$. For $\|X\| < r/\sqrt{2}$, $\left|\sum_{k=1}^p (r\alpha_k - X_k)^2\right| \geqq r^2/2 - r'^2$, which suffices to prove the Lemma. □

Corollary 3.23. i) *If $h(x)$ is harmonic in $\mathbb{R}^p$, then its complexification $f(X)$ in $\mathbb{C}^p$ is an entire function and its growth in $\mathbb{C}^p$ satisfies $m_c(r) \leqq C_p m(2r)$.*

ii) *If $|h(x)| \leqq C\|x\|^s$, $x \in \mathbb{R}^p$, then $|f(X)| \leqq C'\|X\|^s$ with $C' = 5 \cdot 2^{s+p-2}C$.*

iii) *If $h(x)$ is harmonic in $\mathbb{R}^p$ and verifies $\|h(x)\| \leqq C\|x\|^s$, then h is a polynomial of degree $q \leqq s$. If $s < 1$, then $h(x) \equiv 0$.*

Proof. i) and ii) follows from (3,22) and iii) from Corollary 1.7. □

Returning to the form $\theta' = i/\pi\ (\partial\bar{\partial} I_q - \theta)$, we see that we can write

$$\theta' = i/\pi \sum_{p,q} A_{p,k}(z) dz_p \wedge d\bar{z}_k, \tag{3,19}$$

where $A_{p,k}(z)$ is harmonic and is zero at the origin as well as all of its derivatives up to order $q-2$ if $q \geqq 2$, since θ vanishes in a neighborhood of the origin and I_q is zero at the origin as well as all of its derivatives up to order q. Thus, we obtain:

Proposition 3.24. $\theta' = i/\pi(\partial\bar{\partial} I_q - \theta)$ *is a* (1,1) *form with harmonic coefficients* $A_{p,k}$. *Moreover for* $q \geqq 2$, *their derivatives up to order* $(q-2)$ *are zero at the origin (where q is the genus of θ).*

We shall show in fact that $A_{p,k}(z) \equiv 0$.

Proposition 3.25. *There exist constants $\tilde{C}_{p,k}$ and $\tilde{C}'_{p,k}$ such that*

$$|A_{p,k}(z)| \leqq r^{-2}[\tilde{C}_{p,k} \nu(2r) + \tilde{C}'_{p,k} M(2r)] \tag{3,20}$$

where $M(r) = \sup_{\|z\| \leqq r} I_q(z)$.

Proof. Let $\alpha_\varepsilon(z)$ be the function constructed in Proposition 3.21. Since $A_{p,k}(z)$ is harmonic, $A_{p,k}(z) = A_{p,k} \star \alpha_\varepsilon(z)$, and we have

$$A_{p,k}(z) = \left(\frac{\partial^2}{\partial z_p \partial \bar{z}_k} I_q(z)\right) \star \alpha_\varepsilon - \theta_{p,k} \star \alpha_\varepsilon = S_1(z) - S_2(z).$$

i) $S_1(z)$ will be estimated from $\left|\dfrac{\partial^2 \alpha_\varepsilon}{\partial z_p \partial \bar{z}_k}\right| \leqq \dfrac{M_{p,k}}{\varepsilon^{2n+2}}$, where $M_{p,k} = \sup\left|\dfrac{\partial^2}{\partial z_p \partial \bar{z}_k} \alpha_1\right|$, from which we have

$$|S_1(z)| \leqq M_{p,k} \varepsilon^{-2n-2} \int_{\|u\| \leqq \varepsilon} |I_q(z+u)| d\tau(u).$$

By letting $\varepsilon = r$, we obtain

$$\begin{aligned} |S_1(z)| &\leqq M_{p,q} r^{-2n-2} \int_{\|u\| < r} |I_q(z+u)| \beta_n(u) \\ &\leqq M_{p,k} r^{-2n-2} 2M(2r) \tau_{2n} r^{2n} \\ &\leqq r^{-2} \tilde{C}'_{p,k} M(2r) \end{aligned}$$

where $\tilde{C}'_{p,k} = 2^{2n+1} \tau_{2n} M_{p,k}$.

ii) From Theorem 2.16, we know that the coefficients $\theta_{p,k}$ satisfy $\|\theta_{p,k}\|_K \leqq 2\|\sigma\|_K$ for every compact set K. Thus, if $M_0 = \sup|\alpha_1|$,

$$|S_2(z)| = |\theta_{p,k} \star \alpha_\varepsilon| \leqq 2 \int \alpha_\varepsilon(z-u)\,d\sigma(u) \leqq 2\|\sigma\|_{B(z,\varepsilon)} M_0 \varepsilon^{-2n},$$

and if $\varepsilon = r$, we have

$$|S_2(z)| \leqq M_0 r^{-2n} \sigma(2r) \leqq r^{-2} C_{p,k} \nu_\theta(2r). \qquad \square$$

Theorem 3.26. *The canonical potential $I_q(z)$ with respect to a positive closed current θ of degree 1 whose support does not contain the origin and which is of finite order ρ and genus q is plurisubharmonic in $\mathbb{C}^n$ and satisfies equation* (3,10).

Proof. It is sufficient to show that $A_{p,k}(z) \equiv 0$ for all p, k. First for all, we note that Corollary 3.23 and Proposition 3.25 show that for $\nu(t)$ of finite order $A_{p,k}(z)$ is a polynomial. We shall consider several cases:

i) if $\rho < 2$, Proposition 3.24 shows that $|A_{p,k}(z)|$ tends to zero when $\|z\|$ tends to ∞, and hence Corollary 3.23 imples that $A_{p,k} \equiv 0$;

ii) if $\rho > 2$ is not integral, then the genus q satisfies $q < \rho < q+1$ and $A_{p,k}(z)$ is a polynomial of degree $M \leqq q-2$ by Proposition 3.25 and Corollary 3.23. But the derivatives of $A_{p,k}(z)$ of order up to and including $(q-2)$ are zero at the origin, so $A_{p,k}(z) \equiv 0$;

iii) if $\rho \geqq 2$ is an integer and if $\rho = q$, then $A_{p,k}$ is or degree at most $(q-2)$ and the conclusion follows as in (ii). In particular, if $\rho = q = 2$, $|A_{p,k}(z)|$ is bounded and zero at the origin;

iv) if $\rho \geqq 2$ is an integer and $q = \rho - 1$, then $\nu_\theta(t)$ is of minimal type of order ρ and, from Theorem 3.18, $M(r) = \sup\limits_{\|z\| \leqq r} I_q(z)$ is also of minimal type. Thus, from (3,20) Proposition 3.25, $|A_{p,k}(z)| \leqq \varepsilon(r) r^{\rho-2}$ where $\lim\limits_{r \to \infty} \varepsilon(r) = 0$. Hence, by Corollary 3.23, $A_{p,k}$ is a polynomial of degree $\rho - 3$ at most, if $\rho \geqq 3$, and if $\rho = 2$, $A_{p,k} \equiv 0$. Since for $\rho \geqq 3$, all derivatives of order $q-2 = \rho - 3$ or less are zero at the origin, so $A_{p,k} \equiv 0$. $\square$

Let $V(z)$ be a solution of (3,10) in $\mathbb{C}^n$. Then $i\partial\bar{\partial}(V - I_q) = 0$ and hence $V = I_q + H$ where H is pluriharmonic in $\mathbb{C}^n$, thus the real part of an entire function $\varphi(z)$ (*cf. Proposition* 2.29). *If* $V(z)$ is of finite order, then $H = \operatorname{Re}\varphi$ is of polynomial growth and φ is a polynomial by Corollary 3.23. We distinguish two cases:

i) the order ρ of θ is not an integer, in which case φ is a polynomial of degree m with order $V = \sup(\rho, m)$;

ii) ρ is an integer.

This leads to:

Proposition 3.27. *If the positive closed* (1,1) *current* θ *has an indicator* $v_\theta(r)$ *of finite order* $\rho \geqq 0$, *then there exists a real finite dimensional vector space* E_ρ *of all the solutions of minimal order* ρ *of the equation* $i\partial\bar\partial V = \theta$:

i) *if* ρ *is not an integer* $\rho - 1 < q < \rho$ *then*

$$V(z) = V(0) + 2\,\mathrm{Re}\left\{\sum_{j=1}^{q} \frac{1}{j!}\frac{\partial^j}{\partial t^j}[V(tz)]_{t=0}\right\} + I_q(z), \tag{3,21}$$

where $V(z) = P_q(z) + I_q(z)$ *and* $P_q(z)$ *is determined by the value of* V *and its derivatives up to order* q *at the origin; we then have order* $P_q \leqq q < \rho =$ *order* I_q.

ii) *if* ρ *is an integer and* $q = \rho$, *the solutions of order* ρ *are all given by* (3,21); *if* $q = \rho - 1$, *then* $V(z) = P_{\rho-1}(z) + I_q(z) + P_\rho(z)$ *where* P_ρ *is the real part of a homogeneous polynomial of degree* ρ *in* z.

In conclusion, we resume:

Theorem 3.28. *If* θ *is a positive closed* (1,1) *current of finite order* ρ *in* $\mathbb{C}^n$ *such that* $0 \notin \mathrm{supp}\,\theta$, *the canonical solution of* (3,10) *is a solution of smallest order; its order is the order of the indicator* $v_\theta(r)$. *The solutions* V *of* (3,10) *of finite order* ρ' *are obtained by adding to* I_q *the real part of a polynomial* $P(z)$ *of degree at most* ρ'. *An arbitrary solution of* (3,10) *is obtained by the addition of the real part of an entire function.*

§6. The Case of a Cousin Data

Let $X = (f_j, U_j)$ be a Cousin data such that $0 \notin Y(X)'$ and let q be the genus of $v_X(t)$ and $\sigma_X = \theta_X \wedge \beta_{n-1}$. Then for

$$I_q(z) = k_{2n-2}^{-1} \int_{r_0}^{\infty} e_{2n-2}(a, z, q)\, d\sigma_X(a),$$

we have

$$i\partial\bar\partial(I_q - \log|f_j|) = 0 \quad \text{in } U_j \tag{3,22}$$

since in U_j, $\theta_X = i\partial\bar\partial \log|f_j|$ and $i\partial\bar\partial(I_q - \theta_X) = 0$.

Proposition 3.29. *There exists an entire function* $F_0(z)$ *such that*

$$\log|F_0(z)| = I_q(z). \tag{3,22}$$

Proof. Let B be a ball such that $B \cap Y(X) = \emptyset$. If the Proposition is true, $F_0(z)$ is non-zero in B and hence $G(z) = \log F_0(z) = A_1(z) + iA_2(z)$ is holomorphic in B. Then $\partial\bar G = \bar\partial G = 0$, or equivalently $\bar\partial A_1 - i\bar\partial A_2 = 0$, hence $dG = \partial A_1 + i\partial A_2 = 2\partial A_1$. We also have $I_q = \log|F_0| = A_1$ so $dG = d\log F_0 = 2\partial A_1$

$=2\partial I_q$ and thus

$$G=\log F_0(z)=\log F(0)+2\int_0^z \partial I_q(\xi), \tag{3,23}$$

where the integral is taken over a polygonal path from 0 to z compact in $\mathbb{C}^n - Y(X)$.

Let us show that (3,23) defines the logarithm of a non-zero holomorphic function. If we replace the path γ by the path γ', we must verify that we obtain a multiple of $2\pi i$ over the closed path $\gamma_0=(\gamma, \gamma')$. The open set $\mathbb{C}^n - Y(X)$ is locally arc connected. It is enough thus to prove the result when γ_0 is in U_j and is the boundary of a manifold $\tilde{\gamma}_0$. By Stokes' Theorem and the equation $\partial\bar{\partial} I_q=\partial\bar{\partial}\log|f_j|$ in U_j, we write successively

$$2\int_{\gamma_0} \partial I_q(z)=2\int_{\tilde{\gamma}_0} d\partial I_q(z)=-2\int_{\tilde{\gamma}_0} \partial\bar{\partial} I_q=-2\int_{\tilde{\gamma}_0} \partial\bar{\partial}\log|f_j|$$

$$2\int_{\gamma_0} \partial I_q(z)=2\int_{\tilde{\gamma}_0} d\partial\log|f_j|=2\int_{\gamma_0} \partial\log|f_j|=\int_{\gamma_0} d\log f_j=2\pi i N.$$

This shows that $G=\log F_0$ is determined by (3,23) and $F_0=e^G$ is well defined in all $\mathbb{C}^n$.

To prove that $F_0(z)$ is holomorphic in $\mathbb{C}^n$, we choose $z_0 \notin Y(X)$. Then by the definition of G in (3,23), dG is a (1,0) form in a neighborhood of z_0; $dG=2\partial I_q$ implies $\bar{\partial} G=0$. Then $G=\log F_0(z)$ is holomorphic in z_0; moreover $\log|F_0|=I_q$ is locally bounded above on compect subsets, and hence $|F_0|$ is locally bounded and can be extended as a holomorphic function to all of $\mathbb{C}^n$ by Riemann's Theorem (cf. Corollary I.23).

In U_j, we have $\frac{i}{\pi}\partial\bar{\partial}\log|F_0(z)|=\theta=\frac{i}{\pi}\partial\bar{\partial}\log|f_j|$, which shows that $F_0 f_j^{-1}$ and $f_j F_0^{-1}$ are never zero in U_j. Hence, we obtain $F_0=f_j\cdot\varphi_j$ where φ_j is holomorphic in U_j, $\varphi_j \neq 0$. Furthermore, if F is an entire function vanishing on the Cousin's data X, then $FF_0^{-1}=(Ff_j^{-1})\varphi_j^{-1}$, and so $g=F\cdot F_0^{-1}$ is holomorphic in every U_j, hence entire, with $F=gF_0$. If F is a solution of Cousin's Second Problem for the data X, then $F=gF_0$, with $g=e^h \neq 0$ in $\mathbb{C}^n$, $h\in\mathscr{H}(\mathbb{C}^n)$. □

Using Theorem 3.26, we obtain:

Theorem 3.30. *Let $X=(f_j, U_j)$ be a Cousin data such that $0\notin Y(X)$ and such that θ_X is of finite order ρ. Then there exists an entire function $F_0(z)$ which has exactly X as its zero set (that is, which solves Cousin's Second Problem with data X) such that*

i) $$\log|F_0(z)|=I_q(z)=k_{2n-2}^{-1}\int e_{2n-2}(a, z, q)\,d\sigma_X(a)$$

and

$$\log F_0(z)=2k_{2n-2}^{-1}\int_0^z \partial I_q(\xi)=2k_{2n-2}^{-1}\int d\sigma_X(a)\int_0^z \partial e_{2n-2}(a, \xi, q),$$

where $(0, z)$ *is any compact polygonal path in* $\mathbb{C}^n - Y(X)$ *and* q *is the genus of* X;

ii) $$\log|F_0(z)| \leqq A(n, q) r^q \left[\int_{r_0}^{r} t^{-q-1} v_X(t)\,dt + \int_{r}^{\infty} t^{-q-2} v_X(t)\,dt \right],$$

where $v_X(t) = [\tau_{2n-2} t^{2n-2}]^{-1} \sigma_X(t)$ *is the projective indicator of* X *and* $B(0, r_0) \cap Y(X) = \emptyset$;

iii) F_0 *is of the same order as* X *and if* $v_X(t)$ *is of normal type with respect to a proximate order* $\rho(t)$ *and* ρ *is non-integral, then* F_0 *is also of normal type with respect to the proximate order* $\rho(t)$;

iv) F_0 *divides every entire function which is zero on* X;

v) *Among the set of entire functions which are zero on* X *and only on* X *(with the given multiplicities),* F_0 *is the unique function which has order equal to the order of* $v_X(t)$ *and such that* $F_0(0) = 1$ *and all derivatives up to and including* $q =$ *genus* $v_X(t)$ *are zero at the origin;*

vi) *any entire function which has properties* i)–iv) *can be written as* $F_0(z) \exp P(z)$, *where* $P(z)$ *is any polynomial of degree at most* ρ.

Corollary 3.31. *An entire function* F *of finite order* ρ *is determined by the set* X *of its zeros and its value at a finite number of points in* $\mathbb{C}^n$ *equal to* $\dim E_\rho$ *(cf. Proposition* 3.27).

Definition 3.32. The genus q' of an entire function f of finite order will be defined by $q' = \sup(q, p)$, where q is the genus of the zero set of f, and p is the degree of the polynomial P such that $f(z) = F_0(z) \exp P(z)$.

The genus of f is at most equal to its order ρ and is strictly smaller if ρ is not an integer.

§7. Slowly Increasing Cousin Data: the Genus $q=0$; the Algebraic Case

Estimates of the growth of $F_0(z)$ for a Cousin data X of finite order in $\mathbb{C}^n$ depend on a constant $C(2n-2, q)$ (Proposition 3.15), which in its turn depends upon two constants $C_1(2n-2, q)$ and $C_2(2n-2, q)$ (Proposition 3.14). For $q=0$, we have $C_2(2n-2, 0) = 1$, which is independant of the dimension n, but $C_1(2n-2, 0)$ depends on n. If $\tau = \left(\dfrac{2n-2}{2n-1}\right)^{2n-2}$, for $n \geqq 2$ then $0 < e^{-1} < \tau \leqq \frac{4}{9}$.

By using the remark following Theorem 3.17, we obtain:

$$\log|F_0(z)| \leqq \int_{r_0}^{\sigma r} v_X(t) t^{-1}\,dt + \left(\frac{2n-1}{2n-2}\right) C_1(2n-2, 0) r \int_{\sigma r}^{\infty} v_X(t) t^{-2}\,dt$$
$$+ (2n-2)^{-1} v_X(\sigma r)$$

where $\sigma=\tau^{-1}$. It is interesting to consider the case where the second integral on the right is negligible with respect to the first – in this case, we obtain an estimate independant of the dimension n of the space. This will be the case if $\nu_X(t)$ satisfies an estimate of the type

$$\nu_X(t) \leqq C(\log^+ t)^s + C' \qquad C>0,\ C'>0,\ s \geqq 0. \tag{3,24}$$

Let $I_s = \int\limits_r^\infty (\log t)^s t^{-2} dt$. An integration by parts shows that

$$r I_s(r) = s r I_{s-1}(r) + (\log r)^s \quad \text{and } r I_0(r) = 1.$$

Thus, we obtain

$$\begin{aligned} M_0(r) &= \sup_{\|z\| \leqq r} \log |F_0(z)| \\ &\leqq C(s+1)^{-1} (\log^+ r)^{s+1} + A_{n,s} (\log^+ r)^s (1+\varepsilon_r) \end{aligned} \tag{3,25}$$

where $\varepsilon_r = o(r)$. We resume these results as follows (cf. Theorem 1.6).

Theorem 3.33. *If X is a Cousin data such that $\nu_X(t)$ satisfies* (3,24), *then*

$$\limsup_{r\to\infty} \frac{M(r)}{(\log r)^{s+1}} \leqq (s+1)^{-1} \limsup_{r\to\infty} \frac{\nu_X(r)}{(\log r)^s} \tag{3,26}$$

holds for the canonical solution of Cousin's Second Problem. If $s=0$, that is if $0<\nu_X(t)<\nu_\infty<\infty$, we obtain

$$\limsup_{r\to\infty} \frac{M_0(r)}{\log r} = \lim_{r\to\infty} \nu_X(r) = \nu_\infty, \tag{3,27}$$

and in this case F_0 is a polynomial of degree ν_∞, which is an integer.

Proof. (3,26) follows directly from (3,25). If $s=0$, $\nu_X(t)$ is bounded and hence $\nu_\infty = \lim_{t\to\infty} \nu_X(t)$ exists. Then (3,25) gives

$$\limsup_{r\to\infty} \frac{M_0(r)}{\log r} = \nu_\infty. \tag{3,28}$$

Since $M_0(r)$ is an increasing convex function of $\log r$, the limit on the left hand side of (3,28) exists. What is more, since $\nu_X(r) = \dfrac{\partial}{\partial \log r} \lambda(0, r, \log |F_0|)$ by Gauss' Formula and since the two functions are increasing and convex in $\log r$, we have

$$\nu_\infty = \lim_{r\to\infty} \frac{\lambda(0, r, \log |F_0|)}{\log r} \leqq \lim_{r\to\infty} \frac{M_0(r)}{\log r} \leqq \nu_\infty.$$

Again, by the convexity of $M_0(r)$ with respect to $\log r$, we see that $M_0(r) \leqq M_0(1) + \nu_\infty \log r$ or $|F_0(z)| \leqq C \|z\|^{\nu_\infty}$. Hence $F_0(z)$ is a polynomial of degree ν_∞ (cf. Theorem 1.6). □

Theorem 3.34. *Let $F(z)$ be an entire function and let* $\lim_{r\to\infty}(\log r)^{-1}M(r)=a$ *and* $\lim_{r\to\infty} v_X(r)=b$, *where* $v_X(r)$ is the indicator of $\frac{i}{\pi}\partial\bar\partial\log|F|$ and $v_X(r)=\frac{\partial}{\partial\log r}\lambda(0,r,\log|F|)$. *Then*

i) *if* $a=0$, $F(z)=F(0)$ *is constant and* $b=0$;

ii) *if* $0<a<\infty$, $a=b$ *and* F *is a polynomial of degree* a; *if* $F(0)\neq 0$, *we have*

$$\log F(z)=\log F(0)+2k_{2n-2}^{-1}\int_0^z\int\partial[-\|a-z\|^{2-2n}]\,d\sigma_X(a)$$

$$\log|F(z)|=\log F(0)+k_{2n-2}^{-1}\int[-\|a-z\|^{2-2n}+\|a\|^{2-2n}]\,d\sigma_X(a);$$

iii) *if* $a=+\infty$ *and if* $b\neq+\infty$, *then* b *is an integer and* $\liminf_{r\to\infty}\frac{M(r)}{r}>0$.

Proof. i) and ii) follow from Theorems 3.33 and 3.30 respectively. To prove iii), we remark that $a=+\infty$, the value of b and the conclusion do not depend on the choice of the origin. Thus we can suppose that $F(0)\neq 0$. Then there exists a polynomial P_0 of degree b (an integer) such that $P(0)=1$ and $F(z)=P_0(z)g(z)$ where $g(z)$ has no zeros in $\mathbb{C}^n$, Thus, we can write $g(z)=\exp g_1(z)$ for $g_1(z)$ an entire function, $g_1(z)=A(z)+iB(z)$. Suppose now that there exists an increasing sequence $r_m\to\infty$ such that $\lim_{m\to\infty} r_m^{-1}M(r_m)=0$. Since

$$\log|F(z)|=\log|P_0(z)|+A(z)$$

$$\text{and}\quad b\log r\geqq\lambda(0,r,(\log(|P|)^+)-\lambda(0,r,(\log P)^-)\geqq 0,$$

we see that $b\log r\geqq\lambda(0,r,(\log|P|)^-)$ and $\lim_{m\to\infty} r_m^{-1}\lambda(0,r_m,A^+)=0$. Since $A(z)$ is an harmonic function, its complexification $\tilde A(Z)$ is an entire function of $Z\in\mathbb{C}^{2n}$. Moreover, $A(0)=\lambda(0,r,A^+)-\lambda(0,r,A^-)$ from which it follows that $\lambda(0,r,|A|)\geqq 2\lambda(0,r,A^+)-A(0)$. Thus $\lim_{m\to\infty} r_m^{-1}\lambda(0,r_m,|A|)=0$, and by Lemma 3.22, there exists a constant c_n such that $m_c(c_n r_m)=\sup_{\|z\|=c_n r_m}|\tilde A(Z)|$ satisfies $\lim_{m\to\infty} r_m^{-1}m_c(c_n r_m)=0$. It then follows from Corollary 1.7 that $\tilde A\equiv\tilde A(0)$, hence A is constant. By the Cauchy-Riemann equations, for $z_k=x_k+iy_k$, we have

$$\frac{\partial A}{\partial x_k}=\frac{\partial B}{\partial y_k}\quad\text{and}\quad\frac{\partial A}{\partial y_k}=-\frac{\partial B}{\partial x_k}.$$

Thus $B(z)$ is also identically constant. Hence $g(z)$ is constant and $F(z)=CP_0(z)$. But then $a=b<+\infty$, which contradicts the hypotheses. □

Corollary 3.35. *(Characterization of algebraic sets of co-dimension 1.) Let X be a Cousin data in $\mathbb{C}^n$, that is $\sum m_k Y_k(X)$ for $Y_k(X)$ irreducible branches of $Y(X)$ and positive integers m_k. Then X is algebraic if and only if $v_X(r)$ is bounded, and in this case, k is finite and $v_X(\infty)=\sum_k m_k v_k(\infty)$ is the degree of X and of the polynomial $F_0(z)$.*

§ 8. The Case of Integral Order: Extension of a Theorem of Lindelöf

We have not yet given a complete treatment of the comparison between the growth of $\nu_X(r)$ and $M_0(r) = \sup_{\|z\| \leqq r} |F_0(z)|$ for the case of ρ an integer. For $n=1$, the necessary information is given by the quantity $S_\rho(r) = \sum_{|a_n| \leqq r} a_n^{-\rho}$ (introduced by E. Lindelöf). This quantity takes into account not only the moduli of the zeros but their argument as well. Let us recall the basic results (with respect to ρ constant).

i) if ρ is an integer, then F_0 is of minimal or normal type if and only if $|S_\rho(r)|$ remains bounded for all r;

ii) F_0 is of minimal type with respect to ρ if $q=\rho-1$ or if $q=\rho$ and $\lim_{r \to +\infty} S_\rho(r) = 0$.

In order to treat the case $n \geqq 2$, we will replace $S_\rho(r)$ by a family of homogeneous polynomials of degree ρ which will be harmonic in $\mathbb{R}^{2n}$. Let

(3,29) $$\Phi_{R,\rho}(z) = k_{2n-2}^{-1} \int_{\|a\| \leqq R} \|a\|^{2-2n} P_\rho(a, z)\, d\sigma_X(a)$$

with $P_\rho(a, z) = \|a\|^{2n-2} \dfrac{1}{q!} \dfrac{\partial^q h(a, tz)}{\partial t^q}\bigg|_{t=0}$ the polynomial of degree $\rho = q$ in the canonical kernel $e_{2n-2}(a, z, q)$.

Definition 3.36. A family $\{P_t(x)\}_{t \geqq a > 0}$ of polynomials of degree at most μ in the variable $x \in \mathbb{R}^p$ will be said to be *bounded* (respectively *to go to zero at infinity*) if there exists an open set $\omega \subset \mathbb{R}^p$ such that $M_t = \sup_{x \in \omega} |P_t(x)| \leqq C$ (respectively $\lim_{t \to \infty} M_t = 0$).

Proposition 3.37. *For a family $\{P_t(x)\}$ with $\deg P_t = \mu < +\infty$, the following properties are equivalent:*

i) *the family $\{P_t(x)\}_{t \geqq a > 0}$ is bounded in modulus independent of t (resp. goes to zero uniformly when t goes infinity) for every bounded set of $\mathbb{C}^p$;*

ii) *the complexification $\{P_t(X)\}_{t \geqq a \geqq 0}$ is bounded in modulus independently of t (resp. goes to zero uniformly when t goes to infinity) for $\|X\| \leqq 1$;*

iii) *the coefficients $a_{\alpha,t}$ of P_t are bounded in absolute value (resp. go to zero) when t goes to infinity.*

Proof. i) ⇒ ii). ω contains a cube $\Delta_r = \{x\colon |x_k - x_k^{(0)}| < r\}$ for some $r > 0$. Let $\varphi_t(x) = P_t(x - x^{(0)})$. Then $\varphi_t(x_k) \leqq M_t$ for $|x_k| < r$, and it follows from Lemma 3.22 that for $|X| < r/\sqrt{2}$; $|\varphi_t(X)| \leqq C_p M_t$. It follows from the Cauchy Integral Formula that $|a_{\alpha,t}| \leqq C'_p M_t$, where $a_{\alpha,t}$ are the coefficients of $\varphi_t(X)$. Thus $|P_t(X)| \leqq C_{p,\mu,r} M_t (1 + \|X - x^{(0)}\|)^\mu$. The implications ii) ⇒ iii) ⇒ i) are trivial. □

Proposition 3.38. *Let $\{P_t(x)\}$ be a family of real valued harmonic polynomials in $\mathbb{R}^p$ homogeneous of degree μ. If $\sup_{\|x\|=1} P_t(x)=m_t$ (resp. $\lambda(0,1,P_t^+)$) is bounded above, then $\{P_t(x)\}$ is a family of bounded polynomials. If m_t tends to zero (resp. $\lambda(0,1,P_t^+)$ tends to zero) as t tends to infinity, then $\{P_t(x)\}$ goes to zero as t goes to infinity.*

Proof. If $\deg P_t=0$, there is nothing to prove. If $\deg P_t \geqq 1$, then $P_t(0) =\lambda(0,1,P_t)=0$; hence, $\lambda(0,1,|P_t|)=2\lambda(0,1,P_t^+)\leqq 2m_t$. Thus, from Lemma 3.22, for $\|X\|\leqq 1/2$

(3,30) $$|P_t(X)|\leqq 5\cdot 2^{p-2}\lambda(0,1,|P_t|)\leqq 5\cdot 2^{p-1} m_t,$$

and the result now follows from (iii) of Proposition 3.37. □

Now we consider the case where $v_X(t)$ is of order ρ an integer and of genus q. Let

(3,31) $$D_R(z)=I_q(z)-\Phi_{R,\rho}(z)=k_{2n-2}^{-1}\int_{\|a\|\leqq R} e_{2n-2}(a,z,q-1)d\sigma_X(a)$$
$$+k_{2n-2}^{-1}\int_{\|a\|>R} e_{2n-2}(a,z,q)d\sigma_X(a)=I_1+I_2.$$

Then $D_R(z)$ is a subharmonic function in $\mathbb{R}^{2n}$. We set

(3,32) $$M_D(R)=\sup_{\|z\|\leqq R} D_R(z).$$

We use (3,9) to estimate successively the two integrals I_1 and I_2;

$$I_1\leqq A_1\|z\|^\rho\int_{r_0}^{R}\frac{d\sigma_X(t)}{(t+\|z\|)t^{2n+\rho-3}}\leqq A_2 v_X(R)+A_3R^\rho\int_{r_0}^{R}\frac{v_X(t)}{t^\rho(t+R)}dt$$

$$I_2\leqq A_4R^{\rho+1}\int_R^\infty\frac{v_X(t)dt}{(t+R)t^{\rho+1}}.$$

Let (3,33)

$$\gamma_v=\limsup_{t\to\infty} t^{-\rho}v_X(t) \quad\text{and}\quad \gamma_0=\limsup_{t\to\infty} M_0(t)\,t^{-\rho}$$

where $M_0(r)=\sup_{\|z\|\leqq r}\log|F_0(z)|$. Since $v_X(t)<(\gamma_v+\varepsilon)\,t^\rho$ for $t>R_\varepsilon$, we have

$$I_1\leqq A_2\,v_X(R)+A_3\,R^\rho\int_{r_0}^{R_\varepsilon} v_X(t)\,t^{-\rho}(t+R)^{-1}dt+A_3\,R^\rho(\gamma_v+\varepsilon).$$

Hence

$$R^{-\rho}M_D(R)\leqq(\gamma_v+\varepsilon)\,[A_2+A_3+A_4]+\frac{A_3}{R}\int_{r_0}^{R_\varepsilon} v_X(t)\,t^{-\rho}\,dt,$$

and

(3,34) $$\limsup_{R\to\infty} R^{-\rho}M_D(R)\leqq A_5\gamma_v,$$

where A_5 depends only on n and q. We thus arrive at the following statement:

Proposition 3.39. *The quantity* $M_D(R) = \sup_{\|z\|=R} [I_q(z) - \Phi_{R,\rho}(z)]$ *is of order at most* ρ *and at most of normal type if* $v_X(R)$ *is of normal type with respect to the order* ρ.

Proof. If $q=\rho$, $\rho \geqq 1$, then this follows from (3,34). If $q=\rho=0$, then it still remains true, since the integral I_1 reduces to

$$-k_{2n-2}^{-1} \int_{\|a\| \leqq R} h_{2n-2}(a, z) d\sigma_X(a) \leqq 0$$

and hence $I_1 \leqq 0$ and the estimates of I_2 are still valid. Finally, if $q=\rho-1$, then $\gamma_v=0$ and (3,34) still holds, which shows that $M_D(r)$ is of minimal type. □

Let $\xi(r) = \sup_{\|z\| \leqq r} \Phi_{R,\rho}(z) = r^{-\rho} \sup_{\|z\| \leqq r} \Phi_{r,\rho}(z)$ and $\xi = \limsup_{r \to \infty} \xi(r)$.

Proposition 3.40. *We have* $\gamma_0 \leqq A_5 \gamma_v + \xi$.

Proof. From (3,31), we obtain $\log|F_0(z)| = \Phi_{R,\rho}(z) + D_R(z)$, and hence $r^{-\rho} M_0(r) \leqq \xi(r) + A_5 \gamma_2 + o(r)$, from which the conclusion follows. □

Proposition 3.40 gives an upper bound for the type γ_0 of $F_0(z)$ if ξ is finite. If $\xi=0$ and $v_X(t)$ is of minimal type, then $\gamma_0=0$, and $F_0(z)$ is of minimal type. It remains to prove the inverse: to give an upper bound for ξ in terms of γ_v and γ_0. This will be a simple consequence of the fact that $\Phi_{R,\rho}(z)$ is harmonic and homogeneous of order ρ. Proceeding as in the proof of Proposition 3.39, we see that

$$\lambda(0, R, \Phi_{R,\rho}^-) = \lambda(0, R, \Phi_{R,\rho}^+) = 1/2 \lambda(0, R, |\Phi_{R,\rho}|)$$

and $D_R(z) = I_q(z) - \Phi_{R,\rho}(z)$, so $\lambda(0, R, \Phi_{R,\rho}^-) \leqq \lambda(0, R, I_q^-) + \lambda(0, R, D_R^+)$.

Furthermore, since I_q is $\mathbb{R}^{2n}$-subharmonic and zero at the origin, $\lambda(0, R, I_q^-) \leqq \lambda(0, R, I_q^+) \leqq M_0(R)$. Hence,

$$\lambda(0, R, \Phi_{R,\rho}^+) \leqq \lambda(0, R, I_q^+) + \lambda(0, R, D_R^+) \leqq M_0(R) + M_D(R).$$

It follows from the Poisson Integral Formula that

$$\sup_{\|z\| \leqq R/2} \Phi_{R,\rho}(z) \leqq A_6 \lambda(0, R, \Phi_{R,\rho}^+).$$

This gives, by the homogeneity of Φ:

$$\xi(R) = R^{-\rho} \sup_{\|z\| \leqq R} \Phi_{R,\rho}(z) \leqq A_6 [M_0(R) + M_D(R)] R^{-\rho},$$

or

$$\xi \leqq A_6 [\gamma_0 + A_5 \gamma_v]. \tag{3,35}$$

We resume this as follows:

Theorem 3.41. *Let* γ_v *be the type of the indicator* $v_X(t)$, γ_0 *that of the canonical potential* $I_q(z)$, *and* ξ *that of the family of harmonic polynomials*

$\Phi_{R,\rho}(z)$ defined by (3,29). Then there exist constants C_i, $i=1,\ldots,5$ (depending only on n and $q=\rho$) such that

i) $\gamma_v \leqq e^{\rho}\gamma_0$
ii) $\gamma_0 \leqq \xi + C_1\gamma_v$
iii) $\xi \leqq C_2[\gamma_0 + C_3\gamma_v]$
iv) $C_4 \sup(\xi, C_3\gamma_v) \leqq \gamma_0 \leqq C_5 \sup(\xi, C_3\gamma_v)$.

Corollary 3.42. *If the order ρ of a Cousin data X is an integer, if the type of $v_X(t)$ with respect to ρ is zero and if $\Phi_{R,\rho}(z)$ converges to zero when R goes to ∞, then F_0 is also of minimal type of the order ρ. This is always the case when the genus of X is equal to $\rho-1$. If $\sup(\xi, \gamma_v)$ is finite and different from zero, F_0 is of normal type, and if $\sup(\xi, \gamma_v) = +\infty$, F_0 is of maximal type.*

Let us suppose that $v_X(r)$ is of normal type with respect to a proximate order $\rho(r)$. As before, we let

$$\gamma_v = \limsup_{t\to\infty} v_X(t)t^{-\rho(t)} \quad \text{and} \quad \gamma_0 = \limsup_{t\to\infty} M_0(t)t^{-\rho(t)},$$

and we let $\xi' = \limsup_{R\to\infty} \dfrac{\xi(R)}{R^{\rho(R)-\rho}}$.

Theorem 3.43. *There exist constants C_i (the same as in Theorem 3.41) such that*

i) $\gamma_v \leqq e^{\rho}\gamma_0$
ii) $\gamma_0 \leqq \xi' + C_1\gamma_v$
iii) $\xi' \leqq C_2[\gamma_0 + C_3\gamma_v]$
iv) $C_4 \sup(\xi', C_3\gamma_2) \leqq \gamma_0 \leqq C_5 \sup(\xi', C_3\gamma_v)$.

Proof. Since $v_X(t) \leqq (\gamma_v + \varepsilon)t^{\rho(t)}$ for $t > R_\varepsilon$, we obtain

$$R^{-\rho(R)}M_D(R) \leqq (\gamma_v+\varepsilon)\Big[A_2 + A_3R^{\rho-1}\int_{r_0}^{R} t^{\rho(t)-\rho}dt$$
$$+ A_4R^{\rho+1}\int_{R}^{\infty} t^{\rho(t)-\rho-2}dt\Big] + A_3R^{\rho}.$$

We use Lemma 3.20 to evaluate the two integrals and find

$$R^{-\rho(R)}M_D(R) \leqq (\gamma_v+\varepsilon)[A_2+A_3+A_4+o(R)] + C[R^{\rho-\rho(R)-1}].$$

Thus

$$\limsup_{R\to\infty} R^{-\rho(R)}M_D(R) \leqq [A_2+A_3+A_4]\gamma_v.$$

On the other hand from

$$M_D(R) = \sup_{\|z\|=R} [I_q(z) - \Phi_{R,\rho}(z)], \qquad I_q(z) = \Phi_{R,\rho}(z) + D_R(z),$$

we have

$$R^{-\rho(R)}M_0(R)\leqq R^{-\rho(R)}M_D(R)+R^{\rho-\rho(R)}\xi(R).$$

Thus

$$\gamma_0\leqq(A_2+A_3+A_4)(\gamma_v+\varepsilon)+\xi(R)R^{\rho-\rho(R)},$$

and hence $\gamma_0\leqq C_1\gamma_v+\xi'$. Furthermore, proceeding exactly as in Theorem 3.41, from

$$\xi(R)=R^{-\rho}\sup_{\|z\|=R}\Phi_{R,\rho}(z)\leqq A_6[M_0(R)+M_D(R)]R^{-\rho}$$

we find that

$$\xi(R)R^{\rho-\rho(R)}\leqq A_6[M_0(R)R^{-\rho(R)}+M_D(R)R^{-\rho(R)}]$$

and

$$\xi'\leqq A_6[\gamma_0+A_5\gamma_v].$$

□

Remark. Using Proposition 3.37, it is easy to replace in both cases the conditions on $\Phi_{R,\rho}(z)$ by a finite number of conditions that the integrals

$$\int_{\|a\|\leqq R}\|a\|^{2-2n+\rho}m_\lambda(a)d\sigma_X(a)$$

remain bounded independently of R, where the $m_\lambda(a)$ are a finite set of monomials in $a_i, \bar{a}_j$ of degree ρ.

§9. Trace of a Cousin Data on Complex Lines

Let X be a Cousin data of finite order ρ such that $0\notin Y(X)$ and let $I_q(z)$ be the canonical function constructed in Section 7. Then we will say that the trace of X on the complex line $u\cdot z$ through the origin is the set of zeros $a_n(z)$ of $F_0(u\cdot z)$ (counted with multiplicities). We set

$$n_z(r)=\operatorname{card}\{a_n(z)\colon |a_n(z)|\leqq r\}\quad\text{and}\quad N_z(r)=\int_0^r\frac{n_z(t)dt}{t},$$

the integrated counting function. It follows from Jensen's Theorem that

$$N_z(r)=\frac{1}{2\pi}\int_0^{2\pi}\log|F(r\,e^{i\theta}\,z)|\,d\theta$$

and hence $N_z(r)$, for r fixed, is a plurisubharmonic function of z (cf. Proposition I.14). Furthermore,

$$\omega_{2n}^{-1}\int_{\|a\|=1}N_\alpha(r)d\omega_{2n}(\alpha)=\omega_{2n}^{-1}\int_{\|\alpha\|=1}\log|F(r\alpha)|\,d\omega_{2n}(\alpha)=\int_0^r\nu_X(t)t^{-1}dt.$$

By Proposition 1.35 and Corollary 1.43, we see that the order of $N_z(r)$ is exactly ρ except perhaps for z in a pluripolar cone in $\mathbb{C}^n$.

We now study a particular property of the restrictions of entire functions of finite order to one of the variables. We shall use this result to study the restriction of an entire function on the set of complex lines through the origin.

Theorem 3.44. *Let $F(z', u)$ be an entire function of $z=(z', u)\in\mathbb{C}^n$ $z'\in\mathbb{C}^{n-1}$ such that F has finite order $\rho(z')$ with respect to u for fixed z' (see Definition 1.42). Let $E_s\subset\mathbb{C}^{n-1}$ be the set of z' such that the function $F(z', u)$ has at most s zeros as a function of u. Let $A=\{z'\in\mathbb{C}^{n-1}: F(z', u)\equiv 0,\ u\in\mathbb{C}\}$. Then $A\cup E_s$ is closed and is either all of $\mathbb{C}^{n-1}$ or is contained in an analytic subvariety M_s.*

Proof. Given $z'_0\notin(E_s\cup A)$, we prove that $z'\notin(E_s\cup A)$ in a neighborhood of z'_0. Let $\gamma\subset\mathbb{C}_u$ be a curve which contains a least $s+1$ (isolated) zeros of $F(z'_0, u)$ and such that $F\neq 0$ on γ. Then

$$\frac{1}{2i\pi}\int_\gamma [F(z', u)]^{-1}\frac{\partial F}{\partial u}(z', u)\,du\geqq s+1$$

and by the continuity of the integral, this will hold in a neighborhood U of z'_0, so for $z'\in U$, $n_{z'}(r)\geqq s+1$ for $r>\sup[|u|; u\in\gamma]$. Thus $E_s\cup A$ is a closed set in $\mathbb{C}^{n-1}$.

It is sufficient to prove the theorem in a ball $B(0, m)$. By Theorem 1.41, $\rho_m=\sup\limits_{\|z'\|\leqq m}\rho(z')$ is finite. Let ρ be the greatest integer less than or equal to ρ_m. If $F(z, u)\equiv 0$, then $E_s=\emptyset$, so we suppose that $F(z, u)\not\equiv 0$; then $Y=[(z, u)\in\mathbb{C}^n: F(z, u)=0]$ is an analytic subvariety of $\mathbb{C}^n$ and the analytic set $A\times\mathbb{C}_u$ is the intersection of Y with its translates $Y_v=[(z, u): (z, u-v)\in Y]$. Now we define an analytic set $M_s\subset\mathbb{C}^n$, invariant by such translations, and containing $(A\times\mathbb{C}_u)\cup(E_s\times\mathbb{C}_u)$. First we define $\tilde{M}_s$ in a neighborhood Δ of (z'_0, u_s) such that $\Delta\cap Y=\emptyset$, $\Delta=[(z', u): z'\in U,\ |u-u_s|<r_0]$, where U is a neighborhood of z'_0.

The function

$$G(z', u)=F^{-1}(z', u)\frac{\partial F}{\partial u}(z', u)=\sum_{q=0}^{\infty} a_q(u-u_s)^q$$

is holomorphic in Δ and the coefficients

$$a_q=\frac{1}{q!}\frac{\partial^q}{\partial u^q}\left[F^{-1}\frac{\partial F}{\partial u}\right]$$

are holomorphic functions in $\mathbb{C}^n\setminus Y$. By an easy calculation, we find that

$$a_q=F^{-q-1}a'_q$$

and a'_q is a polynomial in $F(z', u), \ldots, \dfrac{\partial^{q+1}}{\partial u^{q+1}}F(z', u)$, hence it is an entire function in $\mathbb{C}^n$. For $z'\in E_s\cap U$, we obtain

$$F(z', u)=\prod_1^s (u-\alpha_j)\exp\cdot P(u).$$

The α_j and the coefficients of the polynomial $P(u)$ depend on z', and degree $P \leqq \rho$. Then for $z' \in E_s \cap U$:

$$G(z', u) = \sum_{j=1}^{s} \frac{1}{u - \alpha_j} P'(u).$$

Set

$$Q_{z'}(u) = \prod_{j=1}^{s} (u - \alpha_j) = \sum_{p=0}^{s} b_j (u - u_0)^p.$$

Then

$$G(z', u) Q_{z'}(u) = R_{z'}(u)$$

for a polynomial $R_{z'}(u)$ of degree at most $s + \rho - 1$ in u, where the coefficients depend on $z' \in E_s \cap U$. Then for $v \geqq s + \rho = k$, and $(z', u_0) \in \Delta$

$$\sum_{\xi=0}^{s} a_{v-\xi}(z', u_0) b_\xi = 0. \tag{3,36}$$

As a consequence, all the determinants

$$D_v(z', u) = \begin{pmatrix} a_v & \dots & a_{v-s} \\ a_{v+s} & \dots & a_v \end{pmatrix} \tag{3,37}$$

vanish for $|u - u_0| < r_0$ and $z' \in E_s \cap U$. If we chose Δ' another neighborhood of (z', u') in $\mathbb{C}^n \setminus Y$, and calculate a'_v and D'_v as above, we obtain

$$D_v(z', u) = F^{-\sigma} D'_v(z', u), \qquad \sigma = (s+1)(v+1)$$

where in D'_v we have to write a'_v instead of a_v; then D'_v is an entire function of $(z', u) \in \mathbb{C}^n$. The equations $D'_v(z', u) = 0$ define an analytic set $\tilde{M}_s$ in $\mathbb{C}^n$ or $\tilde{M}^s = \mathbb{C}^n$; moreover, $\tilde{M}_s$ is invariant by the translation $u \to u - v$, $v \in \mathbb{C}$. Obviously $\tilde{M}_s$ contains $A \times \mathbb{C}_u$. Set $M_s = \tilde{M}_s \cap [u = 0]$. For $z'_0 \in M_s$, $z'_0 \notin A$, there exists u_0 such that $F(z'_0, u_0) \neq 0$, and a neighborhood Δ of (z'_0, u_0) such that $\Delta \cap Y \neq 0$. Then there exist solutions l_ξ of (3,36), $l_\xi = l_\xi(z'_0, u_0)$ and $F^{-1}(z'_0, u) \dfrac{\partial F}{\partial u}(z'_0, u)$ for $u \in \mathbb{C}$ has at most s poles for $u \in \mathbb{C}$; therefore $F(z'_0, u)$ has at most s zeros and $z'_0 \in E_s$. As a consequence, if $\tilde{M}_s \neq \mathbb{C}^n$, $A \cup E_s$ and E_s are contained in the analytic set M_s, and if $M_s = \mathbb{C}^{n-1}$, and $F \not\equiv 0$, $E_s \cup A$ is all of $\mathbb{C}^{n-1}$.

Corollary 3.45. *Let X be a Cousin data of finite order ρ such that $0 \notin Y(X)$. Let $E_s = [L \in P(\mathbb{C}^n)\colon \operatorname{card}(L \cap X) \leqq s]$. Then E_s is closed and if E_s is not contained in the set $[z\colon Q(z) = 0]$ for some homogeneous polynomial Q in $\mathbb{C}^n$ then X is algebraic. In particular, if $\bigcup_{s=0}^{\infty} E_s$ is not a countable union of algebraic sets, X is algebraic.*

Proof. Let $F_0(z)$ be the canonical function which defines X. Let Δ be a neighborhood of the origin such that $F_0(z) \neq 0$ for $z \in \Delta$. Let $F(z_1, \dots, z_n, u) = F(z_1 u, \dots, z_n u)$, and $z' = (z_1, \dots, z_n)$. Then there exists R and r_0 such that $(z', u) \in \Delta$ for $\|z'\| < R$ and $\|u\| \leqq r_0$.

The coefficients $a'_\nu(z')$ defined in the proof of Theorem 3.44 are homogeneous polynomials of the variable z' and hence $D'_{(\nu)}$ is a homogeneous polynomial of z'. The set A is empty by the hypothesis $0 \notin X$. Thus E_s is an analytic subvariety of the projective space $P(\mathbb{C}^n)$ or $D_{(\nu)} \equiv 0$ for $\nu \geq k$. But then $E_s = P(\mathbb{C}^n)$ and $\nu_X(t) < \infty$ and hence $Y(X)$ is algebraic. □

§ 10. The Case of a Cousin Data of Infinite Order

In Sections 4–6, we have solved the Cousin Problem for the case of a divisor of finite order by the construction of the canonical solution. In this section, we shall study the problem where the indicator $\nu_X(t)$ of X is of infinite order, that is when $\limsup_{t \to \infty} \frac{\log \nu_X(t)}{\log t} = +\infty$.

In this case too we can find a solution of the Cousin Problem F an entire function such that the growth of $\log|F|$ is similar to that of ν_X, but the relationship between the growth will be less precise than in the case of finite order.

The technique we shall follow is as above the resolution of the equation $\frac{i}{\pi} \partial \bar{\partial} V = \theta_X$, where θ_X is the current associated with the Cousin data X. We first solve the equation $dv = \theta_X$ using a homotopy formula and then resolve the operators ∂ and $\bar{\partial}$ using the L^2-estimates of L. Hörmander (cf. Appendix III), which will allow us to control the growth of the solution. The results will hold only under the assumption that θ is a positive closed current of degree 1. In the case of a Cousin data, it will be easy to show, as above, that $V = \log|F|$ for F entire, which then gives the desired solution.

Let $\alpha(z) \in \mathscr{C}_0^\infty(B(0,1))$ such that $\alpha \geq 0$, α depends only on $\|z\|$ and $\int \alpha(z) d\tau(t) = 1$, and set $\alpha_\varepsilon(z) = \varepsilon^{-2n} \alpha\left(\frac{z}{\varepsilon}\right)$. We then define the positive closed current

(3,39) $$\theta^\varepsilon = \theta \star \alpha_\varepsilon,$$

whose coefficients are now $\mathscr{C}^\infty$ functions. Furthermore, we let

(3,40) $$\sigma_\varepsilon(z) = \int_{B(z,\varepsilon)} \Delta\theta \cdot \beta_n$$

and

(3,41) $$\nu_\varepsilon(z) = [\tau_{2n} \varepsilon^{2n-2}]^{-1} \sigma_\varepsilon(z)$$

so that $\lim_{\varepsilon \to 0} \nu_\varepsilon(z) = \nu_X(z)$, the Lelong number of the current θ_X (that is, the multiplicity at z of the Cousin data X).

Before proving the main result of this section, we first pause to prove lemmas that we shall need.

Lemma 3.46. *Let Ω ba a domain in $\mathbb{C}^n$ and φ a function defined in Ω such that $\varphi \geqq 0$ and $\varphi \not\equiv 0$. Then $\log \varphi \in \mathrm{PSH}(\Omega)$ if and only if for every $\alpha \in \mathbb{C}^n$, $\varphi_\alpha(z) = \varphi(z) \exp \operatorname{Re}\left(\sum_{i=1}^{n} \alpha_i z_i\right) \in \mathrm{PSH}(\Omega)$.*

Proof. We first prove the sufficiency. Suppose that $\varphi \in \mathscr{C}^2(\Omega)$. Then we set $\varphi_m = \varphi + m$ for $m > 0$ and prove first the sufficiency for φ_m. A simple calculation shows that

$$\frac{\partial^2 \varphi_{\alpha,m}(z)}{\partial z_j \partial \bar{z}_k} = \exp \operatorname{Re}\left(\sum_{i=1}^{n} \alpha_i z_i\right) \left[\frac{\partial^2 \varphi_m(z)}{\partial z_j \partial \bar{z}_k} + \frac{1}{2}\left(\alpha_j \frac{\partial \varphi_m(z)}{\partial \bar{z}_k} + \bar{\alpha}_k \frac{\partial \varphi_m(z)}{\partial z_j}\right) + \frac{1}{4} \alpha_j \bar{\alpha}_k \varphi_m(z)\right]$$

where $\varphi_{\alpha,m}(z) = \varphi_m(z) \exp \operatorname{Re}\left(\sum_{i=1}^{n} \alpha_i z_i\right)$. For fixed $z \in \Omega$, let α be such that $\alpha_k = -\frac{2}{\varphi_m} \frac{\partial \varphi_m(z)}{\partial z_k}$. Then

$$\begin{aligned}\frac{\partial^2 \varphi_{\alpha,m}(z)}{\partial z_j \partial \bar{z}_k} &= \exp \operatorname{Re}\left(\sum_{i=1}^{n} \alpha_i z_i\right) \left[\frac{\partial^2 \varphi_m(z)}{\partial z_j \partial \bar{z}_k} - \frac{1}{\varphi_m(z)} \frac{\partial \varphi_m(z)}{\partial z_j} \frac{\partial \varphi_m(z)}{\partial \bar{z}_k}\right] \\ &= \varphi_m(z) \exp \operatorname{Re}\left(\sum_{i=1}^{n} \alpha_i z_i\right) \frac{\partial^2 \log \varphi_m(z)}{\partial z_j \partial \bar{z}_k}.\end{aligned}$$

Since $\varphi_\alpha \in \mathrm{PSH}(\Omega)$ for all α, $\varphi_{\alpha,m} \in \mathrm{PSH}(\Omega)$ for all α, which shows that the form $\sum_{j,k} \frac{\partial^2 \log \varphi_m(z)}{\partial z_j \partial \bar{z}_k} w_j \bar{w}_k \geqq 0$ for all $w \in \mathbb{C}^n$. Hence $\log \varphi_m \in \mathrm{PSH}(\Omega)$. Since $\log \varphi_m$ decreases to $\log \varphi$ and $\varphi \not\equiv 0$, it follows from Proposition I.3 that $\log \varphi \in \mathrm{PSH}(\Omega)$.

To treat the general case, we choose $\alpha(z) \in \mathscr{C}_0^\infty(B(0,1))$ such that $\int \alpha(z) d\tau(z) = 1$ and α depends only on z and set $\alpha_\varepsilon(z) = \alpha\left(\frac{z}{\varepsilon}\right) \varepsilon^{-2n}$. Then $\varphi_\varepsilon = \varphi \star \alpha_\varepsilon \in \mathrm{PSH}(\Omega_\varepsilon)$, where $\Omega_\varepsilon = \{z : d_\Omega(z) > \varepsilon\}$, $\varphi_\varepsilon \in \mathscr{C}^\infty(\Omega_\varepsilon)$ and φ_ε decreases to φ. Then

$$\exp \operatorname{Re}\left(\sum_{i=1}^{n} \alpha_i z_i\right) \varphi_\varepsilon(z) = \int \varphi_\alpha(z+z') \alpha_\varepsilon(z') \exp\left[-\operatorname{Re}\left(\sum_{i=1}^{n} \alpha_i z'\right)\right] d\tau(z')$$

is in $\mathrm{PSH}(\Omega_\varepsilon)$ by Proposition I.14, so $\log \varphi_\varepsilon \in \mathrm{PSH}(\Omega_\varepsilon)$ by the above, and since φ_ε decreases to φ and $\varphi \not\equiv 0$, $\log \varphi \in \mathrm{PSH}(\Omega)$ by Proposition I.3.

To see the necessity, we note that if $\log \varphi \in \mathrm{PSH}(\Omega)$, then for every α, $\log \varphi(z) + \operatorname{Re}\left(\sum_{i=1}^{n} \alpha_i z_i\right) \in \mathrm{PSH}(\Omega)$. Since $\varphi(t) = e^t$ is an increasing convex function, the result now follows from Proposition I.24. □

Lemma 3.47. *Suppose that $V(z) \in \mathrm{PSH}(\mathbb{C}^n)$ and $\varphi(z)$ is a real valued function such that $\int |V(z)|^2 \exp -2\varphi(z) d\tau(z) < C$. Then there exists a constant $\tilde{C}(n.\varepsilon)$ such that*

$$V(z) \leqq C(n, \varepsilon) \cdot C \exp[\sup_{\|z'-z\| \leqq \varepsilon} \varphi(z')].$$

Proof. By the Inequality of the Mean for subharmonic functions, we have $V(z) \leqq \hat{C}(n, \varepsilon) \int_{\|z'\| < \varepsilon} V(z+z') d\tau(z')$. We now obtain by the Schwarz Inequality

$$\begin{aligned} V(z) &\leqq \hat{C}(n, \varepsilon) [\int_{\|z'\| \leqq \varepsilon} |V(z+z')|^2 \exp -2\varphi(z+z')\, d\tau(z')]^{1/2} \\ &\quad \times [\int_{\|z'\| \leqq \varepsilon} \exp 2\varphi(z+z') d\tau(z')]^{1/2} \\ &\leqq \tilde{C}(n, \varepsilon) \cdot C \cdot \exp[\sup_{\|z'-z\| \leqq \varepsilon} \varphi(z')]. \end{aligned}$$

□

Theorem 3.48. *Let θ be a positive closed (1,1) current such that its trace σ satisfies*

$$\sigma_\varepsilon(z) \leqq C \exp[\Phi(z)] \tag{3,42}$$

where $\Phi(z)$ is a plurisubharmonic function and $C > 0$. Then for every $\alpha > 0$, there exists a plurisubharmonic function V solution of (3,10) in $\mathbb{C}^n$ such that

$$\int [V^+(z)]^2 C(1+\|z\|^2)^{-n-3-\alpha} \exp -2\psi(z) d\tau(z) < C(n) C \tag{3,43}$$

and

$$V(z) \leqq C \cdot \tilde{C}(\alpha, \varepsilon)(1+\|z\|^2)^{n+3+\alpha} \exp \chi(z) \tag{3,44}$$

where $V^+ = \sup(V, 0)$, $\psi(z) = \log \int_0^1 t \exp \Phi(tz) dt$ and

$$\chi(z) = \tfrac{1}{2} \log \int_{\|z'-z\| \leqq \varepsilon} \exp 2\psi(z') d\tau(z').$$

Proof. Let $\theta = i \sum \theta_{jk} dz_j \wedge d\bar{z}_k$. Then $\theta^\varepsilon_{jk} = \theta_{jk} \star \alpha_\varepsilon$. Theorem 2.16 gives an estimate of the coefficients of θ^ε in terms of the trace σ_θ of θ:

$$\begin{aligned} |\theta^\varepsilon_{jk}(z)| &\leqq |\theta_{jk}| \star \alpha_\varepsilon \leqq \tfrac{1}{2} \int \alpha_\varepsilon(z-a) d\sigma(a) \\ &\leqq \tfrac{1}{2} M_\varepsilon \sigma_\varepsilon(z) \leqq C \cdot C(\varepsilon) \exp \Phi(z) \end{aligned} \tag{3,45}$$

where $M_\varepsilon = \sup \alpha_\varepsilon = \varepsilon^{-2n} \sup \alpha(z)$.

Suppose now that V is any solution of (3,10) (cf. Corollary 2.30) and let $W_\varepsilon = V \star \alpha_\varepsilon$. Then it follows from the Mean Value Property that $V \leqq W_\varepsilon$ and what is more $i/\pi \partial \bar{\partial} W_\varepsilon = i/\pi \partial \bar{\partial}(V \star \alpha_\varepsilon) = \theta \star \alpha_\varepsilon = \theta^\varepsilon$. Thus every solution of the equation (3,10) is majorized by a solution W_ε of the equation $i/\pi\, \partial \bar{\partial} W^\varepsilon = \theta^\varepsilon$, whose coefficients satisfy (3,45).

First we solve the equation $id\omega=\theta^{\varepsilon}$. As in Theorem 2.28 the solution is given explicitely by the formula

$$(3,46) \qquad v=\sum_{j,k}\left[\int_0^1 t\theta^{\varepsilon}_{jk}(tz)dt\right]z_j d\bar{z}_k$$
$$-\sum_{j,k}\left[\int_0^1 t\theta^{\varepsilon}_{jk}(tz)dt\right]\bar{z}_k dz_j=v_2-v_1.$$

Let $A^{\varepsilon}_{jk}=\int_0^1 t\theta^{\varepsilon}_{jk}(tz)dt$, which are $\mathscr{C}^{\infty}$ functions. The forms $\Phi_k=\sum_{j=1}^{n}\theta^{\varepsilon}_{jk}dz_j$ are ∂-closed and hence the forms $\sum_{j=1}^{n}A^{\varepsilon}_{jk}dz_j$ are also ∂-closed, so v_1 is ∂-closed, and in a similar manner, v_2 is $\bar{\partial}$-closed. Then

$$\bar{\partial}v_1=\sum_{j,k}\left(-A_{jk}-\sum_s \bar{z}_s\frac{\partial A_{js}}{\partial\bar{z}_k}\right)dz_j\wedge d\bar{z}_k$$
$$=\sum_{j,k}\left(-A_{jk}-\sum_s \bar{z}_s\frac{\partial A_{js}}{\partial\bar{z}_s}\right)dz_j\wedge d\bar{z}_k$$

and

$$\partial v_2=\sum_{j,k}\left(A_{jk}+\sum_s z_s\frac{\partial A_{js}}{\partial z_k}\right)dz_j\wedge d\bar{z}_k$$
$$=\sum_{j,k}\left(A_{jk}+\sum_s z_s\frac{\partial A_{jk}}{\partial z_s}\right)dz_j\wedge d\bar{z}_k.$$

Since $\sum_s\left(\frac{\partial A_{jk}}{\partial z_s}z_s+\frac{\partial A_{jk}}{\partial\bar{z}_s}\bar{z}_s\right)=\int_0^1 t^2\partial\frac{\theta^{\varepsilon}_{jk}(tz)}{\partial t}$, we see that

$$dv=\partial v_2-\bar{\partial}v_1=\left[2\int_0^1\theta^{\varepsilon}_{jk}(tz)dt+\int_0^1 t^2\frac{\partial\theta^{\varepsilon}_{jk}}{\partial t}dt\right]dz_j\wedge d\bar{z}_k=-i\theta^{\varepsilon},$$

which shows that we indeed have $idv=\theta^{\varepsilon}$.

From (3,46) we obtain the estimates

$$|v_j(z)|^2\leqq C\cdot C(\varepsilon)\|z\|^2\exp[2\psi(z)] \qquad j=1,2$$

and hence for every $\alpha>0$ we have

$$(3,47) \quad \int_{\mathbb{C}^n}|v_j(z)|^2(1+\|z\|^2)^{-n-1-\alpha}\exp-2\psi(z)d\tau(z)<C\cdot C(\varepsilon,\alpha), \qquad j=1,2.$$

The function $\Phi(z)$ is plurisubharmonic. To see this, it is enough to show by Lemma 3.44 that the positive function $h(z)=\int_0^1 t\exp\Phi(tz)dt$ generates a plurisubharmonic function $h_{\alpha}(z)=h(z)\exp|\langle\alpha,z\rangle|$ for every vector α. Since $\Phi(z)$ is plurisubharmonic, so is $h(z)$ for every z, and

$$h(x)\exp|\langle\alpha,z\rangle|=\int_0^1\exp(\Phi(tz)+\operatorname{Re}\langle a,z\rangle+\log t)dt$$

is plurisubharmonic.

We now resolve the two equations $\partial u_1 = v_1$ and $\bar{\partial} u_2 = v_1$ with bounds. Indeed, there exist (cf. Appendix III) two functions u_1 and u_2 as above such that

$$\int_{\mathbb{C}^n} |u_j(z)|(1+\|z\|^2)^{-n-3-\alpha} \exp -\psi(z) d\tau(z) < C\cdot\tilde{C}(n,\alpha,\varepsilon), \qquad j=1,2.$$

Then $\theta^\varepsilon = i\partial\bar{\partial}(u_1+u_2)$, and since $\theta^\varepsilon = \bar{\theta}^\varepsilon$, if we let $W^\varepsilon = \mathrm{Re}(u_1+u_2)$, we obtain the estimate (3,43). The estimate (3,44) follows from Lemma 3.47. □

Remark 1. Since $v_\varepsilon(z) = (\tau_{2n-2}\varepsilon^{2n-2})^{-1}\sigma_\varepsilon(z)$, we can replace the estimates in Theorem 3.48 in terms of $\sigma_\varepsilon(r)$ by estimates in terms of $v_\varepsilon(r)$.

Remark 2. The estimates in Theorem 3.48 remain valid if we replace $\psi(z)$ by $\tilde{\psi}(z) = \sup\limits_{0\leqq t\leqq 1} \Phi(tz)$ and $\chi(z)$ by $\tilde{\chi}(t) = \sup\limits_{\|z'\|\leqq\varepsilon} \psi(z+z')$.

Remark 3. If we have only radial estimates, that is $\sigma(r) \leqq C\exp\Phi(r)$ with $r=\|z\|$, where $\Phi(r)$ is plurisubharmonic (that is an increasing convex function of $\log r$), we see that for $\alpha>0$ and $\varepsilon>0$ there exists a solution V of (3.10) such that

$$V(z) \leqq C\cdot C(\varepsilon,n,\alpha)(1+r)^{n+3+\alpha}\exp\Phi(r+\varepsilon). \tag{3,47}$$

Remark 4. The solution we obtain does not require the hypothesis $0\notin\operatorname{supp}\theta$.

Remark 5. For the case of finite order ρ, the estimate given by (3,47) is less precise than that given by Theorem 3.30. We obtain $\sup\limits_{\|z\|\leqq r} V(z) \leqq C(\varepsilon,\alpha)(1+r)^\lambda$ with $\lambda = 3n+1+\alpha+\rho$.

We now apply Theorem 3.48 to the solution of the Cousin Problem.

Theorem 3.49. *Let X be a Cousin data in $\mathbb{C}^n$ and let σ be the trace of the positive closed current associated with X. Let $\sigma_\varepsilon(z)$ be the mass carried by the ball $\overline{B(z,\varepsilon)}$ and suppose that $\sigma_\varepsilon(z) \leqq C\exp\Phi(z)$, where $C>0$ and $\Phi(z)$ is a plurisubharmonic function. Then for every $\alpha>0$, there exists an entire function $F(z)$ which solves the Cousin Problem and such that*

$$\begin{cases} \int_{\mathbb{C}^n} (\log^+|F(z)|)^2(1+\|z\|^2)^{-n-3-\alpha}\exp(-2\psi(z))d\tau(z) < C\cdot C(\varepsilon,\alpha) \\ \log|F(z)| \leqq C\cdot\tilde{C}(\varepsilon,\alpha)(1+\|z\|)^{n+\varepsilon+\alpha}\exp\chi(z) \end{cases} \tag{3,48}$$

where ψ and χ are as in Theorem 3.48. In particular, if Φ depends only on $\|z\|$ and grows so fast that

$$r^{n+3+\alpha}\exp\Phi(r) \leqq \exp\Phi(r+\varepsilon) \qquad \textit{for } r>R_\varepsilon, \tag{3,49}$$

then $\log|F(z)| \leqq C\cdot C(\varepsilon,\alpha)\exp\Phi(r+\varepsilon)$, $r=\|z\|$.

Remark 1. The condition (3,49) is valid if $\dfrac{\partial\Phi}{\partial r}\Big/\log r$ increases to infinity.

Historical Notes

The development of the n dimensional canonical function is due to P. Lelong (cf. the notes of 1953 and [11]) who proved the publisubharmonicity of the potential I_q for a Cousin data of finite order. Other representations are due to Stoll [2] and to Ronkin [9]. Theorem 3.33 for slowly increasing data is due to Avanissian [3]. Theorem 3.44 given here is a generalization of two results, one is due to Sibony and Wong [1] (Corollary 3.35 and the other is an earlier result of P. Lelong [3] who proved the property that the set of the z_2's for which a function $F(z_1, z_2)$ of finite order with respect to z_2 has no zeros as a function of z_2 for fixed z_1, was contained in an analytic set. The extension to n variables of the Theorem of Lindelöf was given by P. Lelong in [11]. The proof of Theorem 3.42 gives a simplification even in the classical case $n=1$. The result of § 10 are due to H. Skoda [1], who used the resolution of the $\bar{\partial}$ equation. For Cousin data of finite order the canonical function gives more precise results, whereas for data of infinite order the opposite is true.

Chapter 4. Functions of Regular Growth

We have seen in Chapter 3 that there is a relationship between the asymptotic growth of the quantity $M_f(r)$ for an entire function f and the area of the zero set of f. In certain cases, however, much more can be said. We shall prove here the fundamental principle for functions of finite order and of regular growth, which, paraphrased a little crudely, states that an entire function of finite order has its zero set "regularly distributed" asymptotically if and only if it has "regular growth" asymptotically along all rays. An equivalent formulation, as we shall see, is to say that for an entire function f, $r^{-\rho(r)}\log|f(rz)|$ converges (as a distribution) in $L^1_{\mathrm{loc}}(\mathbb{C}^n)$ to $h_f^\star(z)$ if and only if $r^{-\rho(r)}\Delta\log|f(rz)|$ converges as a distribution to $\Delta h_f^\star(z)$. These ideas of regularity will be made more precise below.

The technics that we shall use will be simple potential theoretic properties of subharmonic functions as well as the canonical representation of entire functions of finite order developped in Chapter 3. Since we are interested in emphasizing the use of potential theory in this context, we shall develop the subject in a greater generality than that of entire functions. In the first place, the domains we shall consider will be cones in $\mathbb{R}^m$ (for convenience, we shall always assume that the vertex is at the origin; thus if Γ is such a cone, Γ is open and connected, and $tx\in\Gamma$ whenever $x\in\Gamma$ and $t>0$). In the second place, we shall consider subharmonic functions of finite order in these cones; however, the reader should bear in mind that the main application will be to entire functions of finite order in $\mathbb{C}^n=\mathbb{R}^{2n}$.

Let Γ be an open connected cone in $\mathbb{R}^m$ with vertex at the origin, and let $\rho(r)$ be a proximate order. We shall denote by $\mathrm{SH}^{\rho(r)}(\Gamma)$ the family of functions u subharmonic in Γ such that there exist constants A_0 and A_1 (depending on u) with $u(x)\leqq A_0+A_1\|x\|^{\rho(\|x\|)}$. These are the functions of finite type with respect to the order $\rho(r)$. For such a function, we define

$$\hat{h}_u(x)=\limsup_{r\to\infty}\frac{u(rx)}{r^{\rho(r)}}\quad\text{and}\quad \hat{h}_u^\star(x)=\limsup_{\substack{x'\to x\\ x\in\Gamma}}\hat{h}_u(x')$$

$\hat{h}_u^\star(x)$ is the indicator of growth function of u, and it is subharmonic and positively homogeneous of order ρ in Γ (cf. Proposition 1.30). We note that if $\Gamma=\mathbb{C}^n$ and $u=\log|f|$ for an entire function f, then $\hat{h}_u^\star(x)$ is just the radial indicator of growth function as defined by Definition 1.29.

In contrast to the case of entire functions of one complex variable, entire functions of several complex variables can have irregular growth on a small set of rays without affecting the global asymptotic behavior of the function or its zero set. For instance, the function $z_2 \exp z_1$ has regular growth except when $z_2=0$, when it is identically zero. Similarily, the function $z_2+\exp z_1$ has regular growth except when $z_2=0$, $\operatorname{Re} z_1<0$. The reason is, of course that $\hat{h}_u$ and $\hat{h}_u^\star$ may be different on a small set of rays. Since it is $\hat{h}_u^\star$ which describes the asymptotic behavior of u, it is not natural to study the behavior of u along individual rays but rather on smaller and smaller neighborhoods of a ray. With this in mind, we are led to the following: let

$$\begin{aligned} I_u^r(x,\delta) &= (\tau_m r^m \delta^m)^{-1} r^{-\rho(r)} \int_{B(0,r\delta)} u(rx+y)\,d\tau(y) \\ &= r^{-\rho(r)} A(rx, r\delta) \qquad \text{for } \delta < d(x, \complement\Gamma). \end{aligned}$$

Since u is subharmonic, $I_u^r(x,\delta)$ is an increasing function of δ for r fixed.

Definition 4.1. A function $u \in \mathrm{SH}^{\rho(r)}(\Gamma)$ will be said to be of *regular growth* for the ray rx, $r>0$, if $\liminf\limits_{\delta\to 0} \liminf\limits_{r\to\infty} I_u^r(x,\delta) = \hat{h}_u^\star(x)$, $x \notin E_u = \{x\in\Gamma:\ \hat{h}_u^\star(x) = -\infty\}$; u will be said to be of regular growth in a set D if u is of regular growth for all $x\in D$, $x\notin E_u$ and $D \not\subset E_u$.

Remark. It follows from the definition and Theorem 1.18, the the set of x for which u is of regular growth is a cone with vertex at the origin.

§ 1. General Properties of Functions of Regular Growth

Lemma 4.2. *Suppose $u\in \mathrm{SH}^{\rho(r)}(\Gamma)$ and u is of regular growth along the ray rx, $r>0$ (resp. there exists a sequence r_n increasing to infinity such that $\lim\limits_{n\to\infty} \dfrac{u(r_n x)}{r_n^{\rho(r_n)}} \geqq C_0\Big)$.*

Then if $\gamma = d_\Gamma(x)$ and $\delta \leqq \frac{1}{4}\gamma$, there exists R (resp n_0) and a constant C_δ (depending only on δ, C_0 and A_1, where $u(x) \leqq A_0 + A_1 \|x\|^{\rho(\|x\|)}$ such that for $x', x'' \in B(x, \delta/4)$ and $r>R$ (resp. $n \geqq n_0$)

$$\text{(a)}\quad |I_u^r(x',\delta) - I_u^r(x'',\delta)| \leqq C_\delta \|x'-x''\|$$

$$\text{(}\textit{resp.}\text{ (b)}\quad |I_u^{r_n}(x',\delta) - I_u^{r_n}(x'',\delta)| \leqq C_\delta \|x'-x''\|\text{)}.$$

Proof. By Theorem 1.31, there exists R_1 and a constant C_1 such that $u(ry)r^{-\rho(r)} \leqq C_1$ for $y\in B(x, \frac{3}{4}\gamma)$ and $r>R_1$. Let $\eta = \|x'-x''\|$. Then for $r>R_1$, by Theorem 1.18,

$$I_u^r(x',\delta) \leqq I_u^r(x',\delta+\eta) \leqq \frac{\delta^m}{(\delta+\eta)^m} I_u^r(x'',\delta) + \frac{(\delta+\eta)^m - \delta^m}{(\delta+\eta)^m} (2\gamma)^\rho C_1.$$

For $r>R_2$, $I_u^r(x,\frac{3}{4}\gamma)\geqq \hat{h}_u^\star(x)-1=C_0'$, so for $y\in B(x,\frac{1}{4}\gamma)$

$$I_u^r(y,\delta)\geqq\left(\frac{3\gamma}{4\delta}\right)^m C_0'-(\tau_m r^m\delta^m)^{-1}r^{-\rho(r)}\int\limits_{B\left(rx,\frac{3\gamma}{4}r\right)-B(ry,\delta r)} u(w)\,d\tau(w).$$

Thus, for $r>\sup(R_1,R_2)$,

$$-I_u^r(y,\delta)\leqq-\left(\frac{3\gamma}{4\delta}\right)^m C_0'+\delta^{-m}\left[\left(\frac{3\gamma}{4}\right)^m-\delta^m\right](2\gamma)^\rho C_1$$

and so we obtain

$$I_u^r(x',\delta)-I_u^r(x'',\delta)\leqq\left[\frac{\delta^m}{(\delta+\eta)^m}-1\right]I_u^r(x'',\delta)+\frac{(\delta+\eta)^m-\delta^m}{(\delta+\eta)^m}(2\gamma)^\rho C_1\leqq C_\delta\eta. \tag{4,1}$$

Since the estimate (4,1) is independent of x', x'', by reversing the roles, we obtain

$$|I_u^r(x',\delta)-I_u^r(x'',\delta)|\leqq C_\delta\|x'-x''\|.$$

To prove (b), it suffices to write r_n instead of r and $n>n_0$ instead of $r>\sup(R_1,R_2)$. □

Theorem 4.3. *Suppose that $u,v\in\mathrm{SH}^{\rho(r)}(\Gamma)$. If u is of regular growth for the ray rx_0, then $\hat{h}_{u+v}^\star(x_0)=\hat{h}_u^\star(x_0)+\hat{h}_v^\star(x_0)$.*

Proof. Clearly $\hat{h}_{u+v}^\star(x_0)\leqq\hat{h}_u^\star(x_0)+\hat{h}_v^\star(x_0)$, so we prove the converse. If $x_0\in E_v$, we are through, so suppose $x_0\notin E_v$. Let $\varepsilon>0$ be given. Since $\hat{h}_v^\star(x_0)=\limsup\limits_{x'\to x_0}\hat{h}_v(x')$, for every $\eta>0$, there exists x' with $\|x'-x_0\|<\eta$ and $|\hat{h}_v^\star(x')-\hat{h}_v^\star(x_0)|<\varepsilon/8$. Let r_n be an increasing sequence such that $\lim\limits_{n\to\infty}\frac{v(r_n x')}{r_n^{\rho(r_n)}}=\hat{h}_v(x')$. It follows from Theorem 1.31 that for δ and η sufficiently small,

$$\frac{u(rx)}{r^{\rho(r)}}\leqq\hat{h}_u^\star(x_0)+\varepsilon/8,\qquad\frac{v(rx)}{r^{\rho(r)}}\leqq\hat{h}_v^\star(x_0)+\varepsilon/8 \tag{4,2}$$

for $\|x-x_0\|<\eta+\delta$ and $r>R_1(\varepsilon)$. By subharmonicity and the upper semi-continuity of $\hat{h}_u^\star$, for δ sufficiently small (depending on ε) there exists N_0 such that for $n>N_0$

$$I_u^{r_n}(x',\delta)\leqq h_u^\star(x_0)+\varepsilon/8;\qquad\frac{\varepsilon}{8}-\hat{h}_v(x')\leqq I_v^{r_n}(x',\delta)\leqq\hat{h}_v^\star(x')+\varepsilon/8. \tag{4,3}$$

Let $Q_n=\left\{y\colon\|y-x'\|<\delta,\ \frac{v(r_n y)}{r_n^{\rho(r_n)}}\leqq\hat{h}_v^\star(x_0)-\varepsilon/2\right\}$. Then for n sufficiently large, it follows from (4,2) and (4,3) that meas. $(Q_n)\leqq\frac{1}{4}\tau_m\delta^m$. Furthermore, for n sufficiently large and η sufficiently small by Lemma 4.2.

$$I_u^{r_n}(x',\delta)\geqq I_u^{r_n}(x_0,\delta)-\varepsilon/8\geqq\hat{h}_u^\star(x_0)-\varepsilon/4. \tag{4,4}$$

Let $S_n=\left\{y\colon \|y-x'\|<\delta, \frac{u(r_n y)}{r_n^{\rho(r_n)}}\leqq \hat{h}_u^\star(x_0)-\varepsilon/2\right\}$ and set $L_n=Q_n\cup S_n$. It follows from (4,2) and (4,4) that meas. $(S_n)\leqq \tau_m\delta^m/2$ for n sufficiently large.

Suppose that $w\in\bigcap_{n=1}^{\infty}\left(\bigcup_{\mu=n}^{\infty}\complement L_\mu\right)$, which is non-empty since by the Bounded Convergence Theorem meas. $\left(\bigcap_{n=1}^{\infty}\bigcup_{\mu=n}^{\infty}\complement L_\mu\right)=\lim_{n\to\infty}$ meas. $\bigcup_{\mu=n}^{\infty}(\complement L_\mu)>0$.

Then there exists an increasing subsequence $r_{n_j}=s_j$ tending to infinity such that $\frac{u(s_j w)}{s_j^{\rho(s_j)}}+\frac{v(s_j w)}{s_j^{\rho(s_j)}}\geqq \hat{h}_u^\star(x_0)+\hat{h}_v^\star(x_0)-\varepsilon$ with $\|w-x_0\|<\eta$. Since this is true for arbitrarily small η, we obtain $\hat{h}_{u+v}^\star(x_0)\geqq \hat{h}_u^\star(x_0)+\hat{h}_v^\star(x_0)-\varepsilon$, for each $\varepsilon>0$, and hence $\hat{h}_{u+v}^\star(x_0)\geqq \hat{h}_u^\star(x_0)+\hat{h}_v^\star(x_0)$. □

Theorem 4.4. *Let Γ be a convex cone. If $u\in \mathrm{SH}^{\rho(r)}(\Gamma)$ is of regular growth along a ray rx_0, $r>0$, then $v_y(x)=u(x+y)$, $y\in\Gamma$, is of regular growth along the ray rx_0, $r>0$.*

Proof. Let $\varepsilon>0$ be given. Then by Theorems 1.31 and 1.18 there exists $R(\varepsilon)$ and $\delta_0(\varepsilon)$ such that for $r>R(\varepsilon)$ and $\delta<\delta_0(\varepsilon)$, the following inequalities hold:

$$\frac{u(rx)}{r^{\rho(r)}}-\hat{h}_u^\star(x_0)-\frac{\varepsilon}{4}<-\frac{\varepsilon}{8};\qquad \frac{u(rx+y)}{r^{\rho(r)}}-\hat{h}_u^\star(x_0)-\frac{\varepsilon}{4}<-\frac{\varepsilon}{8}$$

for all x such that $\|x-x_0\|<2\delta_0$. Furthermore, by hypothesis, there exists $\delta\leqq\delta_0$ such that for $r\geqq R(\varepsilon,\delta)$

$$-\frac{\varepsilon}{8}>(\tau_m r^m\delta^m)^{-1}\int_{B(0,r\delta)}\left[\frac{u(rx_0+w)}{r^{\rho(r)}}-\hat{h}_u^\star(x_0)-\frac{\varepsilon}{4}\right]d\tau(w)>-\frac{3\varepsilon}{8}.$$

Since this is an increasing function of δ and the integrand is negative, for η sufficiently small,

$$0>(\tau_m r^m\delta^m)^{-1}\int_{B(0,(1+\eta)r\delta)}\left[\frac{u(rx_0+w)}{r^{\rho(r)}}-\hat{h}_u^\star(x_0)-\frac{\varepsilon}{4}\right]d\tau(w)>-\frac{\varepsilon}{2}$$

holds independently of r, and since the integrand is negative, for r sufficiently large $B(rx_0+y,r\delta)\subset B(rx_0,r\delta(1+\eta))$ so

$$0>(\tau_m r^m\delta^m)^{-1}\int_{B(0,r\delta)}\left[\frac{u(rx_0+y+w)}{r^{\rho(r)}}-\hat{h}_u^\star(x_0)-\frac{\varepsilon}{4}\right]d\tau(w)>-\frac{\varepsilon}{2}.$$ □

Theorem 4.5. (i) *Let $u\in\mathrm{SH}^{\rho(r)}(\Gamma)$ be of regular growth on a set $D-S$, where D is open and S is of Lebesgue measure zero. Then u is of regular growth in D.*

(ii) *If D is open in Γ and $\hat{h}_u^\star$ is continuous in D, then the set of points $x\in D$ for which u is of regular growth is closed.*

Proof. Let $A\subset D$ be the set of points for which u is of regular growth, and let $\varepsilon>0$ be given. Suppose $x_0\in\bar{A}$, $x_0\notin E_u$. It follows from Theorem 1.31 that there exists $\xi > 0$ and R_ε such that for $r > R_\varepsilon$ and $\|x - x_0\| < \xi$,

$$\frac{u(rx)}{r^{\rho(r)}} \leqq \hat{h}_u^\star(x_0) + \varepsilon/6. \tag{4,5}$$

It follows from the Mean Value Property for subharmonic functions in (i) and from the continuity of $h_u^\star$ in (ii) that for every $\eta > 0$, $\eta < \xi$, there exists $x' \in A$, $\|x' - x_0\| < \eta$ such that $|\hat{h}_u^\star(x') - \hat{h}_u^\star(x_0)| < \frac{\varepsilon}{6}$. Since $x' \in A$, given δ_0 there exists $\delta < \delta_0$ and $R_1(x')$ such that for $r > R_1(x')$, $I_u^r(x', \delta) \geqq \hat{h}_u^\star(x') - \frac{\varepsilon}{6}$.

Thus, if η is sufficiently small, by (4,5) and Lemma 4.2

$$I_u^r(x_0, \delta) \geqq I_u^r(x', \delta) - \frac{\varepsilon}{6} \geqq \hat{h}_u^\star(x') - \frac{\varepsilon}{3} \geqq h^\star(x_0) - \frac{\varepsilon}{2}$$

for r sufficiently large. □

Remark. For $f(z)$ entire in $\mathbb{C}$, $h_f^\star$ is always continuous (cf. Levin [*D*]), but for $n \geqq 2$, $h_f^\star(z)$ need not be continuous (cf. Lelong [13]).

Theorem 4.6. *Let D be an open connected set in Γ and suppose that $u \in \mathrm{SH}^{\rho(r)}(\Gamma)$ is such that $\hat{h}_u^\star$ is harmonic in D. If u is of regular growth for one point in D, then u is of regular growth in D.*

Proof. A harmonic function is always continuous, so it follows from Theorem 4.5 that the set of points A for which u is of regular growth is closed in D. We shall show that A is open in D, from which it follows that $A = D$.

Let $x_0 \in A$ and $t < d_D(x)/3$. If $\zeta > 0$, by Theorem 1.31, there exists $t(\zeta)$ and $R_1(\zeta)$ such that $I_u^r(x', \delta) \leqq \hat{h}_u^\star(x') + \zeta/4$ for $\|x' - x_0\| < 2t$ and $r > R_1(\zeta)$. Since u is of regular growth for x_0, given $\tilde{\delta}_2 > 0$, there exists $\tilde{\delta}_1(\zeta) < \tilde{\delta}_2$ such that $|I_u^r(x_0, \tilde{\delta}_1) - \hat{h}_u^\star(x_0)| < \frac{\zeta}{4}$ for $r > R_2(\zeta)$.

Since $I_u^r(x, \tilde{\delta}_1)$ is subharmonic in x and $\hat{h}_u^\star(x)$ is harmonic, for $r > \sup(R_1(\zeta), R_2(\zeta)) = R_3(\zeta)$, we have

$$\begin{aligned} 0 &\geqq \tau_m^{-1} t^{-m} \int_{B(x_0, t)} [I_u^r(x, \tilde{\delta}_1) - \hat{h}_u^\star(x) - \zeta/4] \, d\tau(x) \\ &\geqq I_u^r(x_0, \tilde{\delta}_1) - \hat{h}_u^\star(x_0) - \frac{\zeta}{4} \geqq -\frac{\zeta}{2}, \end{aligned}$$

and since the integrand is negative, for $r > R_3(\zeta)$, we have

$$\tau_m^{-1} t^{-m} \int_{B(x_0, t)} |I_u^r(x, \tilde{\delta}_1) - \hat{h}_u^\star(x)| \, d\tau(x) < \frac{3\zeta}{4}. \tag{4,6}$$

Now let $\varepsilon > 0$, $\delta_0 > 0$ be given. Then, by Corollary 1.32 there exists $R_1(\varepsilon)$ and $\delta_1 \leqq \delta_0$ such that for $r > R_1(\varepsilon)$ and $\|x - x_0\| < \delta_1$, $I_u^r(x, \delta_1) \leqq \hat{h}_u^\star(x) + \varepsilon$. Let

$A_r=\{x\in B(x_0,t)\colon I_u^r(x,\delta_1)\leqq \hat{h}_u^\star(x)-\varepsilon\}$. Then for $r>R(\varepsilon,\delta)=\sup(R_3(\zeta),R_1(\varepsilon))$, it follows from (4,6) that

$$\text{(4,7)} \qquad \operatorname{meas.}(A_r)\leqq\frac{\zeta}{\varepsilon}\tau_m t^m \quad \text{if } \delta_1\leqq\tilde{\delta}_1.$$

By Lemma 4.2, there exists $\eta>0$ and $R_2(\varepsilon)$ such that

$$\text{(4,8)} \qquad |I_u^r(x,\delta_1)-I_u^r(x',\delta_1)|<\frac{\varepsilon}{4} \quad \text{for } x,x'\in B(x_0,\eta_1) \text{ for } r>R_2(\varepsilon),$$

and since $\hat{h}_u^\star(x)$ is continuous, there exists η_2 such that

$$\text{(4,9)} \qquad |\hat{h}_u^\star(x)-\hat{h}_u^\star(x')|<\frac{\varepsilon}{4} \quad \text{for } x,x'\in B(x_0,\eta_2).$$

Let $\eta_3=\inf(\eta_1,\eta_2)$. There exists $\delta_1\leqq\delta_0$ such that (4,7) holds for δ_1 and $\zeta=\varepsilon^2\dfrac{\eta_3^m}{2t^m}$.

Since $I_u^r(z,\delta)$ increases with δ,

$$B_r=\{x\in B(x_0,t/2)\colon I_u^r(x,\tilde{\delta}_1)\leqq\hat{h}_u^\star(x)-\varepsilon\}\subset A_r.$$

Thus for $r>\sup(R_3(\zeta),R_2(\varepsilon))$, B_r is empty, since $\operatorname{meas.}(B_r)\leqq\frac{\varepsilon}{2}\tau_m\eta_3^m$ and (4,8) and (4,9) imply that if $y\in B_r$ and $x'\in B(y,\eta_3)$, $I_u^r(x',\tilde{\delta}_1)\leqq\hat{h}_u^\star(x')-\frac{\varepsilon}{2}$, which contradicts (4,6) with $\zeta=\frac{\varepsilon}{2}$. Thus, u is of regular growth in $B\left(x,\frac{t}{2}\right)$. □

Theorem 4.7. *Let $u\in \mathrm{SH}^{\rho(r)}(\Gamma)$ be of regular growth in Γ. Then, setting $u_t(x)=u(tx)t^{-\rho(t)}$, for any Lebesgue measurable set K relatively compact in Γ, $\lim\limits_{t\to\infty}\int_K|u_t(x)-\hat{h}_u^\star(x)|\,d\tau(x)=0$.*

Proof. It is sufficient to prove that for given $x_0\in\Gamma$, the result is true for a ball $K=B(x_0,\alpha)$, for $\alpha>0$. Let $\mu=d_\Gamma(x_0)$ and $\alpha=\frac{\mu}{8}$. Suppose $\varepsilon>0$ is given. If $0<\delta<\alpha$ and $x\in B(x_0,\alpha)$, $x\notin E_u$, then there exists an $R_x(\varepsilon)$ such that $I_u^r(x,\delta)\geqq\hat{h}_u^\star(x)-\frac{\varepsilon}{6}$ for $r>R_x(\varepsilon)$. Furthermore, by Lemma 4.2, there exists η_1 such that $|I_u^r(x,\delta)-I_u^r(x',\delta)|<\frac{\varepsilon}{12}$ for $\|x-x'\|<\eta_1$, and there exists η_2 such that $\hat{h}_u^\star(x')\leqq\hat{h}_u^\star(x)+\frac{\varepsilon}{12}$ for $\|x'-x\|<\eta_2$ by the upper semicontinuity of the indicator function. Thus, for $\eta_3=\inf(\eta_1,\eta_2)$ and $x'\in B(x,\eta_3)$, $r>\tilde{R}_x(\varepsilon)$,

$$I_u^r(x',\delta)\geqq\hat{h}_u^\star(x')-\frac{\varepsilon}{2}.$$

Since $\overline{B(x_0, \alpha)}$ is compact, there exists a finite number of balls $B(x_i, \eta_3^i)$ which cover $\overline{B(x_0, \alpha)}$, and if $R_1(\varepsilon)=\sup R_{x_i}(\varepsilon)$, then for $r \geqq R_1(\varepsilon)$

(4,10) $$I_u^r(x', \delta) \geqq \hat{h}_u^\star(x') - \frac{\varepsilon}{4}.$$

Furthermore, since $I_u^r(x', \delta)$ is locally bounded above (by Theorem 1.31), by Fatou's Lemma

$$\begin{aligned}
&\lim_{\delta\to 0} \limsup_{r\to\infty} \tau_m^{-1} \alpha^{-m} \int_{B(x_0, \alpha)} I_u^r(x', \delta)\, d\tau(x') \\
&\leqq \lim_{\delta\to 0} \tau_m^{-1} \alpha^{-m} \int_{B(x_0, \alpha)} \limsup_{r\to\infty} I_u^r(x', \delta) d\tau(x') \\
&\leqq \tau_m^{-1} \alpha^{-m} \int_{B(x_0, \alpha)} \lim_{\delta\to 0} \limsup_{r\to\infty} I_u^r(x', \delta) d\tau(x') \\
&\leqq \tau_m^{-1} \alpha^{-m} \int_{B(x_0, \alpha)} \hat{h}_u^\star(x') d\tau(x'),
\end{aligned}$$

where at the last step, we invoke the upper semi-continuity of $\hat{h}_u^\star$ and Theorem 1.31. Thus, there exists $\delta_1 < \alpha = \frac{\mu}{8}$ and $R(\delta_1, \varepsilon)$ such that for $r > R(\delta_1, \varepsilon)$,

$$\tau_m^{-1} \alpha^{-m} \int_{B(x_0, \alpha)} (I_u^r(x', \delta) - \hat{h}_u(x')) d\tau(x') < \frac{\varepsilon}{4} \quad \text{for all } \delta < \delta_1.$$

Coupled with (4,10), this imples that for $\delta < \delta_1$, there exists $R(\varepsilon, \delta)$ such that for $r > R(\varepsilon, \delta)$

(4,11) $$\tau_m^{-1} \alpha^{-m} \int_{B(x_0, \alpha)} |I_u^r(x', \delta) - \hat{h}_u^\star(x')|\, d\tau(x') < \frac{3\varepsilon}{4}.$$

Furthermore, it follows from Lemma 4.2 that

$$\tau_m^{-1} \alpha^{-m} \left| \int_{B(x', \alpha)} \frac{u(r\xi)}{r^{\rho(r)}} d\tau(\xi) - \int_{B(x_0, \alpha)} \frac{u(r\xi)}{r^{\rho(r)}} d\tau(\xi) \right| < \frac{\varepsilon}{4}$$

for $x' \in B(x_0, \tilde{\delta}_\varepsilon)$ and $r > \tilde{R}_\varepsilon$. Thus since $I_u^r(x, \delta) \geqq \frac{u(rx)}{r^{\rho(r)}}$, for $\delta \leqq \tilde{\delta}_\varepsilon$, we obtain

$$\begin{aligned}
&\tau_m^{-1} \alpha^{-m} \int_{B(x_0, \alpha)} \left(I_u^r(x', \delta) - \frac{u(rx')}{r^{\rho(r)}} \right) d\tau(x') \\
&= (\tau_m^{-1} \alpha^{-m}) \int_{B(x_0, \delta)} \Big(\int_{B(x', \alpha)} u(r\xi) r^{-\rho(r)} d\tau(\xi) \\
&\quad - \int_{B(x_0, \alpha)} u(r\xi) r^{-\rho(r)} d\tau(\xi) \Big) d\tau(x') \leqq \frac{\varepsilon}{4}.
\end{aligned}$$

Thus for $r > \sup(\tilde{R}_\varepsilon, R(\varepsilon, \tilde{\delta}_\varepsilon))$

$$\tau_m^{-1} 8^m \mu^{-m} \int_{B(x_0, \alpha)} \left| \frac{u(rx)}{r^{\rho(r)}} - \hat{h}_u^\star(x) \right| d\tau(x) < \varepsilon. \qquad \square$$

An alternative formulation of the above result would be to say that if $u_r(x) = u(rx)\cdot r^{-\rho(r)}$, then $\lim_{r\to\infty} u_r(x) = \hat{h}_u^\star(x)$ in the space of distributions $\mathscr{D}(\Gamma)$.

Theorem 4.8. *Let $\rho(r)$ be a strong proximate order. Then $u \in \mathrm{SH}^{\rho(r)}(\Gamma)$ is of regular growth in Γ if and only if $\hat{h}_{u+v}^\star(x) = \hat{h}_u^\star(x) + \hat{h}_v^\star(x)$ for every $v \in \mathrm{SH}^{\rho(r)}(\Gamma)$.*

Proof. It suffices to show that if u is not of regular growth in Γ, then there exists a $v \in \mathrm{SH}^{\rho(r)}(\Gamma)$ such that the above equality does not hold for at least one $x_0 \in \Gamma$, since the necessity follows from Theorem 4.3.

If u is not of regular growth in Γ, there exists x_0, $\varepsilon > 0$ and $\delta > 0$ and a sequence r_n increasing to infinity such that $I_u^{r_n}(x_0, \delta) \leqq \hat{h}_u^\star(x_0) - \varepsilon$. By choosing a subsequence, if necessary, we can assume that $r_{n+1} \geqq 2r_n$. By Lemma 4.2, there exists N_1 and $\eta > 0$ such that $I_u^{r_n}(x', \delta) \leqq \hat{h}_u^\star(x_0) - \frac{\varepsilon}{2}$ for $n \geqq N_1$ and $\|x' - x_0\| < \eta_1$, and since $I_u^{r_n}(x', \delta) \geqq u(r_n x')\cdot r_n^{-\rho(r_n)}$,

$$\frac{u(r_n x')}{r_n^{\rho(r_n)}} \leqq \hat{h}_u^\star(x_0) - \varepsilon/2 \quad \text{for } \|x' - x_0\| < \eta_1 \text{ and } n > N_1. \tag{4,12}$$

Let ψ be a $\mathscr{C}^\infty$ function of the variable $t \in \mathbb{R}$ with support in the interval $(-1, 1)$, $0 \leqq \psi \leqq 1$ and $\psi \equiv 1$ on the interval $(-\frac{1}{2}, +\frac{1}{2})$ and let

$$v(x) = \sup\left(\|x\|^{\rho(\|x\|)} + \sum_n \xi r_n^{\rho(r_n)} \psi\left(\frac{x - r_n x_0}{\eta_2 r_n}\right), A_0\right)$$

where $\eta_2 \leqq \eta_1$ will be fixed later.

By Proposition 1.22, for $\|x\|$ sufficiently large,

$$\begin{aligned} \Delta \|x\|^{\rho(\|x\|)} &= \sum_{i=1}^m \frac{\partial^2 \|x\|^{\rho(\|x\|)}}{\partial(\log\|x\|)^2}\left(\frac{\partial \log\|x\|}{\partial x_i}\right)^2 \\ &\quad + \sum_{i=1}^m \frac{\partial \|x\|^{\rho(\|x\|)}}{\partial(\log\|x\|)} \frac{\partial^2 \log\|x\|}{\partial x_i^2} \\ &\geqq \frac{\rho^2}{2}\|x\|^{\rho(\|x\|)-2} \end{aligned}$$

and so for A_0 sufficiently large, and ξ sufficiently small $v(x) \in \mathrm{SH}^{\rho(r)}(\Gamma)$ by Theorem 1.18. Furthermore, $\hat{h}_v^\star(x_0) = (\|x_0\|\rho + \zeta)$. We now choose η_2 sufficiently small, so that setting $\gamma = \dfrac{(1+\eta_2)}{(1-\eta_2)}$

$$2\gamma^\rho\left(\hat{h}_u^\star(x_0) - \frac{\varepsilon}{2}\right) \leqq \hat{h}_u^\star(x_0) - \frac{\varepsilon}{4}. \tag{4,13}$$

Suppose that $\|x' - x_0\| < \eta_2$. If $rx' \in \bigcup_n B(r_n x_0, \eta_2 r_n x_0)$, then $r_n(\gamma^{-1}) \leqq r \leqq r_n(\gamma)$ and

$$\frac{u(rx')}{r^{\rho(r)}}+\frac{v(rx')}{r^{\rho(r)}}\leqq\frac{r_n^{\rho(r_n)}}{r^{\rho(r)}}\left[\frac{u(r_nx')}{r_n^{\rho(r_n)}}+\frac{v(r_nx')}{r_n^{\rho(r_n)}}\right]$$

$$\leqq\hat{h}_u^\star(x_0)-\frac{\varepsilon}{4}+(\|x'\|^\rho+\xi)$$

$$\leqq\hat{h}_u^\star(x_0)+\hat{h}_v^\star(x_0)-\frac{\varepsilon}{8}$$

for ξ sufficiently small and r sufficiently large, by (4,12), (4,13) and Theorem 1.18.

If $rx'\notin\bigcup_n B(r_nx_0,\eta_2r_nx_0)$, then

$$\frac{u(rx')}{r^{\rho(r)}}+\frac{v(rx')}{r^{\rho(r)}}\leqq\hat{h}_u^\star(x_0)+\left(\|x_0\|^\rho+\frac{\xi}{2}\right)$$

for r sufficiently large by Theorem 1.31, so $\hat{h}_{u+v}^\star \mp h_u^\star(x_0)+h_v^\star(x_0)$. □

If $\Gamma\subset\mathbb{C}^n$ and if f is holomorphic in Γ with $\log|f|\in\mathrm{SH}^{\rho(r)}(\Gamma)$, then we shall say that f is of regular grouwth in Γ if $\log|f|$ is of regular growth in Γ, as defined in Definition 4.1. In this case, we can relax the conditions in Theorem 4.8.

Theorem 4.9. *The holomorphic function f is of regular growth in the convex cone Γ (for the strong proximate order $\rho(r)$) if and only if for every holomorphic function g of finite type with respect to $\rho(r)$ in Γ, $h_{fg}^\star(z)=h_f^\star(z)+h_g^\star(z)$.*

The necessity follows from Theorem 4.3, hence we shall prove only the sufficiency. We begin with a lemma that we shall need.

Lemma 4.10. *Let r_m be an increasing sequence of real numbers such that $r_m\geqq e^2r_{m-1}$ and $r_1\geqq1$, and let $z_0\in\Gamma$ be such that $\|z_0\|=1$. If $\varphi(z)=\sum_{m=1}^{\infty}\log\frac{\|z-r_mz_0\|}{r_m}$, then given $\varepsilon>0$, there exists R_ε such that $\varphi(z)\leqq\|z\|^\varepsilon$ for $\|z\|>R_\varepsilon$, and for $1/2\leqq\|z-r_mz_0\|\leqq1$, there exists a constant C such that $\varphi(z)+\log(1+\|z\|)>C$.*

Proof. Let z be such that $e^{j-1}\leqq\|z\|\leqq e^j$. If $m\leqq j^2$, then

$$\log\frac{\|z-r_mz_0\|}{r_m}\leqq\log\left(1+\frac{\|z\|}{r_m}\right)\leqq2j,$$

so

$$\sum_{m=1}^{j^2}\log\frac{\|z-r_mz_0\|}{r_m}\leqq2j^3\leqq8(\log\|z\|)^3\qquad\text{for }\|z\|\geqq e^2\ (i.e.\ j\geqq3).$$

For $m\geqq j^2$, using the Taylor series development for $\log(1+x)$, we obtain

$$\sum_{m=j^2}^{\infty}\log\frac{\|z-r_mz_0\|}{r_m}\leqq\sum_{m=j^2}^{\infty}\log\left(1+\frac{\|z\|}{r_m}\right)\leqq C_1\sum_{m=j^2}^{\infty}\frac{\|z\|}{e^{2m}}=C_1\|z\|\frac{e^{-2j^2}}{1-e^2}\leqq C_1',$$

which proves the first assertion. Suppose that for some m, $\frac{1}{2} \leqq \|z - r_m z_0\| \leqq 1$. Then for $q \leqq m-1$,

$$\log \frac{\|z - r_q z_0\|}{r_q} = \log \left\| \frac{z}{r_q} - z_0 \right\| \geqq \log \left(\frac{r_{m-1}}{r_q} - 1 \right) \geqq \log (c^2 - 2) \geqq 0,$$

and for $q \geqq m+1$, using the Taylor series development for $\log(1-x)$, we obtain

$$\log \frac{\|z - r_q z_0\|}{r_q} = \log \left\| \frac{z}{r_q} - z_0 \right\| \geqq \log \left(1 - \frac{\|z\|}{r_q}\right) \geqq \log \left(1 - \frac{(r_{m-1})}{r_q}\right)$$

$$\geqq \tilde{C}_1 \sum_{m=1}^{\infty} e^{-2m} \geqq \tilde{C}_1'.$$

Furthermore, for r_m

$$\log \frac{\|z - r_m z_0\|}{r_m} \geqq \log \tfrac{1}{2} - \log r_m \geqq \log \tfrac{1}{2} - \log (1 + \|z\|). \qquad \square$$

Proof of Theorem 4.9. As in the proof of Theorem 4.8, we find $z_0 \in \Gamma$, $\varepsilon > 0$, $\eta_1 > 0$ and an increasing sequence r_m tending to infinity such that $h_f^\star(z_0) \neq -\infty$ and

$$\frac{\log |f(r_m z')|}{r_m^{\rho(r_m)}} \leqq h_f^\star(z_0) - \frac{\varepsilon}{2} \qquad \text{for } \|z' - z_0\| < \eta_1.$$

We suppose without loss of generality that $\|z_0\| = 1$ and $\{r_m\}$ satisfies the hypotheses of Lemma 4.10. Let ψ be in $\mathscr{C}_0^\infty(\mathbb{R})$, $0 \leqq \psi \leqq 1$, $\operatorname{supp} \psi \subset (-1, +1)$, and $\psi \equiv 1$ for $t < \frac{1}{2}$; set

$$A(z) = \|z\|^{\rho(\|z\|)} + \sum_m \zeta r_m^{\rho(r_m)} \psi \left(\frac{z - r_m z_0}{\eta_2 r_m} \right),$$

where $0 < \eta_2 < \eta_1$ and η_2 is so small that $(1 - 2\eta_2)^{-\rho} \left(h_f^\star(z_0) - \frac{\varepsilon}{2} \right) \leqq h_f^\star(z_0) - \frac{\varepsilon}{4}$. By Proposition 1.22, for r sufficiently large

$$\sum_{j,k} \frac{\partial^2 r^{\rho(r)}}{\partial z_j \partial \bar{z}_k} w_j \bar{w}_k = \frac{d^2 r^{\rho(r)}}{d(\log r^2)^2} \frac{|\langle z, w \rangle|^2}{r^4} + \frac{d r^{\rho(r)}}{d(\log r^2)} \frac{(r^2 \|w\|^2 - |\langle z, w \rangle|^2)}{r^4}$$

$$\geqq \inf \left(\frac{\rho}{4}, \frac{\rho^2}{4} \right) r^{\rho(r) - 2} \|w\|^2,$$

and so for $\zeta > 0$ sufficiently small (depending on ψ and η_2) and A_0 sufficiently large, $V_1(z) = \sup(A(z), A_0)$ is a plurisubharmonic function. Let

$$\alpha(z) = \sum_{m=1}^{\infty} \psi(z - r_m z_0) \exp V_1(r_m z_0) \quad \text{and} \quad \varphi(z) = \sum_{m=1}^{\infty} \log \frac{\|z - r_m z_0\|}{r_m}.$$

If $\|z' - r_m z_0\| \leqq 1$ and $\eta_2 r_m > 2$, then by the Mean Value Theorem, $\|A(z') - A(r_m z_0)\| \leqq C_0 \|z'\|^{\rho'}$ for $\rho > \rho' > \inf(0, \rho - 1)$ by Proposition 1.19. Let

$$V_2(z) = V_1(z) + 2n\varphi(z) + 3n \log(1 + \|z\|) + C_0 \|z\|^{\rho'}.$$

Then for $\beta = \bar{\partial}\alpha$, $\operatorname{supp}\beta \subset \bigcup_m \{z: \frac{1}{2} \leqq \|z - r_m z_0\| \leqq 1\}$, so by Lemma 4.10, we have

$$\int_\Gamma |\beta|^2 \exp -2V_2(z)\,d\tau(z) \leqq C \sum_{m=1}^{\infty} \exp -\log(1 + r_{m-1}) \leqq C \sum_{m=1}^{\infty} e^{1-2m} \leqq C'.$$

Now, we apply the resolution of the $\bar{\partial}$-equation (see Appendix III). There exists γ such that $\bar{\partial}\gamma = \beta$ and

$$\int_\Gamma |\gamma|^2 \exp -2V_3(z)\,d\tau(z) < +\infty,$$

where

$$V_3(z) = V_2(z) + \frac{3n}{2} \log(1 + \|z\|^2).$$

Then $g(z) = \alpha(z) - \gamma(z)$ is holomorphic, and since $\exp -2n\varphi(z)$ is non-integrable in a neighboorhood of the point $r_m z_0$, we must have $\gamma(r_m z_0) = 0$ and $g(r_m z_0) = \alpha(r_m z_0) = \exp V_1(r_m z_0)$. Thus $h_g^\star(z_0) \geqq h_g(z_0) \geqq (1 + \zeta)$.

If $z' \in B\left(z_0, \frac{\eta_2}{2}\right)$ and r is such that $rz' \in \bigcup_m B\left(r_m z_0, \frac{3\eta_2 r_m}{2}\right)$, then

$$\frac{(1 + \eta_2/2)}{(1 - \eta_2)} r_m \leqq r \leqq r_m \frac{(1 + \eta_2)}{(1 - \eta_2/2)}$$

for some m, and hence there exists $z'' \in B(z_0, \eta_2)$ such that $r_m z'' = rz'$. Since, by Lemma 3.47, for $\|z\|$ sufficiently large

$$\begin{aligned} |g(z)| &\leqq C'' \exp \sup_{\|z' - z\|} V_3(z) \\ &\leqq C''' \exp\left(\|z\|^{\rho(\|z\|)} + \zeta \|z\|^{\rho(\|z\|)} + \frac{\varepsilon}{8} \|z\|^{\rho(\|z\|)}\right), \end{aligned}$$

for η_2 sufficiently small and r sufficiently large, we have

$$\begin{aligned} \frac{\log|f(rz')|}{r^{\rho(r)}} + \frac{\log|g(rz')|}{r^{\rho(r)}} &\leqq \frac{r_m^{\rho(r_m)}}{r^{\rho(r)}} \left[\frac{\log|f(r_m z'')}{r_m^{\rho(r_m)}} + \frac{\log|g(r_m z'')|}{r_m^{\rho(r_m)}}\right] \\ &\leqq h_f^\star(z_0) - \frac{\varepsilon}{4} + (1 + \zeta) + \frac{\varepsilon}{8} \\ &\leqq h_f^\star(z_0) + h_g^\star(z_0) - \frac{\varepsilon}{16}. \end{aligned}$$

Furthermore, if $rz' \notin \bigcup_m B(r_m z_0, 3\eta_2 r_m/2)$, it then follows from Lemma 3.47 and Theorem 1.31 that for $\|z' - z_0\|$ sufficiently small and r sufficiently large that

$$\frac{\log|f(rz')|}{r^{\rho(r)}} + \frac{\log|g(rz')|}{r^{\rho(r)}} \leqq h_f^\star(z_0) + 1.$$

Thus, $h_{fg}^\star(z_0) < h_f^\star(z_0) + h_g^\star(z_0)$. □

§2. Distribution of the Zeros of Functions of Regular Growth

In this section, we shall prove the fundamental principle for entire functions of regular growth, which relates the regularity of growth of the function to the regularity of the distribution of its zero set.

Let $\mu = \Delta u(x)$ and $\mu_t = t^{-\rho(t)} \Delta u(tx)$ for $u \in \mathrm{SH}^{\rho(r)}(\Gamma)$. These are positive measures in Γ. We shall say that the cone Γ' is compactly included in Γ if $\overline{\Gamma' \cap bd B(0,1)} \subset \Gamma$. If Γ' is compactly contained in Γ, we set $\mu_{\Gamma'}(r) = \mu(\Gamma' \cap B(0,r) \cap \complement \overline{B(0,1)})$. (We write $\mu(A)$ and $\Delta h_u^\star(A)$ for the positive mass μ and for the mass of $\Delta h_u^\star$ supported by a set A).

Theorem 4.11. *Suppose that $u \in \mathrm{SH}^{\rho(r)}(\Gamma)$ is of regular growth in Γ and suppose that Γ' compactly included in Γ satisfies $\Delta h_u^\star(bd\Gamma') = 0$. Then*

$$\lim_{r \to \infty} \frac{\mu_{\Gamma'}(r)}{r^{m-2+\rho(r)}} = \Delta h_u^\star(\Gamma' \cap B(0,1)).$$

Proof. It follows from Theorem 4.7 that if $\varphi \in \mathscr{C}_0^\infty(\Gamma)$, then $\lim_{t \to \infty} \int \varphi \, d\mu_t = \int \varphi \Delta h_u^\star$. Let Ω be a bounded open set such that $\bar{\Omega} \subset \Gamma$. If $\{\varphi_n\}$ is a sequence in $\mathscr{C}_0^\infty(\Gamma)$ such that $\varphi_n \uparrow \chi_\Omega$, the characteristic function of Ω, then $\liminf_{t \to \infty} \mu_t(\Omega) \geqq \lim_{t \to \infty} \int \varphi_n \, d\mu_t = \int \varphi_n \Delta h_u^\star$ for all n, and hence $\liminf_{t \to \infty} \mu_t(\Omega) \geqq \Delta h_u^\star(\Omega)$. In exactly the same way, if $\psi_n \in \mathscr{C}_0^\infty(\Gamma)$ is such that $\psi_n \downarrow \chi_{\bar{\Omega}}$,

$$\limsup_{t \to \infty} \mu_t(\bar{\Omega}) \leqq \Delta h_u^\star(\bar{\Omega}).$$

We have $\mu_t(\Omega) = \dfrac{\mu(t\Omega)}{t^{m-2+\rho(t)}}$ $\left(\text{where } t\Omega = \left\{x : \dfrac{x}{t} \in \Omega\right\}\right)$.

Hence, if $\Delta h_u^\star(\partial \Omega) = 0$, then

$$\lim_{t \to \infty} \frac{\mu(t\Omega)}{t^{m-2+\rho(t)}} = \Delta h_u^\star(\Omega). \tag{4,14}$$

By homogeneity, we have

$$\Delta h_u^\star(\Gamma' \cap B(0,1)) = \lim_{r \to \infty} \Delta h_u^\star\left(\Gamma' \cap B(0,1) \cap \complement \overline{B\left(0, \frac{1}{r}\right)}\right),$$

and from (4,14)

$$\frac{\mu_{\Gamma'}(t)}{t^{m-2+\rho(t)}} \geqq \frac{\mu_t(\Gamma' \cap B(0,t) \cap \complement \overline{B(0,t/r)})}{t^{m-2+\rho(t)}} \geqq \Delta h_u^\star(\Gamma' \cap B(0,1)) - \varepsilon$$

for r sufficiently small and t sufficiently large (depending on r and ε). Since ε is arbitrary, $\liminf_{t \to \infty} \dfrac{\mu_{\Gamma'}(t)}{t^{m-2+\rho(t)}} \geqq \Delta h_u^\star(\Gamma' \cap B(0,1))$.

Now, given $\varepsilon>0$, there exists $\sigma(\varepsilon)$ such that

$$\mu(\Gamma'\cap B(0,\sigma r)\cap \complement\,\overline{B(0,1)})\leqq \varepsilon r^{2-m+\rho(r)}$$

for r large enough. To see this, we note that if $\Omega=\Gamma'\cap B(0,2)\cap\complement\,\overline{B(0,1)}$, then by (4,14), there exists t_0 such that $\frac{\mu(t\Omega)}{t^{m-2+\rho(t)}}\leqq \Delta h_u^\star(\Omega)+1$ for $t\geqq t_0$, and if $2^q t_0\leqq r<2^{q+1}t_0$ and $\gamma=2^{-q_0}$, then

$$\frac{\mu_{\Gamma'}(\gamma r)}{r^{m-2+\rho(r)}}\leqq\frac{\mu(\Gamma'\cap B(0,t_0)\cap\complement\,\overline{B(0,1)})}{r^{m-2+\rho(r)}}$$

$$+\frac{(\Delta h_u^\star(\Omega)+1)}{r^{m-2+\rho(r)}}\sum_{k=0}^{q-q_0+1}(2^k t_0)^{m-2+\rho(2^k t_0)}.$$

Thus

$$I_{\gamma r}=r^{2-m-\rho(r)}\mu_{\Gamma'}(\gamma r)\leqq o(r)+C'\sum_{k=0}^{q-q_0+1}2^{(k-q)\left(m-2+\frac{\rho}{2}\right)}$$

for r large enough, since $r^{\rho(r)-\frac{\rho}{2}}$ is an increasing function. It then follows that

$$I_{\gamma r}\leqq o(r)+C'\sum_{q_0-1}^{q}2^{-k\left(m-2+\frac{\rho}{2}\right)}\leqq\varepsilon$$

if q_0 is large enough. Thus given $\varepsilon>0$, for t sufficiently large

$$I_t=\frac{\mu_{\Gamma'}(t)}{t^{m-2+\rho(t)}}\leqq\frac{\varepsilon}{2}+\frac{\mu\left(\Gamma'\cap B(0,t)\cap\complement\,B\left(0,\frac{\gamma\varepsilon t}{2}\right)\right)}{t^{m-2+\rho(t)}}$$

$$\leqq\varepsilon+\Delta h_u^\star\left(\Gamma'\cap B(0,1)\cap\complement\,B\left(0,\frac{\gamma\varepsilon t}{2}\right)\right),$$

and since ε was arbitrary, $\limsup_{t\to\infty}\frac{\mu_{\Gamma'}(t)}{t^{m-2+\rho(t)}}\leqq\Delta h_u^\star(\Gamma'\cap B(0,1))$. □

Corollary 4.12. *Suppose that $u\in \mathrm{SH}^{\rho(r)}(\Gamma)$ is of regular growth in Γ. Then $h_u^\star$ is harmonic in Γ if and only if for every Γ' compactly included in Γ, $\lim_{r\to\infty}\frac{\mu_{\Gamma'}(r)}{r^{m-2+\rho(r)}}=0$. If $\Gamma\subset\mathbb{C}^n$ and u is plurisubharmonic, then $h_u^\star$ is pluriharmonic if and only if for every Γ' compactly included in Γ, $\lim_{r\to\infty}\frac{\mu_{\Gamma'}(r)}{r^{2n-2+\rho(r)}}=0$.*

Proof. If $h_u^\star$ is harmonic in Γ, then $\Delta h_u^\star(\partial\Gamma')=0$ for any Γ' compactly included in Γ. On the other hand, we can find a sequence Γ_n and Γ_n compactly included in Γ_{n+1}, $\bigcup_{n=1}^{\infty}\Gamma_n=\Gamma$ and $\Delta h_u^\star(\partial\Gamma_n)=0$. Then by Theorem 4.11, $\Delta h_u^\star(\partial\Gamma_n)=0$, so $h_u^\star$ is harmonic in Γ. If u is plurisubharmonic, then $h_u^\star$ is harmonic and plurisubharmonic in Γ. Since a harmonic function is $\mathscr{C}^\infty$, we see that, by semi-positivity

$$\left|\frac{\partial^2 h_u^\star}{\partial z_j \partial \bar{z}_k}\right| \leqq \frac{\partial^2 h_u^\star}{\partial z_j \partial \bar{z}_j} + \frac{\partial^2 h_u^\star}{\partial z_k \partial \bar{z}_k} \leqq \Delta h_u^\star = 0,$$

since for a $\mathscr{C}^2$ plurisubharmonic function $\frac{\partial^2 h_u}{\partial z_i \partial \bar{z}_i} \geqq 0$. Thus, $\frac{\partial^2 h_u}{\partial z_j \partial \bar{z}_k} \equiv 0$ and $h_u^\star$ is pluriharmonic. □

Remark. Using Green's Theorem and the positive homogeneity of the function $h_u^\star(x)$, one can express $\Delta h_u^\star(\Gamma \cap B(0,1))$ in terms of $h_u^\star$ on $\Gamma \cap bd B(0,1)$ and the normal derivatives of h_u on $bd\Gamma \cap B(0,1)$. This leads, for instance, to the following generalization of a classical result: if Γ_w^Φ is the circular cone with axis $tw, t \geqq 0$, and angular opening Φ, then

$$\Delta h_u^\star(\Gamma_w^\Phi \cap B(0,1)) = \rho \int_{\Gamma_w^\Phi \cap bd B(0,1)} h_u^\star(z) d\omega_m(z)$$
$$+ (\rho + m - 2)^{-1} \frac{d}{d\Phi} \int_{bd\Gamma_w^\Phi \cap bd B(0,1)} h_u^\star(z) d\tau_{2n-2}(z)$$

where the derivative exists except perhaps for a countable number of Φ (depending on w).

To prove that an entire function whose zeros admit an angular density is of regular growth, we shall use the integral representation formula of Chapter 3. We shall need slightly more information about it. We refer to the notations of Chapter 3, Section 4.

Lemma 4.13. *Given* $\varepsilon > 0$, *there exists* $s(\varepsilon)$ *such that for* $s \geqq s(\varepsilon)$, $\|a\| > 3(2n-2)\|z\|$,

$$|e_{2n-2}(a, z, q+s)| \leqq \frac{\varepsilon \|z\|^{q+1}}{\|a\|^{2n-1+q}}.$$

Proof. We have $|P_s(a,z)| \leqq \frac{\|z\|^s}{\|a\|^{2n-2+s}} b_{n,s}$ with

$$b_{n,s} = \frac{1}{s!}(2n-2)(2n-1)\dots(2n+s-3)$$

by Proposition 3.14, and so

$$|e_{2n-2}(a, z, q+s)| \leqq \sum_{k=s+q+1}^{\infty} \frac{\|z\|^k}{\|a\|^{2n-2+k}} \frac{1}{k!}(2n-2)\dots(2n+k-3)$$
$$\leqq \frac{\|z\|^{q+1}}{\|a\|^{2n-1+q}} \sum_{k=s}^{\infty} \left(\frac{1}{\tau}\right)^k \frac{1}{(k+q)!}(2n-2)\dots(2n+k+q-3)$$

for $\|a\| > \tau \|z\|$.

If $\tau = 3(2n-3)$ and $s > (2n-2)$ is sufficiently large, then

$$|e_{2n-2}(a, z, q+s)| \leqq \frac{\|z\|^{q+1}}{\|a\|^{2n-1+q}} \sum_{k=s}^{\infty} \left(\frac{2}{3}\right)^k \leqq \frac{\varepsilon \|z\|^{q+1}}{\|a\|^{2n-1+q}}. \qquad \square$$

Lemma 4.14. *Suppose* $X = \{z : f(z) = 0\}$ $0 \notin X$ *and* $\sigma_X(r) < C r^{\rho(r)+2n-2}$ (*cf.* *Chapter* 3). *Then there exists an absolute constant* A *such that for* μ *sufficiently small and* $q < \rho$ *and* $\|w\| = 1$, $\mu^{\rho-q} r^{\rho(r)}$

$$|k_{2n-2}^{-1} \int_{\|a\| \leqq \mu r} e_{2n-2}(a, rw, q) d\sigma_X(a)| \leqq A C \mu^{\rho-q} r^{\rho(r)}.$$

Proof. By Proposition 3.14, for μ small enough and $\|w\| = 1$, we have

$$|k_{2n-2}^{-1} e_{2n-2}(a, rw, q)| \leqq \frac{C_2 r^q}{\|a\|^{2n-2+q}},$$

and thus we obtain by an integration by parts:

$$|k_{2n-2}^{-1} \int_{\|a\| \leqq \mu r} e_{2n-2}(a, rw, q) d\sigma_X(a)|$$

$$\leqq C_2 r^q \int_0^{\mu r} \frac{d\sigma_X(t)}{t^{2n-2+q}} \leqq C_2 r^q \left[\frac{\sigma_X(t)}{t^{2n-2+q}}\right]_0^{\mu r} + C_3 r^q \int_0^{\mu r} \frac{\sigma_X(t) dt}{t^{2n-1+q}}$$

$$\leqq A C \mu^{\rho-q} r^{\rho(r)}$$

since

$$r^q \frac{\sigma_X(\mu r)}{(\mu r)^{2n-2+q}} \leqq C \frac{(\mu r)^{\rho(\mu r)}}{\mu^q} \leqq A_1 C \mu^{\rho-q} r^{\rho(r)}$$

by Proposition 1.20 and

$$r^q \int_0^{\mu r} \frac{\sigma_X(t) dt}{t^{2n-1+q}} dt \leqq C r^q \int_0^{\mu r} \frac{(t)^{\rho(t)+2n-2}}{t^{2n-1+q}} dt$$

$$\leqq \mu^{\rho-q} C A_1 r^q \int_0^r t^{\rho(t)-q+1} dt$$

by Proposition 1.20. We now apply Lemma 3.20. $\square$

If w is a vector in $\mathbb{C}^n$, we let Γ_w^Φ be the right circular cone with vertex at the origin, w as axis and angular opening Φ.

Theorem 4.15. *Let* $f(z)$ *be an entire function of non-integral order* ρ *and normal type with respect to the proximate order* $\rho(r)$. *If* Γ *is a cone in* $\mathbb{C}^n$ *(with vertex at the origin), we set* $\sigma(\Gamma, r) = \int_{\Gamma \cap B(0,r)} \Delta \log |f|$. *Then if for every* $w \in S_{2n-1}$, $\lim_{r \to \infty} \frac{\sigma(\Gamma_w^\Phi, r)}{r^{\rho(r)+2n-2}} = A(\Phi)$ *exists except perhaps for a countable set of* Φ *(depending on* w*),* f *is of regular growth in* $\mathbb{C}^n$.

Proof. Let $\varepsilon > 0$ and $\delta_0 > 0$ be given. We suppose without loss of generality that $\lim_{\Phi \to 0} A(\Phi) = 0$, since there are at most a countable number of vectors w_m for which this is not the case, and if f is of regular growth in $\mathbb{C}^n - \bigcup_m \{t w_m : t \geqq 0\}$, then f is of regular growth in $\mathbb{C}^n$ by Theorem 4.5.

Let Φ_0 be such that $\lim_{r \to \infty} \frac{\sigma(\Gamma_w^\Phi, r)}{r^{\rho(r)+2n-2}} \leqq l$ for $0 \leqq \Phi \leqq \Phi_0$ where we fix l later (and l will depend only on δ_0). Set $L_w(\Phi_1, \Phi_2) = \Gamma_w^{\Phi_1} - \Gamma_w^{\Phi_2}$. For a vector $\vec{a}$, let $\vec{a}_\theta$ be the point in the 2-dimensional plane determined by $\vec{a}$ and $\vec{w}$ such that $\|\vec{a}_\theta\| = \|\vec{a}\|$ and $\vec{a}_\theta$ forms an angle θ with $\vec{w}$, and let $P_q^\theta(a, z) = P_q(a_\theta, z)$.

By Proposition 3.14 and Lemma 4.14, there exist μ_0 and λ_0 such that

$$|k_{2n-2}^{-1} \int_{\|a\| \geqq \lambda r} e_{2n-2}(a, rw, q) d\sigma_X(a)| \leqq \frac{\varepsilon}{12} r^{\rho(r)}$$

and

$$|k_{2n-2}^{-1} \int_{\|a\| < \mu r} e_{2n-2}(a, rw, q) d\sigma_X(a)| \leqq \frac{\varepsilon}{12} r^{\rho(r)}$$

for $\lambda \geqq \lambda_0$ and $\mu \leqq \mu_0$.

In addition, given $\xi > 0$, there exists $\eta(\xi, \mu, \lambda) > 0$ such that for $\theta_1, \theta_2 \geqq \Phi_0$, $|\theta_i - \theta_{i+1}| < \eta$ and $a \in L_w(\theta_1, \theta_2)$, $\mu r \leqq \|a\| \leqq \lambda r$,

$$|k_{2n-2}^{-1} e_{2n-2}(a, rw, q) - k_{2n-2}^{-1} (e_{2n-2}(a_{\theta_i}, rw, q)| < \xi r^{2-2n}.$$

We set $\xi = \frac{\varepsilon}{12 d \lambda^{2n-2+\rho}}$, where $d = \lim_{r \to \infty} \frac{\sigma(\Gamma_w^\pi, r)}{r^{2n-2+\rho(r)}}$.

We divide the interval $(\Phi_0, \pi]$ into a finite number of sub-intervals $(\theta_1, \theta_{i+1}]$, $i = 0, 1, \ldots, M$, such that $|\theta_{i+1} - \theta_i| < \eta(\xi, \mu, \lambda)$. We set

$$\sigma_i(r) = \sigma_{\Gamma_w^\theta(i+1)}(r) - \sigma_{\Gamma_w^\theta i}(r), \qquad A_i = A(\theta_{i+1}) - A(\theta_i), \qquad \Gamma' = \complement \Gamma_w^{\Phi_0},$$

$$T_i^r = L_w(\theta_{i+1}, \theta_i) \cap \complement B(0, \mu r) \cap B(0, \lambda r),$$

$$D_\mu^\lambda(r) = \Gamma_w^{\Phi_0} \cap \complement B(0, \mu r) \cap B(0, \lambda r),$$

$$T(z) = k_{2n-2}^{-1} \int e_{2n-2}(a, z, q) d\sigma_X(a).$$

$$J(r) = k_{2n-2}^{-1} \int_{D_\mu^\lambda(r)} e_{2n-2}(a, rw, q) d\sigma_X(a)$$

Then for r sufficiently large,

$$\left| T(rw) - \sum_{i=1}^{M} k_{2n-2}^{-1} \int_{T_i^r} e_{2n-2}(a_{\theta_i}, rw, q) d\sigma_X(a) \right|$$

$$\leqq |T(rw) - J(r)| + \left| J(r) - \sum_{i=1}^{M} k_{2n-2}^{-1} \int_{T_i^r} e_{2n-2}(a_{\theta_i}, rw, q) d\sigma_X(a) \right|$$

$$\leqq \frac{2\varepsilon}{12} r^{\rho(r)} + \xi d \lambda^{2n-2+\rho} r^{\rho(r)} \leqq \frac{\varepsilon}{4} r^{\rho(r)}.$$

Furthermore, we let $S_n^r(\theta, t, q) = k_{2n-2}^{-1} e_{2n-2}(a_\theta, rw, q)$ for $t = \|a_\theta\|$, and $k_{2n-2}^{-1} P_q(a, rw) = \dfrac{r}{\|a\|^{2n-2+q}} B_q(\cos\theta)$ for a polynomial B_q. Then

$$k_{2n-2}^{-1} \int_{T_i^r} e_{2n-2}(a_{\theta_i}, rw, q)\, d\sigma_X(a) = \int_{\mu r}^{\lambda r} S_n^r(\theta_i, t, q)\, d\sigma_i(t).$$

Given $\zeta > 0$, for $t \in (\mu r. \lambda r)$ and r large enough, if $L(t) = t^{\rho(t)-\rho}$

$$\begin{aligned}
(4{,}15)\quad & |\sigma_i(t) - L(r) A_i t^{\rho+2n-2}| \\
& \leqq |\sigma_i(t) - A_i t^{\rho(t)+2n-2}| + |A_i t^{\rho(t)+2n-2} - L(r) A_i t^{\rho+2n-2}| \\
& \leqq \frac{\zeta}{2} L(t) t^{\rho+2n-2} + A_i t^{\rho+2n-2} L(r) \left| \frac{L(t)}{L(r)} - 1 \right| \leqq \zeta L(r) t^{\rho+2n-2}
\end{aligned}$$

by Theorem 1.18, and so, setting $\tau = \dfrac{1}{3(2n-2)}$,

$$\begin{aligned}
&\int_{\mu r}^{\tau r} S_n^r(\theta_i, t, q)\, d\sigma_i(t) \\
&\quad = \int_{\mu r}^{\tau r} \left\{ \frac{-1}{|(t\cos\theta_i - r)^2 + t^2 \sin^2\theta_i|^{n-1}} + \frac{B_1(\cos\theta_i)}{t^{2n-1}} + \dots + \frac{B_q(\cos\theta_i)}{t^{2n-2+q}} \right\} d\sigma_i(t)
\end{aligned}$$

and

$$\begin{aligned}
&\int_{\mu r}^{\tau r} \frac{d\sigma_i(t)}{|(t\cos\theta_i - r)^2 + t^2\sin^2\theta_i|^{n-1}} \\
&\qquad = \left. \frac{\sigma_i(t)}{|(t\cos\theta_i - r)^2 + t^2 \sin^2\theta_i|^{n-1}} \right]_{\mu r}^{\tau r} \\
&\qquad\quad + (2n-2) \int_{\mu r}^{\tau r} \frac{\sigma_i(t)\,(t - r\cos\theta_i)\, dt}{|(t\cos\theta_i - r)^2 + t^2\sin^2\theta_i|^n}.
\end{aligned}$$

It now follows from an integration by parts, Lemma 4.14, and (4,15) that for μ sufficiently small

$$\text{(a)}\quad \left| \left. \frac{\sigma_i(t)}{|(t\cos\theta_i - r)^2 + t^2\sin^2\theta_i|^{n-1}} \right]_{\mu r}^{\tau r} - \frac{A_i \tau^{2n-2+\rho} r^{\rho(r)}}{|(\tau\cos\theta_i - 1)^2 + \tau^2 \sin^2\theta_i|^{n-1}} \right| \leqq \frac{\varepsilon A_i r^{\rho(r)}}{24 d(q+1)}$$

and for r sufficiently large:

$$\text{(b)}\quad \left| \int_{\mu r}^{\tau r} \frac{\sigma_i(t)\,(t - r\cos\theta_i)\, dt}{|(t\cos\theta_i - r)^2 + t^2\sin^2\theta_i|^n} - r^{\rho(r)} A_i \int_0^{\tau} \frac{s^{\rho+2n-2}(s - \cos\theta_i)\, ds}{|(s\cos\theta_i - 1)^2 + s^2\sin^2\theta_i|^n} \right| \leqq \frac{\varepsilon A_i r^{\rho(r)}}{24 d(q+1)}$$

for r sufficiently large. We note in passing that

$$\left| \int_0^{\tau} \frac{t^{\rho+n-2}(t - \cos\theta_i)\, dt}{|(t\cos\theta_i - 1)^2 + t^2 \sin\theta_i|^n} \right| \leqq C_{\Phi_0} \quad \text{for } \theta_i \geqq \Phi_0.$$

Furthermore,

$$r^q \int_{\mu r}^{\tau r} \frac{d\sigma_i(t)}{t^{2n-2+q}} = \left[\frac{\sigma_i(t)}{t^{2n-2+q}}\right]_{\mu r}^{\tau r} + (2n-2+q) \int_{\mu r}^{\tau r} \frac{\sigma_i(t)\,dt}{t^{2n-1+q}}.$$

(c) Thus by (4,15) and Lemma 4.14 for μ small enough and r large enough

$$\left| r^q \int_{\mu r}^{\tau r} \frac{d\sigma_i(t)}{t^{2n-2+q}} - A_i \tau_2^{\rho-q} r^{\rho(r)} + \frac{(2n-2+q)}{(\rho-q)} \tau^{\rho-q-1} \right| \leqq \frac{\varepsilon A_i r^{\rho(r)}}{12d(q+1)}.$$

By Lemma 4.13, for β large enough and $t \geqq \tau r$, we have

$$\left| S_n(\theta_i, t, q) - \sum_{k=q+1}^{q+\beta+1} B_k(\cos\theta_i) \frac{r^k}{t^{2n-2+k}} \right| \leqq \frac{\varepsilon r^{q+1}}{12 A t^{2n-1+q}}$$

for $A > 0$ and $\beta(A, \varepsilon)$. Thus

$$\begin{aligned}
&\left| \sum_{i=1}^{M} \int_{\tau r}^{\lambda r} S_n(\theta_i, t, q)\, d\sigma_i(t) - \int_{\tau r}^{\lambda r} \sum_{i=1}^{M} \sum_{k=q+1}^{q+\beta-1} B_k(\cos\theta_i) \frac{r^k}{t^{2n-2+k}}\, d\sigma_i(t) \right| \\
&\quad \leqq \sum_{i=1}^{M} \frac{\varepsilon r^{q+1}}{12 d A} \int_{\tau r}^{\lambda r} \frac{d\sigma_i(t)}{t^{2n-1+q}} \\
&\quad \leqq \sum_{i=1}^{M} \frac{\varepsilon r^{q+1}}{12 d A} \left\{ \left[\frac{\sigma_i(t)}{t^{2n-1+q}}\right]_{\tau r}^{\lambda r} + (2n-1+q) \int_{\tau r}^{\lambda r} \frac{\sigma_i(t)\,dt}{t^{2n-1+q}} \right\} \\
&\quad \leqq \sum_{i=1}^{M} \frac{\varepsilon A_i r^{\rho(r)}}{12}
\end{aligned}$$

for A and r large enough.

Furthermore,

$$\int_{\tau r}^{\lambda r} r^k \frac{d\sigma_i(t)}{t^{2n-2+k}}\, dt = r^k \left[\frac{\sigma_i(t)}{t^{2n-2+k}}\right]_{\tau r}^{\lambda r} + (2n-2+k) r^k \int_{\tau r}^{\lambda r} \frac{\sigma_i(t)}{t^{2n-1+k}}\, dt,$$

and thus, for r sufficiently large, by Theorem 1.18, there exists a constant A_k^i such that

$$\left| \int_{\tau r}^{\lambda r} \frac{r^k}{t^{2n-2+k}}\, d\sigma_i(t) - A_k^i r^{\rho(r)} \right| < \frac{\varepsilon}{24 M \beta}.$$

Gathering together all the inequalities, (a), (b), (c), (d) we find that for

$$T(z) = k_{2n-2}^{-1} \int_{\Gamma'} e_{2n-2}(a, z, q)\, d\sigma_X(a),$$

there exists a number γ_w that $\lim\limits_{r\to\infty} \dfrac{T(rw)}{r^{\rho(r)}} = \gamma_w$, and since $T(z)$ is subharmonic in z, given $\varepsilon > 0$, we obtain $I_T^r(w, \delta) \geqq \gamma_w - \varepsilon/2$ for r sufficiently large.

We now turn our attention to $Q(z) = k_{2n-2}^{-1} \int_{\Gamma_w^{\Phi_0}} e_{2n-2}(a, z, q)\, d\sigma_X(a)$. It follows from the estimates of Theorem 3.19 that there exists a constant C

such that $Q(z)=CA(\Phi_0)\,\|z\|^{\rho(\|z\|)}$ and since $Q(0)=0$,

$$0\leqq \int_{B(0,r(1+\delta))} Q(z)d\tau(z)=\int_{B(rw,r\delta)} Q(z)d\tau(z)+\int_{B(0,(1+\delta)r)-B(rw,r\delta)} Q(z)d\tau(z)$$

or

$$-\int_{B(0,r(1+\delta)-B(rw,r\delta)} Q(z)d\tau(z)\leqq \int_{B(rw,r\delta)} Q(z)d\tau(z)$$

and

$$0\leqq \int_{B(0,(1+\delta)r)} Q^-(z)\,d\tau(z)\leqq \int_{B(0,(1+\delta)r)} Q^+(z)\,d\tau(z).$$

Using the increasing nature of $r^{\rho(r)}$ we obtain by Theorem 1.18, for r large

$$\begin{aligned}-\int_{B(,r(1+\delta))-B(rw,r\delta)} Q(z)\,d\tau(z) &\geqq -CA(\Phi_0)((1+\delta)^\rho r)^{\rho(1+\delta)r}\\ &\geqq -2\,CA(\Phi_0)\,(1+\delta)^\rho\, r^{\rho(r)}.\end{aligned}$$

Thus $I_Q^r(w,\delta)\geqq -\dfrac{C'A(\Phi_0)(1+\delta)^\rho}{\delta^n}$, and if Φ_0 is sufficiently small (depending on δ) $I_Q^r(w,\delta)\geqq -\varepsilon/2$ for r large enough. But $\log|f|=\mathbb{R}e\,S(z)+I(z)+C_\delta$, where $S(z)$ is a polynomial of order $q<\rho$, so $I_f^r(w,\delta)\geqq\gamma_w-\varepsilon$ for r sufficiently large, which completes the proof. □

The case ρ an integer is more complex and requires a slightly different treatment. Let α,β be multi-indices of positive n-tuples each of which is an integer, $|\alpha|=\sum_{i=1}^n \alpha_i$, $|\beta|=\sum_{i=1}^n \beta_i$. We set $z^\alpha\bar{z}^\beta=z_1^{\alpha_1}\ldots z_n^{\alpha_n}\bar{z}_1^{\beta_1}\ldots\bar{z}_n^{\beta_n}$. Let

$$P_q(a,z)=\sum_{|\alpha|+|\beta|=q} P_{\alpha,\beta}(a)\,z^\alpha\bar{z}^\beta \quad\text{and}\quad S(z)=\sum_{|\alpha|+|\beta|\leqq q} C_{\alpha,\beta}\,z^\alpha\bar{z}^\beta,$$

where $S(z)$ is the pluriharmonic polynomial in the canonical representation of $\log|f(z)|$ (Theorem 3.30).

Theorem 4.16. *If ρ is an integer, then an entire function of order ρ is of regular growth in $\mathbb{C}^n$ if and only if*

i) $\lim\limits_{r\to\infty}\dfrac{\sigma_{\Gamma_w^\Phi(r)}}{r^{\rho(r)+2n-2}}$ *exists (for every w) except perhaps for a countable set of Φ which depends on w),*

ii) $\lim\limits_{r\to\infty}\left\{\int_{\|a\|<r}\dfrac{(P_{\alpha,\beta}(a)d\sigma(a)+C_{\alpha,\beta})}{r^{\rho(r)-\rho}}\right\}$ *exists for all α,β such that $|\alpha|+|\beta|=\rho$.*

Proof. The long and tedious proof resembles very closely that of Theorem 4.15, so we give only a sketch her indicating the variations necessary.

We note that for ρ an integer we can have $q=\rho$ or $q=\rho-1$ in the canonical representation, but if we choose $q=\rho$, the integral always converges. Let

$$\Phi_r(z)=k_{2n-2}^{-1}\int_{\|a\|\leq r} e_{2n-2}(a,z,q-1)\,d\sigma_X(a)$$
$$+k_{2n-2}^{-1}\int_{\|a\|>r} e_{2n-2}(a,z,q)\,d\sigma_X(a), \qquad r=\|z\|.$$

For r fixed, this is a subharmonic function of z, and one shows just as in Theorem 4.15 that given $\varepsilon>0$, $\delta_0>0$, there exists $\delta\leqq\delta_0$ such that for r large enough $\varepsilon\geqq I^r_{\Phi_r}(w,\delta)-\gamma_w\geqq-\varepsilon$ for some number γ_w. Thus, by setting $T_r(z)=k_{2n-2}^{-1}\int_{\|a\|\leq r} P_q(a,z)\,d\sigma_X(a)$, which is harmonic and homogeneous of order ρ in z, we have for $w\in S^{2n-1}$:

$$(4,16)\qquad I^r_f(w,\delta)-I^r_{\Phi_0}(w,\delta)$$
$$=\sum_{|\alpha|+|\beta|=\rho}\frac{w^\alpha\bar{w}^\beta}{r^{\rho(r)-\rho}}\Big[C_{\alpha,\beta}+\int_{\|a\|\leq r} P_{\alpha,\beta}(a)\,d\sigma(a)\Big]+o(r^{\rho(r)}),$$

and thus if (ii) is satisfied, f is of regular growth for w.

On the other hand, if f is of regular growth in $\mathbb{C}^n$, then (i) is satisfied (by Theorem 4.12) and thus by (4,16),

$$A_w=\lim_{r\to\infty}\sum_{|\alpha|+|\beta|=q}\frac{w^\alpha\bar{w}^\beta}{r^{\rho(r)-\rho}}\Big[C_{\alpha,\beta}+\int_{\|a\|\leq r} P_{\alpha,\beta}(a)\,d\sigma_X(a)\Big]$$

exists except perhaps for a countable number of w. Let γ be the number of multiindices such that $|\alpha|+|\beta|=\rho$. Set $A=[w^\alpha_{(j)}\bar{w}^\beta_{(j)}]$, the square matrix of order $\gamma\times\gamma$, where the variable is taken in $\mathbb{R}^{2n\gamma}$-dimensional Euclidean space. Then $\det A$ is an analytic function not identically zero, and thus there exist $\tilde{w}_{(1)},\ldots,\tilde{w}_{(\gamma)}$ not in the countable exceptional set for which $\det A(\tilde{w}^\alpha_{(i)}\bar{\tilde{w}}^\beta_{(i)})\neq 0$. Thus, we can solve for $r^{\rho-\rho(r)}\big[C_{\alpha,\beta}+\int_{\|a\|<r} P_{\alpha,\beta}(a)\,d\sigma_X(a)\big]$ as a linear combination of

$$\sum_{|\alpha|+|\beta|=\rho}\frac{\tilde{w}^\alpha_{(i)}\bar{\tilde{w}}^\beta_{(i)}}{r^{\rho(r)-\rho}}\Big[C_{\alpha,\beta}+\int_{\|a\|\leq r} P_{\alpha,\beta}(a)\,d\sigma_X(a)\Big], \qquad i=1,\ldots,\gamma.$$

Since the latter admit finite limits when r tends to ∞, so do the former. □

Historical Notes

Most of this chapter reformulates in a different context classical results on regular growth applied to entire functions of one variable, and the reader is referred to the very thorough book of B.Ja. Levin [D] for results as well as proofs for $n=1$. Functions of regular growth play an important role in the theory of entire and meromorphic functions of one variable of finite order.

For many problems in Nevanlinna theory, they provide extremal solutions. They also have a number of applications in the theory of differential equations and Fourier transforms. The theory is not yet as extensive for functions of regular growth in $\mathbb{C}^n$. We give some applications of this theory in Chapter 9. Other interesting applications can be found in the recent work of Wiegerinck [1, 2].

The beginning of Sections 1 and 2 were announced in Gruman [7] without proofs. Theorem 3.8 is due to Favarov [1]. Some simplifications in the beginning of Section 2 are due to Berndtsson [1]. Additional results on functions of regular growth are to be found in the works of Agranovič [2] and Agranovič and Ronkin [1].

Chapter 5. Holomorphic Mappings from $\mathbb{C}^n$ to $\mathbb{C}^m$

We shall study four problems here related to entire mappings defined on $\mathbb{C}^n$. If X is a Cousin data in $\mathbb{C}^n$, we have seen in Chapter 3 that we can define X as the zero set of an entire function f whose growth is related to the growth of $v_X(r)$, the indicator of X. Our first task will be to show a similar property for analytic varieties Y of arbitrary co-dimension in $\mathbb{C}^n$. We shall show that we can define Y as $Y=\{z\colon F(z)=0\}$ where $F\colon \mathbb{C}^n \to \mathbb{C}^{n+1}$ and the growth of $\|F\|$ is bounded by $v_Y(r)$, the projective indicator of growth of the current of integration on the analytic set Y (cf. Definition 2.24).

There are three other questions related to entire functions that we shall examine:

i) if f is an entire function, then it follows from Gauss' Formula that if $X=f^{-1}(a)$, $v_X(r)$ is bounded by the growth of $|f|$ (where X is the Cousin data $(X,\mathbb{C}^n)$); in particular, for every $\alpha>1$, there exists C_α such that

$$v_X(r) \leqq C_\alpha \sup_{\|z\| \leqq \alpha r} (\log|f|) + C'_f.$$

Can one estimate the size of $F^{-1}(a)$ by the growth of $\|F\|$ for a holomorphic map?

ii) if Y is an analytic variety of co-dimension 1 in $\mathbb{C}^n$, then we can define Y as $f^{-1}(0)$ for an entire function whose asymptotic growth is related to $v_{[Y]}(r)$, and it follows from Jensen's Theorem for one complex variable that for *any* complex line L such that $L \not\subset X$, the asymptotic growth of $L \cap X \cap B(0,r)$ cannot be greater than the asymptotic growth of $|f|$, hence not much greater than the asymptotic growth of $v_{[Y]}(r)$. If Y is of co-dimension superior to 1 and L is a linear subspace equal to the co-dimension of Y, can one estimate the intersection $L \cap X \cap B(0,r)$ in terms of $v_{[Y]}(r)$?

iii) if X is a Cousin data and f is an entire function such that $X=(f,\mathbb{C}^n)$ and $\log|f|$ grows like $v_X(r)$, then for any complex line L,

$$\int_0^r \frac{1}{t} \operatorname{card}(L \cap X \cap B(0,t))\,dt = \frac{1}{2\pi} \int_{\|a\|=1} \log|f(re^{i\theta}\alpha)|\,d\theta - \log|f(0)|$$

(where the points in the intersection are counted with multiplicity). Since the right hand side is a plurisubharmonic function whose average over

S^{2n-1} is equal to $v_X(r)$, we see that the set of L such that $L \cap X \cap B(0,r)$ "grows much more slowly" than $v_X(r)$ is locally pluripolar in $\mathbb{P}(\mathbb{C}^n)$ (cf. Corollary 1.43). If Y is of arbitrary co-dimension p and L^p is a linear subspace of dimension p, for how small a set of L^p can $L^p \cap Y \cap B(0,r)$ grow "more slowly" than $v_{[Y]}(r)$?

We shall see that the answer to the first two questions is negative; however, we shall show that the set of values for which such estimates are not possible is quite small. What is more, we shall show that the set of L^p such that $L^p \cap Y$ grows more slowly than $v_{[Y]}(r)$ is also quite small.

§1. Representation of an Analytic Variety Y in $\mathbb{C}^n$ as $F^{-1}(0)$

As in Chapter 3, for Y an arbitrary analytic variety in $\mathbb{C}^n$, we are interested in expressing Y as $F^{-1}(0)$, where F is an entire mapping whose growth is related to the growth of $v_Y(r)$. There are, however, two fundamental differences:

i) if X is a Cousin data, $X=(U_k, f_k)$, then X determines an analytic variety $Y(X)$ in $\mathbb{C}^n$ and in addition, for every irreducible branch Y_k of $Y(X)$, a non-negative integer m_k, the multiplicity of X on Y_k, and the solution f has a zero of order m_k on Y_k;

ii) the Cousin data gives an expression for the current of integration, $\theta_X = \sum m_k [X_k]$ in $\mathbb{C}^n$.

In the general case, we define Y as a set in each U_k of an open covering of $\mathbb{C}^n$; $Y \cap U_k = \{z\colon f_{kj}(z)=0, j=1, \ldots, j_k\}$. Then we require only that $Y=F^{-1}(0)$ *as a set*. The construction of the current of integration over $[Y]$ as a positive closed current is carried out in Chapter 2. The basic plan is as follows:

a) using the positive closed current $[Y]$, we construct local potentials with density on Y;

b) using a partition of unity, we construct a global potential with density on Y;

c) by adding a function whose Levi form is strictly positive and whose growth is related to $v_{[Y]}(r)$, we construct a plurisubharmonic function V whose growth is related to $v_{[Y]}(r)$ and such that, if $t' = \frac{i}{\pi} \partial \bar{\partial} V$, then $v_{t'}(z) = v_{[Y]}(z) \geqq 1$, where $v_{t'}(z)$ and $v_{[Y]}(z)$ are respectively the Lelong numbers at x of t' and $[Y]$ the current of integration on the analytic set $[Y]$ (cf. Theorem 2.23);

d) using the existence theorems for the $\bar{\partial}$-operator with growth conditions (cf. Appendix III), we construct an entire mapping F such that $Y=F^{-1}(0)$ with estimates for the growth.

The construction, with modifications, is also valid in any pseudo-convex domain and for any positive closed current t, but we shall not treat it in its full generality. These technics are due to H. Skoda.

§ 2. Local Potentials and the Defect of Plurisubharmonicity

Let $c_p=\omega_{2p}^{-1}$, where ω_{2p} is the area of the unit sphere in $\mathbb{R}^{2p}$. We define the kernels $g_p(a,z)=-c_p\|a-z\|^{-2p}$ for $1\leqq p\leqq n-1$ and $g_0(a,z)=\log\|a-z\|$ for $p=0$; these functions are $\mathbb{R}^{2n}$ subharmonic but plurisubharmonic only for $p=0$. Furthermore, the Laplacian

$$\Delta_z g_p(a,z)=2p(2n-2-2p)c_p\|a-z\|^{-2p-2}\geqq 0$$

is (for a fixed) locally integrable if $p\leqq n-2$. For $p=n-1$, we have $\Delta_z=2\pi\delta(a)$ as a distribution.

Proposition 5.1. *Let σ be a positive measure carried by a ball B_0 of $\mathbb{C}^n$ and $d_{2p}(z)=\lim\limits_{r\to\infty}\dfrac{\sigma(z,r)}{\tau_{2p}r^{2p}}$. Then if $U(z)=\int g_p(a,z)\,d\sigma(a)$ and $\sigma'=\dfrac{1}{2\pi}\Delta U$, we have*

$$\lim_{r\to 0}\frac{\sigma'(z,r)}{\tau_{2n-2}r^{2n-2}}=d'_{2n-2}(z)=d_{2p}(z). \tag{5,1}$$

Proof. If $p=n-1$, then $\sigma'=(2\pi)^{-1}\Delta U=\sigma$, from which (5,1) follows. Suppose then that $1\leqq p<n-1$ and that $z=0$. Then

$$(2\pi)^{-1}\Delta U=c_p c'_p\int\|a-z\|^{-2p-2}\,d\sigma(a)\qquad\text{with } c'_p=2p(n-p-1)\,\pi^{-1}$$

and ΔU is a function. Since $-\|a-z\|^{\lambda}$ is $\mathbb{R}^{2n}$-subharmonic for $0\leqq\lambda\leqq 2n-2$, if $k_{p+1}(t,r)=\int\|a-\tilde{\alpha}r\|^{-2p-2}\,d\omega_{2n}(\tilde{\alpha})$ for fixed $\|a\|=t$, $k_{p+1}(t,r)$ is a decreasing function of r, and for $p\leqq n-2$, $\dfrac{dk_{p+1}}{dt}<0$. We then have for $r>0$, $r\to 0$:

$$\begin{aligned} I(r)=I(0,r,\Delta U)&=c_p c'_p\int_0^R d\sigma(t)k_{p+1}(t,r)\\ &=c_p c'_p\int_0^R\sigma(t)\left(\frac{-\partial k_{p+1}}{\partial t}\right)dt+O(1).\end{aligned}$$

Since the integrand is positive and since by hypothesis

$$\sigma(t)=d_{2p}(0)\tau_{2p}t^{2p}+\alpha(t)\tau_{2p}t^{2p}\qquad\text{for } 0\leqq\alpha(t)<\varepsilon$$

when $t<R_\varepsilon$, we obtain

$$I(0,r,\Delta U)=c_p c'_p(1+\theta\varepsilon)\tau_{2p}\int_0^R d_{2p}(0)t^{2p}\left(-\frac{\partial k_{p+1}}{\partial t}(t,r)\right)dt+O(1)$$

with $0 \leqq \theta \leqq 1$. By integrating by parts, we have

$$\int_0^R t^{2p}\left(-\frac{\partial k_{p+1}}{\partial t}\right) dt = [-t^{2p} k_{p+1}]_0^R + 2p \int_0^R t^{2p-1} k_{p+1}(t,r)\, dt$$

and

$$\int_0^R t^{2p-1} k_{p+1}(t,r) dt = \int_0^\infty t^{2p-1} k_{p+1}(t,r) dt + O(1) = J(r) + O(1),$$

where

$$J(r) = c_n \int_{\mathbb{R}^{2n}} \|a\|^{2p-2n} \|a-z\|^{-2p-2} d\tau(a).$$

This last integral is a convolution of $r^{-\alpha}$ and $r^{-\beta}$ with $\alpha = -2p+2n$ and $\beta = 2p+2$. It is easy to see by a substitution that $J(r) = A_{n,p} c'_p c_n r^{-2}$, and $A_{n,p} = \pi^n [p(n-p-1)(n-2)!]^{-1}$, so

$$I(r) = 2p c_p \tau_{2p} c'_p d_{2p}(0) A_{n,p} c_n r^{-2}(1+\theta\varepsilon), \qquad 0 \leqq \theta \leqq 1.$$

Let $\sigma'(r)$ be the mass of $\sigma' = \dfrac{1}{2\pi} \Delta U$ carried by $B(0,r)$. By Gauss' Formula,

$$\sigma'(r) = c_n^{-1} \int_0^r I(t) t^{-2n+1} dt = d_{2p}(0) A_{n,p} (2n-2)^{-1} c'_p (1+\theta\varepsilon) r^{2n-2}$$

and

$$(\tau_{2n-2} r^{2n-2})^{-1} \sigma'(r) = C(1+\theta\varepsilon) d_{2p}(0)$$

where

$$C = A_{n,p} c'_p \tau_{2n-2}^{-1} (2n-2)^{-1} = 1,$$

which proves (5,1) for $1 \leqq p \leqq n-2$.

For $p=0$, $g_0(a,z) = \log\|a-z\|$ and $\sigma(t) = \delta(0) d_0 + \varepsilon(t)$ with $0 \leqq \varepsilon(t) \leqq \varepsilon$ for $0 \leqq t \leqq R$. This reduces to the case where $U(t)$ is a point mass at the origin in which case (5,1) is evident. □

Let $\eta(z) \in \mathscr{C}_0^\infty(\mathbb{C}^n)$, $0 \leqq \eta(z) \leqq 1$ and $\eta \equiv 1$ on an open set ω. We form the local potential

(5,2) $\quad U(z) = -c_p \int \|a-z\|^{-2p} \eta(a) d\sigma(a) = -c_p \int \|a-z\|^{-2p} \eta t \wedge \beta_p(a)$

where $t \in \tilde{T}_{n-p}^+(\mathbb{C}^n)$ and $\sigma = t \wedge \beta_p$ is the trace of t. Then $d_{2p}(z) = v_t(z)$ is the Lelong number of t at z, and we obtain:

Corollary 5.2. *The local potential $U(z)$ defined by (5,2) is $\mathbb{R}^{2n}$ subharmonic, and the measure $\dfrac{1}{2\pi} \Delta U$ has a density $d'_{2n-2}(z) = v_t(z)$ on $\{z\colon \eta(z)=1\}$.*

We shall now evaluate the defect of plurisubharmonicity of the potential (5,2), that is we shall find a positive measure $\psi(z)$ such that

(5,3) $$L(u,\lambda) = \sum_{p,q} \frac{\partial^2 U}{\partial z_p \partial \bar{z}_q} \lambda_p \bar{\lambda}_q \geqq -C(p,n) \|\lambda\|^2 \psi(z).$$

We shall see that outside the support of $d\eta$, we can take $\psi(z)$ to be $\mathscr{C}^\infty$.

Let $g'_p(z) = -c_p \|z\|^{-2p} \beta_p$, and suppose $0 \leqq p \leqq n-1$.

Lemma 5.3. i) *for* $p = n-1$, *the current*

$$\frac{i}{\pi} \partial \bar{\partial} g'_{n-1} \text{ represents the Dirac measure } \delta(0). \tag{5,4}$$

ii) *for* $0 \leqq p \leqq n-2$, *setting* $\gamma = \frac{i}{2} \partial \|z\|^2 \wedge \bar{\partial} \|z\|^2$, *we have*

$$\frac{i}{\pi} \partial \bar{\partial} g'_p = \alpha^{p+1}, \text{ where } \alpha = \frac{i}{2\pi} \partial \bar{\partial} \log \|z\|^2 = \frac{1}{\pi} \left[\frac{\beta}{\|z\|^2} - \frac{\gamma}{\|z\|^4} \right]. \tag{5,5}$$

Proof. If $p = n-1$, then $\partial \beta_{n-1} = \bar{\partial} \beta_{n-1} = 0$, and so

$$\frac{i}{\pi} \partial \bar{\partial} g'_{n-1} = \frac{i}{\pi} \sum_{p,q} \left[\frac{\partial^2}{\partial z_p \partial \bar{z}_q} (-\|z\|^{-2n+2}) dz_p \wedge d\bar{z}_q \right] c_{n-1} \beta_{n-1}$$

$$= \frac{1}{2\pi} c_{n-1} \Delta (-\|z\|^{-2n+2}) \beta_n = \delta(0),$$

since $c_{n-1} c_n^{-1} = \frac{\pi}{n-1}$. For $0 \leqq p \leqq n-2$, we obtain

$$\left(\frac{\beta}{\|z\|^2} - \frac{\gamma}{\|z\|^4} \right)^{p+1} = \frac{\beta^{p+1}}{\|z\|^{2p+2}} - (p+1) \frac{\gamma \wedge \beta^p}{\|z\|^{2p+4}},$$

since $\gamma \wedge \beta = \beta \wedge \gamma$ and $\gamma \wedge \gamma = 0$. Thus

$$\frac{i}{\pi} \partial \bar{\partial} [-\|z\|^{-2p}] = \frac{2p}{\pi} \left[\frac{\beta}{\|z\|^{2p+2}} - \frac{i(p+1) \partial \|z\|^2 \wedge \bar{\partial} \|z\|^2}{2\|z\|^{2p+4}} \right],$$

from which (5,5) follows. □

Remark. We have $\alpha^n = 0$ outside the origin; (5,4) can be obtained from (5,5) by setting $\alpha^n = \delta(0)$.

Let f be a holomorphic mapping of $\Omega \subset \mathbb{C}^n$ onto $\Omega' \subset \mathbb{C}^m$. If φ is a differential form in Ω', we define $f^\star \varphi$, the *pullback* of φ to Ω, to be the form obtained by substituting in φ the variables $z' \in \Omega'$ and their differentials in terms of z and dz. If t is a current with compact support in Ω, we define a current $f_\star t$ on Ω', called the *image* or *pushforward* of t, by duality: $f_\star t(\varphi) = t(f^\star \varphi) = t(\alpha f^\star \varphi)$, where $\alpha(z)$ is any function in $\mathscr{C}_0^\infty(\Omega)$ such that $\alpha \equiv 1$ on support t. If t does not have compact support, we can still define $f_\star t$ if f is a proper mapping.

In order to obtain the lower estimate (5,3) for the potential (5,2), we shall use the product space $\mathbb{C}^n(a) \times \mathbb{C}^n(z)$, where $E_1 = \mathbb{C}^n(a)$ and $E_2 = \mathbb{C}^n(z)$. We then have the projections:

$$q: (a, z) \to a \qquad q: E_1 \times E_2 \to E_1$$
$$q': (a, z) \to z \qquad q': E_1 \times E_2 \to E_2$$
$$\tau: (a, z) \to a - z \qquad \tau: E_1 \times E_2 \to \mathbb{C}^n.$$

Then $\tau^\star g_p = -c_p \|z-a\|^{-2p}\beta_p(z-a)$, with

$$\beta(z-a) = \frac{i}{2}\sum_{k=1}^{n}(dz_k - da_k)\wedge(d\bar{z}_k - d\bar{a}_k)$$

and $\beta_p = [p!]^{-1}\beta^p$. We first treat the case where t is a positive current with coefficients in $\mathscr{C}_0^\infty$. Then

$$\tau^\star g_p \wedge q^\star t = -c_p \|z-a\|^{-2p}\beta_p(z-a)\wedge t(a),$$

and the potential (5,3) has the form

$$(5,6)\qquad U(z) = -c_p \int \|z-a\|^{-2p}\beta_p(a)\wedge t(a) = \int q'_\star[\tau^\star g_p \wedge q^\star t],$$

where $q^\star t$ is the pullback of t by q and $\tau^\star g_p$ is the pull back of g_p by τ, which is defined on $E_1 \times E_2$. We obtain (5,6) by taking the image of $\tau^\star g_p \wedge q^\star t$ (defined on $E_1 \times E_2$) by q'. This is well defined, since the restriction of q' to the support of the product $\tau^\star g_p \wedge q^\star t$ is a proper map.

We calculate $i\partial\bar{\partial}U(z)$ from (5,6). The operators ∂ and $\bar{\partial}$ commute with the images and pullbacks, and exterior products of forms of even degree are commutative, so for $\partial = \partial_z + \partial_a$ and $\bar{\partial} = \bar{\partial}_z + \bar{\partial}_a$, we have

$$i\partial\bar{\partial}U(z) = \int q'_\star[i\partial\bar{\partial}(\tau^\star g_p \wedge q^\star t)].$$

We apply $\partial\bar{\partial}$ to the product and obtain

$$(5,7)\qquad i\partial\bar{\partial}U = \int q'_\star[\tau^\star(i\partial\bar{\partial}g_p)\wedge q^\star t] + J_1 + J_2 + J_3,$$

where

$$J_1 = \int q'_\star[i\tau^\star(\partial g_p)\wedge q^\star(\bar{\partial}t)], \qquad J_2 = \int -q'_\star[i\tau^\star(\bar{\partial}g_p)\wedge q^*(\partial t)]$$

and $J_3 = \int q'_\star[\tau^\star g_p \wedge q^\star(i\partial\bar{\partial}t)]$.

Lemma 5.4. *If t is a $(n-p,\, n-p)$ positive form with compact support, then $q'_\star[\tau^\star(i\partial\bar{\partial}g_p)\wedge q^\star t]$ is a positive current of type* (1,1).

Proof. If $p = n-1$, Lemma 5.3 implies that $\tau^\star(\partial\bar{\partial}g_p) = \pi\tau^\star\delta$, where $\tau^\star\delta$ is the current of integration on the diagonal Δ of $E_1 \times E_2$. Let $[\Delta]$ represent this current. Then

$$q'_\star[\tau^\star(i\partial\bar{\partial}g_p)\wedge q^\star t] = \pi q'_\star([\Delta]\wedge q^\star t] = \pi t \in \Phi^+(\mathbb{C}^n).$$

For $p < n-1$, we use Lemma 5.3 and (5,5). Then

$$(5,8)\qquad q'_\star[\tau^\star(i\partial\bar{\partial}g_p)\wedge q^\star t] = q'_\star[\tau^\star(\pi\alpha^{p+1})\wedge q^\star t].$$

The forms t, α, and their pull backs $q^\star t$ and $q^\star\alpha$ are positive. The form $\tau^\star\alpha$ is of degree 1, so by Theorem 2.12, the current $(\tau^\star\alpha)^{p+1}\wedge q^\star t$ is a positive current on $E_1 \times E_2$, and the same is true for its image under q', which gives (5,8). □

We shall now estimate $|J_1|$, $|J_2|$, and $|J_3|$ for ηt, where t is a positive closed form in $\mathbb{C}^n$ and $\eta(a)\in\mathscr{C}_0^\infty$, $0\leqq\eta(a)$. We have

$$J_1=\int q'_\star[i\tau^\star(\partial g_p)\wedge q^\star(\bar\partial\eta\wedge t)]$$
$$J_2=\int -q'_\star[i\tau^\star(\bar\partial g_p)\wedge q^\star(\partial\eta\wedge t)]$$
$$J_3=\int q'_\star[\tau^\star g_p\wedge q^\star(i\partial\bar\partial\eta\wedge t)],$$

since t is closed. Let $K_1(a,z)$ be the component of type $(1,1)$ in z and type $(p,p-1)$ in a in the form $i\tau^\star(\partial g_p)$.

Then $J_1(z)=\int\limits_{\mathbb{C}^n} K_1(a,z)\wedge\bar\partial\eta(a)\wedge t(a)$. We have

$$i\partial_a g_p=pc_p\|a\|^{-2p-2}\sum_{j=1}^n \bar a_j da_j\wedge\beta_p(a),$$

hence

$$i\tau^\star(\partial_z g_p)=pc_p\|z-a\|^{-2p-2}\sum_{j=1}^n(\bar a_j-\bar z_j)(da_j-dz_j)\wedge\beta_p(z-a).$$

Thus, the coefficients of $K_1(a,z)$ are bounded in absolute value by $C\|a-z\|^{-2p-1}$, where C depends only on n and p. On the other hand, $\sigma=t'\wedge\beta_p$ gives an estimate of the coefficients of t' (cf. Theorem 2.16). We thus have

$$\|J_1(z)\|\leqq C(p,n)\int\limits_{\mathbb{C}^n}\|a-z\|^{-2p-1}\|\bar\partial\eta(a)\|\,\beta_p(a)\wedge t(a),$$

where

$$|\bar\partial\eta|=\left[\sum_{j=1}^n\left|\frac{\partial\eta}{\partial\bar z_j}\right|^2\right]^{1/2}\quad\text{and}\quad\|J_1(z)\|=\left[\sum_{j,k=1}^n|a_{j,k}(z)|^2\right]^{1/2},$$

with $J_1(t)=\sum a_{jk}(z)dz_j\wedge d\bar z_k$. In exactly the same way, we obtain estimates for J_2 and J_3. This leads to:

Proposition 5.5. *Let t be a positive closed current of degree $n-p$ and $\eta\in\mathscr{C}_0^\infty$, $\eta(z)\geqq 0$. Let U be defined by*

$$U(z)=-c_p\int\limits_{\mathbb{C}^n}\|z-a\|^{-2p}\eta(a)\beta_p(a)\wedge t(a).$$

Then the Levi form of U, $L(U,\lambda)$ satisfies

$$(5,9)\qquad L(U,\lambda)=\sum_{p,q=1}^n\frac{\partial^2 U}{\partial z_p\partial\bar z_q}\lambda_p\lambda_q$$
$$\geqq -C(p,n)\|\lambda\|^2\int\limits_{\mathbb{C}^n}\left[\frac{|\bar\partial\eta|}{\|z-a\|}+|\partial_z\bar\partial_z\eta|\right]\frac{\beta_p(a)\wedge t(a)}{\|z-a\|^{2p}}$$

as a distribution.

Remark 1. The brackets give the corrective term $\psi(z)$ in (5,3). It must be interpreted as a distribution – that is, if $\varphi\geqq 0$, $\varphi\in\mathscr{C}_0^\infty(\mathbb{C}^n)$,

$$L(U,\lambda)(\varphi)\geqq -C(p,n)\|\lambda\|^2\int\limits_{\mathbb{C}^n}\psi(z)\varphi(z)d\tau(z);$$

$\psi(z)$, which is in L^1_{loc}, defines a positive measure which measures the "defect of plurisubharmonicity". Outside the support of $\bar{\partial}\eta$, $\psi(z)$ is a $\mathscr{C}^\infty$ function.

Remark 2. Let ω be an open set such that $\eta \equiv 1$ on ω. Then, by Proposition 5.1, for $z \in \omega$, $d'_{2n-2}(z) = v_t(z)$, where d'_{2n-2} is the density of the measure $(2\pi)^{-1} \Delta U$.

The proof of Proposition 5.5 was given for t a form, but the general case follows by approximating t by positive closed currents with $\mathscr{C}^\infty$ coefficients.

Remark 3. If T_1 and T_2 are two distributions, we will write $T_1 \geqq T_2$ if $T' = T_1 - T_2$ is a positive distribution (that is, for $\varphi \in \mathscr{C}_0^\infty$, $\varphi \geqq 0$, $T'(\varphi) \geqq 0$). In what follows, we shall have T_2 given by a measure, in which case, if $T_1 \geqq T_2$, then T_1 is also a measure.

§3. Global Potentials

We patch together the local potentials to obtain a global potential in $\mathbb{C}^n$, but in such a way as to control the growth.

Let ε be fixed, $0 < \varepsilon < 1$, and $\chi(z) \in \mathscr{C}_0^\infty(\mathbb{C}^n)$, a decreasing function of $\|z\|$ with $0 \leqq \chi(z) \leqq 1$, $\chi(z) \equiv 1$ for $\|z\| \leqq 1$ and $\chi(z) = 0$ for $\|z\| \geqq 1 + \varepsilon$. Set $\chi_j(z) = \chi(z/j)$ for $j \geqq 1$ and $\rho_1 = \chi_1$, $\rho_j = \chi_j - \chi_{j-1}$, $j \geqq 2$. The support of ρ_j is contained in the annulus $j - 1 \leqq \|z\| \leqq (1+\varepsilon) j$ for $j \geqq 2$. Furthermore, $\sum_{j=1}^{\infty} \rho_j = 1$. Let $\eta_j(z) = \chi_j\left(\dfrac{z}{(1+2\varepsilon)j}\right)$; then $\eta_j = 1$ for $\|z\| \leqq (1+2\varepsilon)j$, and in particular, $\eta_j \equiv 1$ on $\operatorname{supp} \rho_j$ and $\eta_j \equiv 0$ for $\|z\| \geqq (1+5\varepsilon)j$. Let

$$U_j(z) = -\int_{\mathbb{C}^n} c_p \|z-a\|^{-2p} \eta_j(a) \beta_p(a) \wedge t'(a)$$

and

$$U(z) = \sum_{j=1}^{\infty} \rho_j U_j(z). \tag{5,10}$$

On any compact set, there exists only a finite number of non-zero ρ_j, so the sum converges. We shall use (5,9) to estimate the defect of plurisubharmonicity of $U(z)$. Let M be a constant such that $M \geqq |\bar{\partial}\chi| + |\partial\bar{\partial}\chi|$. Then $|\bar{\partial}\eta_j(z)| \leqq M(1+2\varepsilon)^{-1} j^{-1}$ and $|\partial\bar{\partial}\eta| \leqq M(1+2\varepsilon)^{-2} j^{-2}$. What is more, for $z \in \operatorname{supp} \rho_j$ and $a \in \operatorname{supp} \bar{\partial}\eta_j$, $\|z-a\| \geqq \varepsilon j$. Thus

$$L(U,\lambda) = \sum_{j=1}^{\infty} \rho_j L(U_j,\lambda) + \sum_{j,k,l} \left(\frac{\partial \rho_j}{\partial z_k}\frac{\partial U_j}{\partial \bar{z}_l} + \frac{\partial \rho_j}{\partial \bar{z}_l}\frac{\partial U_j}{\partial z_k}\right) \lambda_k \bar{\lambda}_l$$
$$+ \sum U_j \cdot L(\rho_j,\lambda) = L_1 + L_2 + L_3.$$

From (5,9), we see that the first term satisfies

$$L_1 \geqq -C(\varepsilon, n, p) M \|\lambda\|^2 \sum_{j=1}^{\infty} \frac{\rho_j(z)}{j^{2p+2}} \sigma_t[(1+\varepsilon)j]. \tag{5,11}$$

If for $z, \rho_j(z) \neq 0$, then $j-1 \leqq \|z\| < (1+\varepsilon)j$ and $\dfrac{1+\|z\|}{2(1+\varepsilon)} \leqq j \leqq 1+\|z\|$, which gives (since $\sum_{j=1}^{\infty} \rho_j \equiv 1$) from (5,11)

$$L_1(U, \lambda) \geqq -C(\varepsilon, n, p) \|\lambda\|^2 \sigma_t[l(1+\|z\|)](1+\|z\|)^{-2p-2},$$

where $l = 1+5\varepsilon$. We obtain similar estimates for the other two terms, from which we obtain, when we replace $\sigma_t(r)$ *by* $v_t(r)$:

Proposition 5.6. *Let* $\varepsilon > 0$. *We can choose a partition of unity* ρ_j *and the associated* η_j *such that for* U *given by* (5,10),

$$L(U, \lambda) \geqq -C(\varepsilon, n, p)(1+r)^{-2} v_t((1+\varepsilon)(1+r)) \|\lambda\|^2.$$

In order to obtain a plurisubharmonic function $V = U + W$, it is enough to construct W such that

$$L(W, \lambda) \geqq C(\varepsilon, n, p)(1+r)^{-2} v_t[(1+\varepsilon)(1+r)] \|\lambda\|^2.$$

We shall obtain W as a continuous function of r.

Let $q(r) = \log(1+r^2) + \frac{1}{2}\log^2(1+r^2)$. Then $L(q, \lambda) \geqq (1+r^2)^{-1} \|\lambda\|^2$. We choose $W = h \circ q$, where $h(r)$ is an increasing convex function of r such that

$$h' \circ q(\|z\|) \geqq C(\varepsilon, n, p) v_t((1+\varepsilon)(1+r)). \tag{5,12}$$

Then we shall have $L(W, \lambda) \geqq (1+r^2)^{-1} h' \circ q(r) \|\lambda\|^2$. Let $q^{-1}(r)$ be the inverse function of $q(r)$. Condition (5,12) will hold if we have

$$h'(r) \geqq C(\varepsilon, n, p) v_t((1+\varepsilon)(1+q^{-1}(r))).$$

Since $v_t(r)$ is increasing, we can take $h(r)$ to be

$$h_0(r) = C(\varepsilon, n, p) \int_0^r v_t((1+\varepsilon)(1+q^{-1}(\xi))) d\xi.$$

Then $W_0(z) \leqq C(\varepsilon, n, p) q(r) v_t((1+\varepsilon)(1+r))$ and

$$W_0(z) \leqq C(\varepsilon, n, p) \log^2 r\, v_t[(1+\varepsilon)(1+r)]. \tag{5,13}$$

If $d > 0$, if in place of $q(r)$, we take the function

$$G(r) = (1+r)^d \int_0^{1+r} \frac{v_t((1+\varepsilon)\xi)}{\xi^{d+1}} d\xi,$$

then $L(G, \lambda) \geqq d/4\, v_t((1+\varepsilon)(1+r))(1+r)^{-2}$, which permits us to choose

$$W_0(z) = C(\varepsilon, d)(1+r)^d \int_1^{1+r} \xi^{-(d+1)} v_t((1+\varepsilon)\xi) d\xi; \tag{5,14}$$

this gives a better estimate when v_t is of finite order. When $v_t(r)$ is of infinite order, we obtain a better growth estimate by using the following partition of unity: we let $\chi(t)\in\mathscr{C}_0^\infty$, $0\leqq\chi(t)\leqq 1$ for $t\leqq\varepsilon$, $\chi(t)=0$ for $t\geqq 2\varepsilon$ and set $\chi_j(z)=\chi(\|z\|-j\varepsilon+\varepsilon)$ for j a positive integer and $\rho_1=\chi_1$, $\rho_j=\chi_j-\chi_{j-1}$, $j\geqq 2$. Then $\operatorname{supp}\rho_j$ is contained in the annulus $(j-1)\varepsilon\leqq\|z\|\leqq(j+1)\varepsilon$. We then set $\eta_j(z)=\chi(\|z\|-j\varepsilon-\varepsilon)=\chi_{j+2}(z)$. If we define $U(z)$ as in (5,10), we obtain the estimate

$$L(U,\lambda)\geqq -C(\varepsilon,p,n)\|\lambda\|^2\sigma_t(r+4\varepsilon).$$

We choose $W_0(z)=h(\|z\|^2)$, where h is an increasing convex function, so that $L(w,\lambda)\geqq h'(\|z\|^2)\|\lambda\|^2$, and thus $W_0(z)=C(\varepsilon,p,n)\int_0^{r^2}\sigma_t(\sqrt{t}+4\varepsilon)dt$, which leads to the estimate

$$W_0(z)\leqq C(\varepsilon,p,n)r^2\sigma_t(r+\varepsilon). \tag{5,15}$$

Now we can replace $W(z)$ by $W_0(z)\in\mathscr{C}^\infty$ with a similar bound.

Theorem 5.7. *Let Y be an analytic variety of pure dimension p in $\mathbb{C}^n$ and t the current of integration on Y or in general a positive closed current of degree p (i.e. type $(n-p,n-p)$). Let $\sigma_t=t\wedge\beta_p$ be the trace of t and $v_t(r)=(\tau_{2p}r^{2p})^{-1}\sigma_t(r)$. Then there exists a plurisubharmonic function V in $\mathbb{C}^n$ such that*

i) *for every compact set $K\subset\mathbb{C}^n$ and ω an open bounded neighborhood of K, $V+c_p\int_\omega\|z-a\|^{-2p}d\sigma_t(a)$ is $\mathscr{C}^\infty$ on K;*

ii) *if $M_V(r)=\sup_{\|z\|\leqq r}V(z)$, we have one of the following*

$$\begin{cases} M_V(r)\leqq C(\varepsilon,n,p)\log^2 r\, v_t((1+\varepsilon)r) & \text{for } r>r_0,\\ M_V(r)\leqq C(\varepsilon,d)(1+r)^d\int_1^{1+r} v_t((1+\varepsilon)\xi)\,\xi^{-d-1}d\xi, & \\ M_V(r)\leqq C(\varepsilon,n,p)r^2\sigma_t(r+\varepsilon). & \end{cases} \tag{5,16}$$

iii) *Let $v'_V(z)$ be the Lelong number of $t'=i/\pi\,\partial\bar\partial V$. Then $v'_V(z)=v_t(z)$.*

Proof. Parts i) and ii) follow from the above construction and (5,13), (5,14) and (5,15), since $V=U+W$ and $U\leqq 0$. Part iii) follows from Proposition 5.1 and Corollary 5.2, since W is $\mathscr{C}^\infty$ and hence its density is identically zero. □

Theorem 5.8. *Let Y be an analytic variety of pure dimension p in $\mathbb{C}^n$. Then there exists a plurisubharmonic function V which can be chosen to verify any one of the three conditions (5,16) and such that $\tilde v_V(z)=v_{[Y]}(z)$ and $\tilde v_V(z)=0$ if $z\notin Y$, where $\tilde v$ is the $(2n-2)$ dimensional density of $(2\pi)^{-1}\Delta V$.*

§4. Construction of a System F of Entire Functions such that $Y=F^{-1}(0)$

Given a plurisubharmonic function V in $\mathbb{C}^n$, we consider the analytic set $E(c, V)=\{z\in\mathbb{C}^n: v_V(z)\geqq c\}$ for $c>0$. We are interested in constructing a representation $E(c, V)=F^{-1}(0)$ for an entire mapping, where we obtain an estimate of the growth of $\|F\|$ in terms of $M_V(r)$. We have already obtained for an analytic set Y a function V such that $Y=E(1, V)$ and an estimate of $M_V(r)$ in terms of $v_{[Y]}(r)$, the indicator of Y. The solution will then give a solution $Y=F^{-1}(0)$ with estimates for the growth of $\|F\|$ in terms of $v_{[Y]}(r)$.

In fact, this problem is more general. We shall see that every analytic set Y in $\mathbb{C}^n$ can be represented as $E(1, V)$ for V a plurisubharmonic function, for if is true for a pure dimensional analytic variety, then $Y=\bigcup_{s=0}^{n-1} Y_s$, where Y_s is of pure dimension s and $Y_s=E(1, V_s)$, $v_{V_s}(z)=0$ for $z\notin Y_s$. Then if $V=\sum_{s=0}^{n-1} V_s$, $Y=\bigcup_s Y_s=E(1, V)$.

Definition 5.9. We say that $c_0>0$ is a number of *complete left stability* for $V\in\mathrm{PSH}(\mathbb{C}^n)$ if $E(c, V)=E(c_0, V)$ for $0<c\leqq c_0$.

Example. For the function V constructed in paragraph 2 from the current $t\in T^+_{n-p}(\mathbb{C}^n)$, we obtain $v'_V(z)=v_t(z)$; if $t=[Y]$, then $v_t(z)=0$ for $z\notin Y$. Thus, 1 is a number of complete left stability.

We now prove some lemmas needed for the construction of F.

Lemma 5.10. *There exists an absolute constant C such that for every plurisubharmonic function Φ defined on a neighborhood of the unit ball $\{z: \|z\|<1\}$ with $\Phi(0)=0$, $\Phi(z)<1$,* $\int_{\|z\|<1/3} \exp-\Phi(z)\, d\tau_{2n}(z)<C$.

Proof. We first consider the case $n=1$. The Riesz Representation Theorem tells us that we can write $\Phi(z)$ as

$$\Phi(z)=\int_{|\xi|<1}\log\left(\frac{|z-\xi|}{|1-z\bar{\xi}|}\right)\Delta\Phi(\xi)\,d\tau_2(\xi)+\frac{1}{2\pi}\int_0^{2\pi}\frac{(1-|z|^2)}{|z-e^{i\theta}|^2}\Phi(e^{i\theta})\,d\theta.$$

Letting $z=0$, we obtain

$$0=\int_{|\xi|<1}\log|\xi|\,\Delta\Phi(\xi)\,d\tau_2(\xi)+\frac{1}{2\pi}\int_0^{2\pi}\Phi(e^{i\theta})\,d\theta$$

or alternatively

$$\int_{|\xi|<1}\log\frac{1}{|\xi|}\,\Delta\Phi(\xi)\,d\tau_2(\xi)+\frac{1}{2\pi}\int_0^{2\pi}(1-\Phi(e^{i\theta}))\,d\theta=1,$$

from which it follows that

$$\int_{|\xi|<1} \log\frac{1}{|\xi|}\, \Delta\Phi(\xi)\, d\tau_2(\xi) \leqq 1, \qquad \frac{1}{2\pi}\int_0^{2\pi} |\Phi(e^{i\theta})|\, d\theta \leqq 2,$$

and hence

$$\left|\frac{1}{2\pi}\int_0^{2\pi} \frac{(1-|z|^2)}{|z-e^{i\theta}|^2}\, \Phi(e^{i\theta})\, d\theta\right| \leqq 4 \qquad \text{for } |z|<1/3.$$

Let $a=\frac{1}{2\pi}\int_{|\xi|<R} \Delta\Phi(z)$ so that if $R<1/e$, $a\leqq\frac{1}{\log R^{-1}}<1$. Since $e<3$, we can choose R so that $e^{-1}>R>1/3$ and

$$\left|\int_{|\xi|>R} \log\left(\frac{|z-\xi|}{|1-\bar{z}\xi|}\right) \Delta\Phi(\xi)\, d\tau_2(\xi)\right| < C_1 \qquad \text{for } |z|<1/3.$$

If $a=0$, $|\Phi(z)|<C_1+4$ for $|z|<1/3$. If $a\neq 0$, by the convexity of the exponential function

$$\begin{aligned}
&\exp\left(-1/2\pi \int_{|\xi|<R} \log\left(\frac{|z-\xi|}{|1-\bar{z}\xi|}\right) \Delta\Phi(\xi)\, d\tau_2(\xi)\right)\\
&\leqq \exp\left(\int_{|\xi|<R} -a\log\left(\frac{|z-\xi|}{|1-\bar{z}\xi|}\,\frac{\Delta\Phi(\xi)}{2\pi a}\right) d\tau_2(\xi)\right)\\
&\leqq \int_{|\xi|<R} \left(\frac{|z-\xi|}{|1-\bar{z}\xi|}\right)^{-a} \frac{\Delta\Phi(\xi)}{2\pi a}\, d\tau_2(\xi),
\end{aligned}$$

and since $a<1$, we have

$$\begin{aligned}
\int_{|z|<1/3} \exp-\Phi(z)\, d\tau(z) &\leqq e^{-C_1-4} \int_{|z|<1/3}\int_{|\xi|<R} \left(\frac{|z-\xi|}{|1-\bar{z}\xi|}\right)^{-a} \frac{\Delta\Phi(\xi)}{2\pi a}\, d\tau_2(z)\\
&\leqq C_2 e^{-C_1-4},
\end{aligned}$$

which establishes the result for $n=1$. For $n>1$, we use polar coordinates to obtain

$$\int_{\|z\|<1/3} \exp(-\Phi(z))\, d\tau(z) = \int_{\|\alpha\|=1} d\omega_{2n}(\alpha) \int_{|t|<1/3} |t|^{2n-2} \exp(-\Phi(t\alpha))\, \frac{d\tau_2(t)}{2\pi},$$

where ω_{2n} is the measure on the unit sphere. □

Corollary 5.11. *If Φ is a plurisubharmonic function in a connected set $\Omega\subset\mathbb{C}^n$, each $z\in\Omega$ not in the pluripolar set $\Phi=-\infty$ has a neighborhood $U_z\subset\Omega$ in which* $\exp(-\Phi)$ *is integrable.*

Proof. Suppose $\Phi(z_0)\neq-\infty$. Then by the upper semi-continuity of Φ, $\Phi(z)<\Phi(z_0)+1$ for $\|z-z_0\|<\delta$. We now apply Lemma 5.10 to $\Phi(z)-\Phi(z_0)$. □

Theorem 5.12. *Let $\Phi\in\mathrm{PSH}(\mathbb{C}^n)$ and $z_0\in\mathbb{C}^n$ such that* $\exp(-\Phi)$ *is integrable in a neighborhood of z_0. Then for $\varepsilon>0$, there exists $f\in\mathscr{H}(\mathbb{C}^n)$ such that*

$$f(z_0)=1, \qquad \int_{\mathbb{C}^n} \frac{|f(z)|^2 \exp-\Phi(z)}{(1+\|z\|^2)^{n+\varepsilon}}\, d\tau(z) < +\infty.$$

Proof. Let ω_0 be a neighborhood of z_0 such that $\int_{\omega_0} \exp[-\Phi(z)]d\tau(z) < +\infty$; such a neighborhood exists by Corollary 5.11. Let $\chi \in \mathscr{C}_0^\infty(\mathbb{C}^n)$ such that $0 \leqq \chi \leqq 1$, $\chi \equiv 1$ on $B(z_0, \delta) \Subset \omega_0$. Set

$$\psi(z) = \Phi(z) + 2n \log\|z - z_0\| + \varepsilon \log(1 + \|z\|^2).$$

Then $\psi \in \mathrm{PSH}(\mathbb{C}^n)$ and

$$\sum_{j,k} \frac{\partial^2 \psi(z)}{\partial z_j \partial \bar{z}_k} w_j w_k \geqq \sum_{j,k} \frac{\partial^2 \varepsilon \log(1 + \|z\|^2)}{\partial z_j \partial \bar{z}_k} w_j \bar{w}_k \geqq \frac{\varepsilon \|w\|^2}{(1 + \|z\|^2)^2}$$

for $w \in \mathbb{C}^n$. Let $\beta = \bar{\partial}\chi$. Then $\frac{1}{\varepsilon}\int \frac{|\beta|^2 \exp -\psi(z)}{(1 + \|z\|^2)^{\varepsilon+2}} d\tau < +\infty$, since $\operatorname{supp}\beta$ is compact and $z_0 \notin \operatorname{supp}\beta$.

Thus (cf. Lemma III.11), we can find u such that $\bar{\partial}u = \beta = \bar{\partial}\chi$ and

$$\int_{\mathbb{C}^n} \frac{|u|^2 \exp -\Phi(z)}{\|z - z_0\|^{2n}(1 + \|z\|^2)^\varepsilon} d\tau(z) < +\infty.$$

Since $\beta \equiv 0$ in $B(z_0, \delta)$, u is holomorphic in $B(z_0, \delta)$ and since $\|z - z_0\|^{-2n}$ has a non-integrable singularity, $u(z_0) = 0$.

Define $f(z) = \chi(z) - u(z) \in \mathscr{H}(\mathbb{C}^n)$; $f(z_0) = 1$, and

$$\int_{\mathbb{C}^n} |f(z)|^2 \frac{\exp -\Phi(z)}{(1 + \|z\|^2)^{n+\varepsilon}} d\tau(z) < +\infty. \qquad \square$$

Note that if $V(z)$ is plurisubharmonic, then

$$H_R(z) = V(z) + c_{2n-2} \int_{\|z\| < R} \frac{d\sigma_V(w)}{\|w - z\|^{2n-2}}$$

is harmonic, hence $\mathscr{C}^\infty$, in $\|w\| < R$ (where $\sigma_V = i\partial\bar{\partial}V \wedge \beta_{n-1}$).

Let $U(z) = V(z) - H(z)$. An integration by parts shows that for $\|z\| = r$ and $\|w\| = t$:

$$\begin{aligned}
U(z) &= c_{2n-2} \int \frac{d\sigma_V(w)}{\|w - z\|^{2n-2}} \geqq c_{2n-2} \int_0^R (r+t)^{-2n+2} d\sigma_V(t) \\
&= c_{2n-2}[\sigma_V(t)(r+t)^{-2n-2}]_0^R + c_{2n-2} \int_0^R (r+t)^{-2n+1} \sigma_V(t)dt \\
&\geqq \int_0^R (r+t)^{-1} v_V(t)dt \geqq v_V(0) \log\left(1 + \frac{R}{\|z\|}\right).
\end{aligned}$$

Then $-V = -U - H_R$ and $\exp(-U) \geqq \left(1 + \frac{R}{\|z\|}\right)^\alpha$ for $\alpha = v_V(0)$. Thus if $v_V(z) \geqq 2n$, the function $\exp(-V)$ is not integrable in a neighborhood of z.

Theorem 5.13. *Let $V \in \mathrm{PSH}(\mathbb{C}^n)$ with $c_0 = 1$ as a number of complete left stability. Then for every $\varepsilon > 0$ and all $\alpha > 0$, there exists $C(n, \varepsilon, \alpha)$ independent* *of r and $(n+1)$ entire functions $F=(f_1, \dots, f_{n+1})$ such that $E(1,V)=F^{-1}(0)$ and*

$$\sup_{\|z\|\leq r} \log \|F\| \leqq n M_V(r+\alpha)+(n+\varepsilon)\log(1+r)+C(n,\varepsilon,\alpha)+C_F. \tag{5,17}$$

Proof. We shall use the L^2-estimates with weight for the solution of the $\bar{\partial}$-equation (cf. Appendix III). Let $\varepsilon>0$ and $\varphi\in \mathrm{PSH}(\mathbb{C}^n)$, z_0 a point such that $e^{-\varphi}$ is integrable in a neighborhood of z_0 (Corollary 5.11). Then, by Theorem 5.12, there exists $f\in\mathscr{H}(\mathbb{C}^n)$ such that $f(z_0)=1$ and

$$\|f\|_\varphi^2=\int_{\mathbb{C}^n} |f(z)|^2 e^{-\varphi(z)}(1+\|z\|^2)^{-n-\varepsilon}\,d\tau(z)<+\infty. \tag{5,18}$$

We let H_φ be the Hilbert space of all $f\in\mathscr{H}(\mathbb{C}^n)$ such that $\|f\|_\varphi<\infty$. The closed set $\eta\subset\mathbb{C}^n$ of points z for which $e^{-\varphi}$ is non integrable in a neighborhood of z is an analytic variety, the set of common zeros of the elements in H_φ. We recall that

$$E(2n,\varphi)\subset\eta. \tag{5,19}$$

On H_φ, a point $z_0\in\mathbb{C}^n$ defines a linear functional $\hat{z}_0$ given by $\hat{z}_0(f)=f(z_0)$, which is zero for $z_0\in\eta$. To see that $\hat{z}_0$ is continuous, we use the Cauchy-Schwarz Inequality and set $\psi(z)=\varphi(z)+(n+\varepsilon)\log(1+\|z\|^2)$ for $\varepsilon>0$: $|f(z_0)|\leqq(\tau_{2n}r^{2n})^{-1}\int_{B(z_0,r)}|f(z)|\,d\tau(z)$ by subharmonicity and

$$|f(z_0)|^2\leqq(\tau_{2n}r^{2n})^{-2}\Big[\int_{B(z_0,r)}|f(z)|^2e^{-\psi(z)}\,d\tau(z)\Big]\Big[\int_{B(z_0,r)}e^{\psi(z)}\,d\tau(z)\Big]$$

so $|\hat{z}_0(f)|=|f(z_0)|\leqq C\|f\|_\varphi$ with $C=(\tau_{2n})^{-1/2}\exp\frac{1}{2}M_\psi(1+\|z_0\|)=C(z_0)$ and $C(z)$ is bounded on every compact subsets of $\mathbb{C}^n$ independantly of $f\in H_\varphi$. The linear form $\hat{z}_0$ thus belongs to the dual space H'_φ and $A_\varphi\colon z\to\hat{z}\in H'_\varphi$ is a mapping of $\mathbb{C}^n$ into H'_φ. Furthermore, $\eta\subset\eta'=\{z\colon \varphi(z)=-\infty\}$, so η is of measure zero in $\mathbb{C}^n$.

Let $z_0\notin\eta'$ and set $\varphi=2nV$. The function $e^{-\varphi}$ is integrable in a neighborhood of z_0 by Corollary 5.11. Thus, we can find $f_1\in H_\varphi$ such that $f_1(z_0)=1$, and from (5,18) and Lemma 3.47, we obtain the estimate

$$\sup_{\|z\|\leqq r}\log|f_1(z)|\leqq nM_v(r+\alpha)+(n+\varepsilon)\log(1+r)+C(n,\varepsilon,\alpha)+C_1$$

with $C(n,\alpha,\varepsilon)=(n+\varepsilon)\log(1+\alpha)-n\log\alpha-1/2\log\tau_{2n}$.

What is more, $E(1,V)\subset f_1^{-1}(0)$, since if $v_V(z)\geqq 1$, for $\varphi=2nV$, we have $v_\varphi(z)\geqq 2n$ and $z\in E(2n,\varphi)$,

$$E(1,V)=E(2n,\varphi) \tag{5,20}$$

with $e^{-\varphi}$ non integrable for every point of $E(1,V)$. This shows that $f_1(z)=0$ for $z\in E(1,V)$.

Let X_j be the irreducible branches of $f_1^{-1}(0)$ which are not contained in $E(1,V)$. For every j, we choose a point $z_j\in X_j\cap\complement E(1,V)$. Since $c_0=1$ is a number of complete stability for V, we have $\nu_V(z_j)=0$ and $\nu_\varphi(z_j)=0$ with $e^{-\varphi}$ integrable in a neighborhood of z_j. We can thus find $f\in H_\varphi$ such that $f(z_j)=1$. Then $\hat{z}_j(f)=0$ defines a proper closed subspace of H_φ. Since a countable union of closed subspaces is of first category in H_φ, there exists $f_2\in H_\varphi$ such that $f_2(z_j)\neq 0$ for every j. Then $E(1,V)\subset f_1^{-1}(0)\cap f_2^{-1}(0)=X_2$. We continue in this way by considering the countable family $X_j^{(2)}$ of the irreducible branches of X_2 not contained in $E(1,V)$. We choose $z_j'\in X_j^{(2)}$, $z_j'\notin E(1,V)$. As before, there exists $f_3\in H_\varphi$ such that $f_3(z_j')\neq 0$ and $E(1,V)\subset X_3=\{z\colon f_k=0,\ n=1,2,3\}$. By iteration, we obtain f_k, $k=1,\dots,n$ such that

$$E(1,V)\subset Z=\{z\colon f_k(z)=0,\ k=1,\dots,n\}$$

and the set of points in $Z\cap \complement E(1,V)$ is discrete. Thus, as before, we find $f_{n+1}\in H_\varphi$ such that $f_{n+1}(z)\neq 0$ for $z\in Z\cap\complement E(1,V)$. Then

$$E(1,V)=\bigcap_{k=1}^{n+1} f_k^{-1}(0)$$

and $\|F\|$ satisfies (5,17), since each f_k does. □

Theorem 5.14. *Let Y be an analytic variety in $\mathbb{C}^n$ of pure dimension p with indicator $\nu(r)$. Then*

$$Y=\{z\colon f_k(z)=0,\ k=1,\dots,n+1\}$$

where the f_k satisfy one of the following estimates:

$$(5,21)\qquad \begin{cases} M_k(r)\leqq C(\varepsilon)\log^2 r\,\nu(r+\varepsilon r),\\ M_k(r)\leqq C(\varepsilon,\alpha)(1+r)^d\displaystyle\int_1^{1+r}\nu(t+\varepsilon t)t^{-d-1}\,dt,\\ M_k(r)\leqq C(\varepsilon)r^2\sigma(r+\varepsilon). \end{cases}$$

Remark 1. Theorems 5.13 and 5.14 show that if $t\in\tilde{T}^+_{n-p}(\mathbb{C}^n)$, $A\subset\operatorname{supp}t$, and $\nu_t(z)\geqq 1$ for $z\in A$ and $\nu_t(z)=0$ for $z\notin A$, then A is an analytic subset of $\mathbb{C}^n$, and we can obtain A as $F^{-1}(0)$ where $\log\|F\|$ satisfies one of the estimates in (5,21).

Remark 2. The entire functions f_j are zero on Y and have no common zero outside Y, but the theorem does not give at $z\in Y$ the value of the integer $\nu_W(z)\geqq 1$, where $W=\frac{1}{2}\log\left(\sum_{j=1}^{n+1}|f_j|^2\right)$.

§ 5. The Case of Slow Growth

The use of a partition of unity in the construction of $U(z)$ means that there is a certain degree of arbitrariness in the behavior of $U(z)$. It is perhaps worth the effort to try to extend the method of canonical potentials, which permit a constructive solution for a Cousin data of finite order, to the case of general analytic varieties. Even in the case of co-dimension 1, the use of canonical potentials loses much of its precision when one treats Cousin data of infinite order. Thus, it is more reasonable to treat only the case of finite order, and we give below an extension of the canonical potential to analytic varieties Y of dimension p such that $v_{[Y]}(r)$ is of finite order. We shall use kernels

$$g'_p(a,z) = -c_p \|a-z\|^{-2p}, \qquad 1 \leqq p \leqq n-1$$

$$g'_p(a,z) = \log \|a-z\|, \qquad p=0$$

and construct as in Chapter 3 the kernels $e_p(a,z,q)$, for q an integer, $q \geqq 0$.

Theorem 5.15. *Let t be a positive closed current of degree $n-p$ and $\sigma_t = t \wedge \beta_p$ the trace of t. Suppose that the indicator $v_t(r)$ satisfies*

(5,22) $$\int_1^\infty v_t(r) r^{-3} dr < +\infty.$$

Then for every n, the canonical potential $I(z) = -c_p \int e_p(a,z,1) d\sigma_t(z)$ is plurisubharmonic.

We shall need the following lemma:

Lemma 5.16. *If $q=0$ or $q=1$, the kernel $e_p(a,z,q)$ differs from $g'_p(a,z)$ by a pluriharmonic function.*

Proof. For $q=1$, $e_p(a,z,1) = g'_p(a,z) + \|a\|^{-2p} - \dfrac{2p}{\|a\|^{2p+2}} \operatorname{Re} \sum \bar{a}_i z_i$ and for $q=0$, $e_p(a,z,0) = g'_0(a,z) + \|a\|^{-2p}$. □

Proof of Theorem 5.15. Suppose first that $0 \notin \operatorname{supp} t$ so that the potential $I_q(z)$ converges. Let $\chi(z)$ be in $\mathscr{C}_0^\infty$ such that $\chi(z)=1$ for $\|z\| \leqq 1$ and $\chi(z)=0$ for $\|z\| \geqq 2$, and let $\chi_j(z) = \chi(z/j)$. There exists a constant $M>0$ such that $M\|z\|^{-1} \geqq |\bar\partial \chi_j|$ and $M\|z\|^{-2} \geqq |\partial\bar\partial \chi_j|$. Since $\bar\partial \chi_j = 0$ for $\|z\| \leqq j$ and for $\|z\| \geqq 2j$, there exists a constant $C(p,n,\chi)$ such that, if

$$I'_j(z) = -c_p \int \|a-z\|^{-2p} \chi_j(a)\, d\sigma_t(a),$$

we have by Proposition 5.5:

(5,23) $$L(I'_j, \lambda) \geqq -C(p,n,\chi) \|\lambda\|^2 \varphi_j(z)$$

with $|\varphi_j(z)| \leqq \int_{\|a\|>j} [\|z-a\|^{-1} |\bar\partial \chi_j(a)| + |\partial\bar\partial \chi_j(a)|]\, \|z-a\|^{-2p} d\sigma_t(a)$, by (5,11). Let $\|z\| \leqq R$ and $j > 2R$ Then for $\|a\| = r$ and $j > 2R$,

(5,24) $$|\varphi_j(z)| \leqq C' \int_j^\infty v_t(r) r^{-3} dr + C'',$$

where C' and C'' depend only on t, R and n but not on $z \in B(0,R)$. Thus, $\varphi_j(z)$ tends uniformly to zero on every compact subset of $\mathbb{C}^n$. Furthermore, by (5,23), if $q=0$ or $q=1$,

$$e_p(a,z,q) = -c_p \|a-z\|^{-2p} + l_q(a,z)$$

where $l_q(q,z)$ is pluriharmonic (by Lemma 5.16). Thus

$$L(I_j,\lambda)=L(I_j',\lambda).$$

Since $L(I,\lambda)=\lim_{j\to\infty} L(I_j,\lambda)\geqq 0$ by (5,22) and (5,23), the theorem is proved. □

Remark 1. The hypothesis $\int_1^\infty v_t(r)r^{-3}dr<+\infty$ implies that the terms J_1, J_2 and J_3 disappear in (5,7), the representation for $i\partial\bar\partial I$. We thus have

i) $i\partial\bar\partial I(z)=q'_\star[\tau^\star(\pi\alpha^{p+1})\wedge q^\star t]$, $0\leqq p\leqq n-2$,

ii) $i\partial\bar\partial I(z)=\pi t$, $p=n-1$.

Remark 2. Theorem 5.15 was proved under the hypothesis that $0\notin\operatorname{supp} t$. If in fact $0\in\operatorname{supp} t$, we subtract from $I(z)$ a pluriharmonic function

$$(5,25)\quad \begin{aligned} I_1(z)&=-c_p\int_{\mathbb{C}^n}[\|z-a\|^{-2p+2}(1+\|a\|^{2p})^{-1}]d\sigma_t(a) \quad \text{if } \int v_t(r)r^{-2}dr<\infty\\ I_1(z)&=-c_p\int_{\mathbb{C}^n}[\|z-a\|^{-2p+2}(1+\|a\|^{2p})^{-1}\\ &\quad +2p(1+\|a\|^{2p+2})^{-1}\mathbb{R}e\langle a,z\rangle]d\sigma_t(a), \end{aligned}$$

if $\int_1^\infty v_t(r)r^{-3}dr<+\infty$. Then Theorem 5.15 remains valid.

Remark 3. For $p=0$, we replace $-\|z\|^{-2p}$ by $\log\|z\|$ and in (5,23), we replace $(1+\|a\|^{2p})^{-1}$ by $\log(1+\|a\|)$ if $\int_1^\infty v_t(r)r^{-2}dr<+\infty$, and if $\int_1^\infty v_t(r)r^{-3}dr<+\infty$, we write

$$(5,25)\quad I_1(z)=c_0\int_{\mathbb{C}^n}\log\left[\|z-a\|-\log(1+\|a\|)-\frac{2p}{1+\|a\|^2}\mathbb{R}e\langle a,z\rangle\right]d\sigma(a).$$

Remark 4. By (5,25) we can calculate $\lambda(I_1,0,r)$ and the indicator $v_\theta(r)$ of the current $\theta=\frac{i}{\pi}\partial\bar\partial I_q$, which has the same Lelong number for each $z\in\mathbb{C}^n$ as the given current t. For $1\leqq p\leqq n-2$, we obtain for $r=\|z\|>0$:

$$v_\theta(r)=r\frac{\partial}{\partial r}\lambda(I_1,0,r)=r\int_0^\infty u^{2p}k_p(u,r)v_t(u)du$$

$k_p(u,r)=(2p)^{-1}\dfrac{\partial^2 h_p}{\partial u\,\partial r}$ and $h_p=h_p(u,r)$ is the mean value of $\|a-z\|^{-2p}$ for a given $\|a\|=u$ and $z\in S^{2n}(0,r)$, $r=\|z\|>0$. If $p=0$, we use $-\log\|a-z\|$ and write $\frac{1}{2}$ instead $(2p)^{-1}$.

The method of the canonical representation allows us, under the hypothesis (5,22), to improve the estimates given in (5,21) for $F^{-1}(0)$ if $[Y]$ is the current of integration on an analytic subvariety of dimension p. First of all, we obtain the equality $V(z)=I_q(z)$ and hence, from (3,14)

$$(5,26)\qquad M_V(r)=M_I(r)\leqq A(p,q)\,r^q\left[\int_0^r v_t(s)s^{-q-1}ds+r\int_r^\infty v_t(s)s^{-q-2}ds\right].$$

Thus, by (5,21), we obtain $Y=F^{-1}(0)$, with $\alpha>0$, $\varepsilon>0$ and

$$M_F(r)\leqq nM_V(r+\alpha)+\varepsilon\log(1+r)+C(n,\varepsilon,0).$$

We resume this result in the following theorem:

Theorem 5.17. *If Y is an analytic subvariety of pure dimension p with indicator $v_{[Y]}(t)$ such that $\int_1^\infty v_{[Y]}(t)t^{-3}dt<+\infty$, then for every $\varepsilon>0$, there exists a representation $Y=F^{-1}(0)$ with*

$$M_F(r)=\sup_{\|z\|\leqq r}\log\|F\|\leqq nM_V((1+\varepsilon)r)+\varepsilon\log(1+r)+C(n,p,\varepsilon)$$

and $M_V(r)$ satisfies (5,26). Thus for $q=0$ or $q=1$

$$(5,27)\qquad \sup_{\|z\|<r}\log\|F\|\leqq nA(p,q)r^q\left[\int_0^r v_Y(t)t^{-q-1}dt+r\int_r^\infty v_Y(t)t^{-q-2}dt\right]$$
$$+\varepsilon\log(1+r)+C(n,p,\varepsilon).$$

Proof. As in the proof of Lemma 3.47, by the Mean Value Property for subharmonic functions, we obtain

$$\sup_{\|z\|<r}\|F\|^2\leqq C(\varepsilon r)^{-2n}\exp[2M_V((1+\varepsilon)r)+2n\log(1+r)]$$

if we choose the ball $B(z,\tau)$ so be of radius εr. □

Thus we obtain a much better control of the growth in this case.

§6. The Algebraic Case

Let Y be an algebraic variety in $\mathbb{C}^n$. It can be defined as the common zero set of $(n+1)$ polynomials $P_j(z)$ and the plurisubharmonic function

$$(5,28)\qquad V=\tfrac{1}{2}\log(\Sigma|P_j|^2)$$

defines Y as the set

$$Y=E(1,V)=\{z\in\mathbb{C}^n\colon v_{[Y]}(z)\geqq 1\}\quad\text{and}\quad v_{[Y]}(z)=\min_j(\text{order }P_j\text{ at }z)$$

and V verifies $\lim\limits_{r\to\infty}\dfrac{M_V(r)}{\log r}=\rho$ with $\rho=\sup(\deg P_j)$.

Definition 5.18. A function $V \in \mathrm{PSH}(\mathbb{C}^n)$ is said to be of the minimal growth class S_a, $a>0$, if $\lim\limits_{r\to\infty} \dfrac{M_V(r)}{\log r} = a$.

Proposition 5.19. *Let Y be an algebraic variety (that is, the common zero set of a family of polynomials in $\mathbb{C}^n$) of pure dimension p. Then Y is the set $E(1, V)$ for V of minimal growth class S_a for $a \leqq m\tilde{C}(p)$, where $\tilde{C}(p)$ is a constant which depends only on p and $m =$ degree $Y = \max \operatorname{card}\{Y \cap \mu\}$, the maximum being taken over all $(n-p)$ dimensional planes μ in $\mathbb{C}^n$ such that $Y \cap \mu$ is discrete.*

Proof. Y is a finite union of irreducible algebraic varieties Y_i of dimension p. A linear subspace μ of dimension $(n-p)$ cuts Y_i in a finite number of isolated points $n_i(\mu)$ (except perhaps for μ belonging to an analytic set $\eta_i \subset G_{n-p}(\mathbb{C}^n)$ - cf. Theorem 2.42). For each Y_i, $m_i(\mu)$ is constant on $G_{n-p}(\mathbb{C}^n) - \eta_i$ and $\sum m_i(\mu) = m$. For $\mu \in G_{n-p}(\mathbb{C}^n)$, we denote by $[\mu]$ the current of integration on the analytic variety μ.

Let $\rho(z) \in \mathscr{C}_0^\infty(B(0,1))$ be such that $0 \leqq \rho(z) \leqq 1$, $\int \rho(z)\, d\tau(z) = 1$ and set $\rho_\varepsilon(z) = \varepsilon^{-2n}\rho(z/\varepsilon)$, $\varepsilon>0$, $\tilde{T}_\varepsilon = [Y] \star \rho_\varepsilon$, where $[Y]$ is the current of integration on Y. Suppose that $0 \notin Y$. Let μ_0 be a fixed subspace of $\mathbb{C}^n$ of dimension $(n-p)$, $U(n)$ the space of unitary matrices, and ω the Haar measure on $U(n)$ normalized so as to have total mass equal to 1. Then from the hypotheses, we obtain the inequality

$$m \geqq \int\limits_{B(0,r)} \int\limits_{U(n)} \tilde{T}_\varepsilon \wedge [\mu_0(\gamma^{-1}(z))]\, d\omega(\gamma).$$

Let $\xi_0 = \int\limits_{U(n)} [\mu_0(\gamma^{-1}(z))]\, d\omega(\gamma)$, which is a positive closed current of type (p, p).

If $\mathbb{P}(\mathbb{C}^n)$ is the projective space and $\pi\colon \mathbb{C}^n - \{0\} \to \mathbb{P}(\mathbb{C}^n)$ is the natural projection, then ξ_0 determines a current $\tilde{\xi}_0$ of type $(p-1, p-1)$ on $\mathbb{P}(\mathbb{C}^n)$: if φ is a form of type $(n-p,\ n-p)$ on $\mathbb{P}(\mathbb{C}^n)$, we set $(\tilde{\xi}_0, \varphi) = (\xi, \varphi(\pi(z)))$. Similarily, if $[\mu_0]$ is the current of integration over μ_0, $[\mu_0]$ determines a current $\tilde{\mu}_0$ of type $(p-1, p-1)$ on $\mathbb{P}(\mathbb{C}^n)$. Let φ be any $(n-p,\ n-p)$ form with $\mathscr{C}^\infty$ coefficients on $\mathbb{P}(\mathbb{C}^n)$, and set $\zeta(\varphi) = \int\limits_{U(n)} \varphi(\gamma(z))\, d\omega(\gamma)$. We shall show that $\zeta(\varphi) = k(\varphi)\, \alpha^{n-p}$ for $\alpha = \dfrac{i}{\pi} \partial\bar{\partial} \log \|z\|^2$.

Since any element of $\mathbb{P}(\mathbb{C}^n)$ can be transformed into any other element by an element of $U(n)$ and both $\zeta(\varphi)$ and α^p are invariant with respect to elements of $U(n)$, it suffices to show $\zeta_z(\varphi) = k(\varphi)\alpha_z^p$ for any arbitrary point $z \in \mathbb{P}(\mathbb{C}^n)$. In particular, we let $z = (z_1, \ldots, z_{n-1}, 1)$. We show that the space of $U(n-1)$ invariant forms of type $(n-p,\ n-p)$ in $\bigwedge \mathbb{C}^{n-1}$ is one dimensional. Let $\zeta_z(\varphi) = \sum e_{IJ} e_I \wedge e_{\bar{J}}$ where $(e_1, \ldots, e_{n-1})$ is a standard orthonormal basis in $\mathbb{C}^{n-1}$, $I = i_1 < \ldots < i_{n-p}$ and $e_I = e_{i_1} \wedge \ldots \wedge e_{i_p}$. Suppose that $I \neq J$; if $\hat{\gamma} \in U(n-1)$ is the element which multiplies the ith coordinate by -1 and leaves the others unchanged, then since $\zeta_z(\varphi)(\gamma(z)) = \zeta_z(\varphi)$, we see that $C_{IJ} = 0$.

Similarly, by considering a permutation of the coordinates, we see that $C_{II}=C_{JJ}$. Hence $\zeta_z(\varphi)=k(\varphi)\sum e_I\wedge e_{\bar{I}}$. We then have

$$k(\varphi)=\int_{\mathbb{P}(\mathbb{C}^n)}\alpha^{p-1}\wedge\xi=\int_{\mathbb{P}(\mathbb{C}^n)}\int_{U(n)}\alpha^{p-1}\wedge\varphi(\gamma(z))d\omega(\gamma)=\int_{\mathbb{P}(\mathbb{C}^n)}\alpha^{p-1}\wedge\varphi,$$

since $\int_{\mathbb{P}(\mathbb{C}^n)}\alpha^{n-1}=1$ and α is invariant with respect to elements of $U(n)$. Furthermore,

$$\begin{aligned}(\tilde{\xi}_0,\varphi)&=\int_{U(n)}([\mu_0(\gamma^{-1}(z))],\varphi(\pi(z)))d\omega(\gamma)\\&=\int_{U(n)}[\eta_0],\varphi(\pi(\gamma(z)))\omega(\gamma)=(\tilde{\mu}_0,\zeta(\varphi))\end{aligned}$$

since $\pi(\gamma(z))=\gamma(\pi(z))$. Thus

$$(\tilde{\xi}_0,\varphi)=(\tilde{\mu}_0,\zeta(\varphi))=(\tilde{\mu}_0,k(\varphi)\alpha^{n-p})=(\tilde{\mu}_0,\alpha^{n-p})(\alpha^{p-1},\varphi)=(\alpha^{p-1},\varphi)$$

and $\tilde{\xi}_0=\alpha^{p-1}$. Since ξ_0 and α^p are both constant on complex lines and both determine the same current α^{p-1} in $\mathbb{P}(\mathbb{C}^n)$, we have $\xi_0=\alpha^p$.

Returning to our original equation, we thus see that $m\geqq\int_{B(0,r)}\int_{U(n)}\tilde{T}_\varepsilon\wedge\alpha^p$. If we now let ε go to zero, we obtain $m\geqq\nu_{[Y]}(r)$. In particular, the indicator $\nu_{[Y]}(r)$ is of genus 0. Let

$$I(z)=-c_p\int[\|z-a\|^{-2p}-\|a\|^{-2p}]d\sigma_{[Y]}(a)=-c_p\int e_p(a,z,0)d\sigma_{[Y]}(a)$$

be the canonical potential. Then, by (5,22) and Theorem 5.15, we see that $I(z)$ is plurisubharmonic, and $Y=E(V,1)$. By Proposition 3.14, since $C_2(p,0)=1$ for all p, we obtain

$$I(z)\leqq c_p\int_{r_0}^{r}\frac{\nu_{[Y]}(t)\,dt}{t}+c'(p)\,r\int_r^{\infty}\frac{\nu_{[Y]}(t)}{t^2}dt. \tag{5,29}$$

Thus for $\|z\|=r$, $I(z)\leqq m\tilde{c}(p)\log r$, and so $I(z)$ is of minimal growth class at most $S_{m\tilde{c}(p)}$. □

Theorem 5.20. *Let Y be an analytic variety of pure dimension p such that $\nu_{[Y]}(t)\leqq m$. Then Y is an algebraic variety which can be defined by polynomials P_j of degree at most $nm\tilde{c}(p)+1/2$.*

Proof. Since $I(z)=-c_p\int[\|z-a\|^{-2p}-\|a\|^{-2p}]d\sigma_Y(a)$ satisfies (5,29), we obtain an estimate of the functions P_k by Theorem 5.17 such that for every k, $\deg|P_k|\leqq nm\tilde{c}(p)+1/2$ so by Corollary 1.7, P_k is a polynomial of degree $\tilde{c}(p)mn+1/2$ at most. □

Corollary 5.21. *An analytic subvariety Y is algebraic if and only if $\nu_{[Y]}(t)$ is bounded, and degree $Y=\lim_{t\to\infty}\nu_{[Y]}(t)$.*

§7. The Pseudo Algebraic Case

The study of infinite order differential operators has called attention to a class of analytic varieties Y of slow growth of genus 0 such that, if

$$h_2(r)=r\int_r^\infty v_{[Y]}(t)t^{-2}dt$$

and

$$h_1(r)=\int_{r_0}^r v_{[Y]}(t)t^{-1}dt, \qquad \text{then } \lim_{r\to\infty} h_2(r)/h_1(r)=0.$$

This condition is satisfied, for instance, when $v_{[Y]}(t)\leqq C(\log^+ t)^s$. Then $I(z)\leqq\dfrac{C}{s+1}(\log^+ r)^{s+1}\ (1+\varepsilon(r))$, where $\varepsilon(r)\to 0$, when $r\to\infty$. We can then define Y by entire functions $f_k(z)$ (i.e. $Y=\{z\colon f_k(z)=0,\ k=1,\dots,n+1\}$) such that $\lim\limits_{r\to\infty} M_{f_k}(r)(\log r)^{-\lambda}=0$ for $\lambda>s+1$. This class of functions has the following property: for every $\varepsilon>0$, there exists R_ε such that if $|f(z)|<1$ and $\|z\|>R_\varepsilon$, the distance of z to the null set X_f verifies $d(z,X_f)\leqq\varepsilon\|z\|$.

§8. Counterexamples to Uniform Upper Bounds

We now turn our attention to the study of problems (i), (ii), and (iii) outlined in the introduction to Chapter 5. We begin by producing examples which show that we cannot obtain uniform upper bounds for problems (i) and (ii).

We define the entire functions $g_k(z)$ of the variable $z\in\mathbb{C}$ by $g_k(z)=\prod\limits_{i\neq k}(1-z2^{-i})$. For $\varepsilon>0$, let C_ε be a constant such that $\log(1+r)\leqq C_\varepsilon r^\varepsilon$ for $r\geqq 1$. Suppose that z is such that $2^p\leqq|z|<2^{p+1}$. Then

$$\begin{aligned}\log|g_k(z)|&\leqq\sum_{i=1}^\infty \log(1+|z|2^{-i})\\ &\leqq\sum_{i=1}^p \log(1+|z|2^{-i})+\sum_{i=p+1}^\infty \log(1+|z|2^{-i})\\ &\leqq C_\varepsilon\sum_{i=1}^p |z|^\varepsilon 2^{-i\varepsilon}+2\leqq C'_\varepsilon|z|^\varepsilon,\end{aligned}$$

where C'_ε does not depend on k. Thus for every k and $|z|>1$

$$|g_k(z)|\leqq\exp(C'_{\varepsilon/2}|z|^{\varepsilon/2})\leqq C^1_\varepsilon \exp|z|^\varepsilon.$$

Let $P(w,c)=\prod\limits_{j=1}^c\left(w-\dfrac{1}{j}\right)$ for c a positive integer.

If $c_1 < \ldots < c_m$ is an increasing sequence of positive integers, we define $f(z,w)$ by the infinite series

$$f(z,w) = \sum_{m=1}^{\infty} 2^{-c_m^2} g_m(z) P(w, c_m),$$

which converges uniformly on compact subsets of $\mathbb{C}^2$. We obtain the estimate for $\varepsilon > 0$

$$|f(z,w)| \leqq C''_\varepsilon \exp|z|^\varepsilon \sum_{m=1}^{\infty} 2^{-m^2} |w+1|^m \leqq C'''_3 \exp(|z|^\varepsilon + |w+1|^\varepsilon)$$
$$\leqq C^{(iv)}_\varepsilon \exp(|z|+|w|)^{2\varepsilon}$$

since, if

$$h_\varepsilon = \left\{\frac{1}{\varepsilon}\right\}, \quad \sum_{m=1}^{\infty} 2^{-m^2} |w+1|^m \leqq \sum_{m=1}^{h_\varepsilon} 2^{-m} |w+1|^m + \exp|w+1|^\varepsilon.$$

Let $g(z,w) = \prod_{i=1}^{\infty} (1 - z2^{-i})$ and $F(z,w) = (g(z,w), f(z,w))$, $F: \mathbb{C}^2 \to \mathbb{C}^2$. Then $F^{-1}(0) = \left\{\left(2^m, \frac{1}{j}\right): m \in \mathbb{Z}, j = 1, \ldots, c_m\right\}$.

If $S(r)$ is any positive increasing function of r, by setting $r_m = 2^m + 1$ and $C_m = S(r_m)$, we obtain $\limsup_{r\to\infty} \frac{\operatorname{card}(F^{-1}(0) \cap B(0,r))}{S(r)} \geqq 1$. Thus, no upper bound of $\operatorname{card}(F^{-1}(a) \cap B(0,r))$ by a function of $\|a\|$, r, and $\log\|F\|$ is possible (independent of F).

Let $Y \subset \mathbb{C}^3$ be the analytic variety defined by

$$Y = \{(z_1, z_2, z_3): f(z_1, z_2) = g(z_3) = 0\}.$$

If we choose $c_1 = 1$, then

$$f(0,0) = -1 + \sum_{m=2}^{\infty} \frac{2^{-c_m^2}}{c_m!} (-1)^{c_m}, \quad \text{and} \quad \left|\sum_{m=2}^{\infty} \frac{2^{-c_m^2}(-1)^{c_m}}{c_m!}\right| \leqq (e^{1/2} - 1),$$

so $e^{1/2} \geqq |f(0,0)| \geqq 2 - e^{1/2}$. For fixed r, there are at most $(\log 2)^{-1} \log r$ values of z_3 for which $g(z_3) = 0$, $|z_3| < r$. The set $Y_1 = \{(z_1, z_2, z_3): g(z_3) = 0\}$ is a union of hyperplanes $Y_m = \{(z_1, z_2, z_3): z_3 = 2^m\}$. On each hyperplane Y_m, we see by the estimates of Chapter 3 that, setting $\tilde{Y} = (z: f(z) = 0)$,

$$r^{-2} \cdot \sigma_{[Y_m \cap \tilde{Y}]}(r) \leqq C^{(v)}_\varepsilon r^\varepsilon,$$

and so

$$r^{-2} \sigma_{[Y]}(r) \leqq \frac{1}{(\log 2)} C^{(v)}_\varepsilon \log r \cdot r^\varepsilon$$

where the constants depend only on ε but not on the choice of the c_m, $m \geqq 2$, as long as $c_1 = 1$. On the other hand, $Y \cap \{z: z_1 = z_3\} = \{(2^m, j, 2^m): m \in \mathbb{Z}, j = 1, \ldots, c_m\}$ and thus for any positive increasing function $S(r)$, by a proper choice of c_m, we obtain $\limsup_{r\to\infty} \frac{\operatorname{card}(Y\cap\{z: z_1=z_3\}\cap B(0,r))}{S(r)} \geqq 1$. Hence, we cannot find an upper bound for $Y\cap L\cap B(0,r)$ by a function of r depending only on $\sigma_{[Y]}(r)$ and valid for all hyperplanes L in $\mathbb{C}^n$; for the variety Y, there exists no upper bound for the growth of $n(Y,L,r)=\operatorname{card} Y\cap L\cap B(0,r)$ by $\sigma_{[Y]}(r)$.

This leads us to look for estimates valid everywhere outside a small exceptional set. It is in this context that we shall study problems (i), (ii), and (iii).

§9. An Upper Bound for the Area of $F^{-1}(a)$ for a Holomorphic Map

We begin by proving some lemmas which will provide the foundations of our calculations. In what follows, if Ω is a domain given by a $\mathscr{C}^\infty$ function ρ (i.e. $\Omega=\{z: \rho(z)<0\}$ such that $\operatorname{grad}\rho\neq 0$ on $bd\Omega$), we will always assume that $bd\Omega$ is oriented so that Stokes' Theorem holds – that is, $\int_{bd\Omega}\psi=\int_\Omega d\psi$ where Ω is given the orientation inherited from $\mathbb{C}^n$.

Lemma 5.22. *Let* $\rho\in\mathscr{C}^\infty(\mathbb{C}^n)\cap \mathrm{PSH}(\mathbb{C}^n)$ *and suppose that*

$$\Omega_r=\{z: \rho(z)<r\}\Subset\mathbb{C}^n$$

and $\operatorname{grad}\rho\neq 0$ *on* $bd\Omega_r$. *Let* θ *be a positive closed* $(n-p,n-p)$ *current in* $\mathbb{C}^n$ *with* $\mathscr{C}^\infty$ *coefficients. Then*

i) $\theta\wedge\beta^{p-1}\wedge i\bar\partial\rho$ *defines a positive measure on* $bd\Omega_r$,

ii) *if* $V\in\mathscr{C}^2(\bar\Omega)$, *then*

$$\int_{bd\Omega_r} V\theta\wedge\beta^{p-1}\wedge i\bar\partial\rho=\int_{\Omega_r} V\theta\wedge\beta^{p-1}\wedge i\partial\bar\partial\rho + \int_{\Omega_r}(r-\rho)i\partial\bar\partial V\wedge\theta\wedge\beta^{p-1}. \tag{5,30}$$

Proof. For (1), we have to prove that for $z_0\in bd\Omega_r$, there exists a neighborhood U of z_0 such that

$$I(h)=\int_{bd\Omega_r} h\theta\wedge\beta_{p-1}\wedge i\bar\partial\rho\geqq 0$$

for $h\in\mathscr{C}_0^0(U\cap bd\Omega_r)$, $h\geqq 0$. Given a continuous function $\varphi(x_1,\dots,x_m)$ in $\mathbb{R}^m$, we can calculate Rest. $\varphi|_{x_m=0}$ as a limit

$$\text{Rest. }\varphi|_{x_m=0}=\lim_{\varepsilon\to 0}\varepsilon^{-1}\int\varphi(x_1,\dots,x_m)\alpha\left(\frac{x_m}{\varepsilon}\right)dx_m,$$

where $\alpha(t)\in\mathscr{C}^\infty(-1,+1)$, $\int\alpha(t)dt=1$ and $\varepsilon^{-1}\alpha\left(\frac{t}{\varepsilon}\right)$ is an approximation to the Dirac measure $\delta(0)$. Similarily, for a form Ψ,

$$\text{Rest.}\,\psi|_{x_m=0}=\lim_{\varepsilon\to 0}\varepsilon^{-1}\int\alpha\left(\frac{x_m}{\varepsilon}\right)dx_m\wedge\psi.$$

By hypothesis, grad $\rho\neq 0$ on $bd\Omega_r$, so we can choose U such that ρ is a coordinate in U. Then if $\tilde{h}$ is a continuation of h to all of U, $h\geqq 0$, we have

$$\begin{aligned}I(h)&=\lim_{\varepsilon-0}\varepsilon^{-1}\int\alpha\left(\frac{\rho-r}{\varepsilon}\right)d\rho\wedge\tilde{h}\theta\wedge\beta_{p-1}\wedge i\bar{\partial}\rho\\&=\lim_{\varepsilon\to 0}\varepsilon^{-1}\int\alpha\left(\frac{\rho-r}{\varepsilon}\right)\tilde{h}\theta\wedge\beta_{p-1}\wedge i\partial\rho\wedge\bar{\partial}\rho\geqq 0\end{aligned}$$

by the definition of a positive form, since $\tilde{h}\alpha\geqq 0$.

Note that by this procedure, we have given an orientation to $bd\Omega_r$ which is consistent with Stokes' Theorem $\int_{bd\Omega_r}\psi=\int_{\Omega_r}d\psi$.

To prove (ii), we first apply Stokes' Theorem to the left hand side of (5,30). We obtain, since $d(\theta\wedge\beta^{p-1})=0$,

$$\int_{bd\Omega_r}V\theta\wedge\beta^{p-1}\wedge i\bar{\partial}\rho=\int_{\Omega_r}dV\wedge\theta\wedge\beta^{p-1}\wedge i\bar{\partial}\rho+\int_{\Omega_r}V\theta\wedge\beta^{p-1}\wedge i\partial\bar{\partial}\rho.$$

Furthermore, since $(\rho-r)=0$ on $bd\Omega_r$, we obtain by an integration by parts:

$$\begin{aligned}\int_{\Omega_r}dV\wedge\theta\wedge\beta^{p-1}\wedge i\bar{\partial}\rho&=\int_{\Omega_r}\partial V\wedge\theta\wedge\beta^{p-1}\wedge i\bar{\partial}(\rho-r)\\&=\int_{\Omega_r}(\rho-r)i\bar{\partial}\partial V\wedge\theta\wedge\beta^{p-1}\\&=-\int_{\Omega_r}(\rho-r)i\partial\bar{\partial}V\wedge\theta\wedge\beta^{p-1}.\end{aligned}$$

□

Lemma 5.23. *Let Y be an analytic variety of pure dimension p in $\mathbb{C}^n$ and let $\theta_{[Y]}$ be the positive closed current of integration over (the regular points of) Y. Let $V\in\mathscr{C}^\infty(\mathbb{C}^n)\cap\mathrm{PSH}(\mathbb{C}^n)$ with $V\geqq 0$, and set $M_V(r)=\sup_{\|z\|\leqq r}V(z)$. Then for every $\gamma>1$, there exists a constant C depending on γ and on n, p, q, such that*

$$\int_{B(0,\gamma^q r)}\theta_{[Y]}\wedge\beta^{p-q}\wedge(i\partial\bar{\partial}V)^q\leqq Cr^{-2q}\sigma_{[Y]}(r)\,[M_V(r)]^q. \tag{5,31}$$

Proof. Let θ_ν be a sequence of positive closed currents with $\mathscr{C}^\infty$ coefficients such that $\theta_\nu\to\theta_{[Y]}$ for the weak topology (cf. Proposition 2.11). Set

$$T_s=(i\partial\bar{\partial}V)^s\quad\text{and}\quad\sigma_s^\nu(t)=\int_{B(0,t)}T_s\wedge\theta_\nu\wedge\beta^{p-s}.$$

Then by Stokes' Theorem

$$\begin{aligned}
M_V(r) \int_{B(0,\gamma^s r)} \theta_v \wedge T_s \wedge \beta^{p-s} &= \int_{bd B(0,\gamma^s r)} M_V(r) T_s \wedge \beta^{p-s-1} \wedge i\bar{\partial}\|z\|^2 \wedge \theta_v \\
&\geqq \int_{bd B(0,\gamma^s r)} V T_s \wedge \beta^{p-s-1} \wedge i\bar{\partial}\|z\|^2 \wedge \theta_v \\
&\quad \text{by Lemma 5.22 (i)} \\
&\geqq \int_{B(0,\gamma^s r)} (r^2 - \|z\|^2) T_{s+1} \wedge \beta^{p-s-1} \wedge \theta_v \\
&\quad \text{by Lemma 5.22 (ii), since } V \geqq 0 \\
&\geqq \int_0^{\gamma^s r} 2t \sigma_{s+1}^v(t) dt \\
&\quad \text{after an integration by parts} \\
&\geqq \sigma_{s+1}^v(\gamma^{s+1} r) \int_{\gamma^{s+1} r}^{\gamma^s r} 2t dt \\
&\geqq r^2 \sigma_{s+1}^v(\gamma^{s+1} r)[\gamma^{2s} - \gamma^{2(s+1)}].
\end{aligned}$$

We thus obtain by iteration:

$$\int_{B(0,\gamma^q r)} \theta_v \wedge \beta^{p-q} \wedge (i\partial\bar{\partial} V)^q \leqq C r^{-2q} [M_V(r)]^q \int_{B(0,r)} \theta_v \wedge \beta^p.$$

If we now let v tend to infinity, we get

$$\int_{B(0,\gamma^q r)} \theta_{[Y]} \wedge \beta^{p-q} \wedge (i\partial\bar{\partial} V)^q \leqq C r^{-2q} [M_V(r)]^q \int_{\overline{B(0,r)}} \theta_{[Y]} \wedge \beta^p.$$

Since $\sigma_{[Y]}(r)$ is an increasing function of r, it is continuous outside a countable set E. If we apply the above inequality for a sequence $r_{m'} \in \complement E$ such that $r_{m'}$ increases to r, we obtain the conclusion of the Lemma. □

Lemma 5.24. *Let Y be an irreducible analytic variety of dimension p contained in a domain $\Omega \subset \mathbb{C}$ and let Y' and $\tilde{Y}$ be its singular and regular points respectively. Let $F = (f_1, \ldots, f_m)$: $Y \to \mathbb{C}^m$ be a holomorphic map and let $\hat{Y} = \{z \in \tilde{Y}$: $\operatorname{rank}(\partial f_1, \ldots, \partial f_m) < \sup_{z \in \tilde{Y}} \operatorname{rank}(\partial f_1, \ldots, \partial f_m) = k\}$. Then for every $z \in (\tilde{Y} - \hat{Y})$, there exist neighborhoods U_z of z in Y and V_z of $F(z)$ in $\mathbb{C}^m$ such that*

i) *$F(U_z)$ is a complex analytic manifold of dimension k in V_z,*
ii) *$\dim(U_z \cap F^{-1}(a)) = p - k$ for $a \in F(U_z)$.*

Proof. Let $z \in \tilde{Y} - \hat{Y}$. Then there exists a neighborhood W_z of z and a biholomorphic map ξ: $W_z \to B(0, 1) \subset \mathbb{C}^p$, $\xi(z) = 0$. Let

$$\begin{aligned}
\tilde{F}(z) &= F(\xi^{-1}(z)), \qquad z \in B(0,1), \\
\tilde{F}(z) &= (\tilde{f}_1(z), \ldots, \tilde{f}_m(z)): B(0,1) \to \mathbb{C}^m,
\end{aligned}$$

and rank $(\partial\tilde{f}_1, \ldots, \partial\tilde{f}_m) = k$. We suppose for simplicity that

$$\det\left[\frac{\partial\tilde{f}_i}{\partial z_j}(0)\right]_{i=1,\ldots,k}^{j=1,\ldots,k} \neq 0$$

(otherwise, we permute the functions and variables). Thus, by the Inverse Mapping Theorem, there exist neighborhoods T and T' of 0 in $B(0, 1)$ such that the map $\pi: T \to T'$ given by $z' = (\tilde{f}_1(z), \ldots, \tilde{f}_k(z), z_{k+1}, \ldots, z_p)$ is a biholomorphic homeomorphism. Set $f_j^{\#}(z) = \tilde{f}_j \circ \pi^{-1}(z)$. Then $\dfrac{\partial f_j^{\#}}{\partial z_l} \equiv 0$ in T' for $j<k$, $l>k$, for otherwise rank $[\partial f_1^{\#}, \ldots, \partial f_m^{\#}] > k$ for at least one point of T', since $f_j^{\#}(z) = z_j$, $j = 1, \ldots, k$. Thus for $j>k$, $f_j^{\#}$ depends only on the variables $(z_1, \ldots, z_k)$, and if $U_z = \xi^{-1}(T)$,

$$F(U_z) = \tilde{F}(T) = F^{\#}(T') = \{w \in \mathbb{C}^m \colon w_j = f_j^{\#}(w_1, \ldots, w_k), j = k+1, \ldots, \mathrm{m}\}$$

is a complex manifold of dimension k in a neighborhood V_z of $F(z)$. We can choose $f_1^{\#}, \ldots, f_k^{\#}$ as local coordinates on V_z, and hence

$$\dim(U_z \cap F^{-1}(a)) = \dim(T' \cap F^{\#-1}(a)) = p - k. \qquad \square$$

Lemma 5.25. *Let Y be an analytic variety of pure dimension p in a domain $\Omega \subset \mathbb{C}^n$ and let $F\colon Y \to \mathbb{C}^m$ be a holomorphic map. Let Y' be the singular points of Y and $\hat{Y} = \{z \in Y - Y' \colon \operatorname{rank}[\partial f_1, \ldots, \partial f_m] < m\}$. Then there exists an F_σ-set E of Lebesgue measure zero in $\mathbb{C}^m$ such that for $a \notin E$, if $Y \cap F^{-1}(a) \neq \emptyset$, $\dim(Y \cap F^{-1}(a)) = p - m$ and no irreducible branch of $(Y \cap F^{-1}(a))$ is contained in $Y' \cup \hat{Y}$.*

Proof. Let Y_j be the irreducible branches of Y and Y_j' be the singular points of Y_j. Set $k_j = \max\limits_{Y_j - Y_j'} \operatorname{rank}[\partial f_1, \ldots, \partial f_m]$ and

$$\hat{Y}_j = \{z \in Y_j - Y_j' \colon \operatorname{rank}[\partial f_1, \ldots, \partial f_m] < k_j\}.$$

We use induction on the dimension p. If $p = 1$ and $k_j = 0$, then the functions $f_i, i = 1, \ldots, m$ are constant on Y_j and so $F(Y_j) = a_j$. If $k_j = 1$, then $\hat{Y}_j \cap Y_j'$ is a countable set. For $a \notin \bigcup\limits_{k_j=0} F(Y_j) \bigcup\limits_{k_j=1} F(\hat{Y}_j \cup Y_j')$, $\dim(F^{-1}(a) \cap Y) = 1$ and no branch of $F^{-1}(a) \cap Y$ is contained in $\hat{Y} \cup Y'$. We now suppose that p is arbitrary. For any $a \in \mathbb{C}^m$ such that $F^{-1}(a) \cap Y \neq \emptyset$, $\dim(F^{-1}(a) \cap Y) \geqq p - m$. If $k_j < m$, it follows from Lemma 5.24 that $\tilde{E}_j = F(Y_j - (\hat{Y}_j \cup Y_j'))$ is an F_σ-set of Lebesgue measure zero, since a proper complex manifold is an F_σ-set of Lebesgue measure zero. We can find a countable number of open subsets Ω_{ij} of $\mathbb{C}^m$ and analytic varieties W_{ij} in Ω_{ij} such that $\dim W_{ij} \leqq (p-1)$ and $\tilde{Y}_j \cup Y_j = \bigcup\limits_i W_{ij}$. It follows from the induction hypothesis that we can find F_σ-sets E_j' of Lebesgue measure zero in $\mathbb{C}^m$ such that if $F^{-1}(a) \cap W_{ij} \neq \emptyset$, $\dim(F^{-1}(a) \cap W_{ij}) \leqq p - 1 - m$ for $a \notin E_{ij}$. If $k_j = m$, set $E_j = \bigcup\limits_i E_{ij}'$ and if $k_j < m$, set $E_j = \tilde{E}_j \bigcup\limits_i E_{ij}'$. Then for $a \notin \bigcup\limits_j E_j$, if

$F^{-1}(a)\cap Y\neq\emptyset$, $\dim(F^{-1}(a)\cap Y)=p-m$ and no branch of $F^{-1}(a)\cap Y$ is contained in $\bigcup (Y_j\cup Y_j')$. □

Theorem 5.26. *Let Y be an analytic variety of pure dimension p in $\mathbb{C}^n$ and let $F\colon Y\to\mathbb{C}^m$ be a holomorphic map, $m<p$. Set $M_F(r)=\sup_{\|z\|\leqq r}\|F\|$. For $a\in\mathbb{C}^m$, let θ_a be the positive closed current of integration over (the regular points of) $F^{-1}(a)\cap Y$ and set $\sigma_{[Y]}(a;r)=\int_{B(0,r)}\theta_a\wedge\beta_{p-m}$. Let $\varepsilon>0$, $\beta>1$ be given. Then the set*

$$\mathscr{E}=\left\{a\in\mathbb{C}^m\colon \dim F^{-1}(a)\neq p-m \text{ and } \limsup_{r\to\infty}\frac{\sigma_{[Y]}(a;r)}{r^{-2m}(\log r)^{\beta}\sigma_{[Y]}((1+\varepsilon)r)(\log M_F((1+\varepsilon)r))^m}\neq 0\right\}$$

is of Lebesgue measure zero.

Proof. Let $V(z)=\log(1+\|F\|^2)$. Then for $\|z\|\leqq r$, $0\leqq V(z)\leqq\log(1+M_F(r)^2)$. It follows from Lemma 5.23 (5,31) that for $\gamma<1$, there exists a constant C such that

$$C[\log^+ M_F(r)]^m\sigma_{[Y]}(r)\cdot r^{-2m}\geqq\int_{B(0,\gamma^m r)}(i\partial\bar\partial V)^m\wedge\theta_{[Y]}\wedge\beta_{p-m}.$$

It follows from Lemma 5.24 that if $\sup_Y \operatorname{rank} F\neq m$, then $F^{-1}(\mathbb{C}^m)$ is of Lebesgue measure zero.

Let $\eta_\nu\in\mathscr{C}_0^\infty(B(0,r))$ be a sequence of functions such that η_ν increases to the characteristic function of $B(0,\gamma^m r)-(Y'\cup\hat Y)$. For a fixed ν, if $z\in\operatorname{supp}\eta_\nu$, we can find a neighborhood $U_z\Subset B(0,\gamma^m r)-(Y'\cup\hat Y)$ and coordinates $(g_1,\dots,g_m,\dots,g_{m+1},\dots,g_p)$ in U_z such that

$$\{z\colon g_{m+1}=\dots=g_p=0,\ |g_i|<\delta_z,\ i=1,\dots,m\}$$

can be mapped by a holomorphic holomorphism π_z onto a neighborhood V_z of $F(z)$. Since $\operatorname{supp}\eta_\nu$ is compact, we can cover it by a finite number of these neighborhoods, say U_i, $i=1,\dots,N$. Let α_i be a $\mathscr{C}^\infty$ partition of unity subordinate to U_i; that is $\operatorname{supp}\alpha_i\subset U_i$, $\alpha_i\geqq 0$, and $\sum\alpha_i\equiv 1$ on $\operatorname{supp}\eta_\nu$. By Fubini's Theorem, we have

$$\begin{aligned}&\int\alpha_i\eta_\nu(i\partial\bar\partial V)^m\wedge\theta_{[Y]}(z)\wedge\beta_{p-m}\\&\quad=\int[i\partial_w\bar\partial_w\log(1+\|w\|^2)]^m\int\alpha_i\eta_\nu\theta_{[Y]}(w)\wedge\beta_{p-m}(z(w))\end{aligned}$$

and summing over i, we obtain

$$\begin{aligned}&\int\eta_\nu(i\partial\bar\partial V)^m\wedge\theta_{[Y]}\wedge\beta_{p-m}\\&\quad=\int[i\partial_w\bar\partial_w\log(1+\|w\|^2)]^m\int\eta_\nu\theta_{[Y]}(w)\wedge\beta_{p-m}(z(w)).\end{aligned}$$

Thus

$$\begin{aligned}
&\int_{B(0,\gamma^m r)} (i\partial\bar\partial V)^m \wedge \theta_{[Y]} \wedge \beta_{p-m} \\
&\quad = \lim_{\nu\to\infty} \int_{B(0,\gamma^m r)} \eta_\nu (i\partial\bar\partial V)^m \wedge \theta_{[Y]} \wedge \beta_{p-m} \\
&\quad = \lim_{\nu\to\infty} \int [i\partial_w\bar\partial_w \log(1+\|w\|^2)]^m \int_{B(0,\gamma^m r)} \eta_\nu \theta_{[Y]}(w) \wedge \beta_{p-m}(z(w)) \\
&\quad = \int [i\partial_w\bar\partial_w \log(1+\|w\|^2)]^m \lim_{\nu\to\infty} \int_{B(0,\gamma^m r)} \eta_\nu \theta_{[Y]}(w) \wedge \beta_{p-m}(z(w)) \\
&\quad = \int \sigma_{[Y]}(w;\gamma^m r)[i\partial_w\bar\partial_w \log(1+\|w\|^2)]^m
\end{aligned}$$

by the Monotone Convergence Theorem.

Set $\gamma^m=(1+\varepsilon)^{-1/2}$, $r_\tau=(1+\varepsilon)^{\tau/2}$ and

$$\begin{aligned}
F_\tau = \{a\in\mathbb{C}^m\colon\ & \sigma_{[Y]}(a;r_\tau) \\
& \geqq (\log r_\tau)^{\beta'} C\sigma_{[Y]}((1+\varepsilon)^{1/2} r_\tau) r_\tau^{-2m}[\log^+ M_F((1+\varepsilon)^{1/2} r_\tau)]^m\}
\end{aligned}$$

where $\beta'=\left(\dfrac{1+\beta}{2}\right)$. Set $\mu(w)=[i\partial_w\bar\partial_w \log(1+\|w\|^2)]^m$, which is a positive measure. Then by Lemma 5.23,

$$\int \sigma_{[Y]}(a;r_\tau)\,d\mu(a) \leqq C r_\tau^{-2m}[\log^+ M_F((1+\varepsilon)^{1/2} r_\tau)]^m \sigma_{[Y]}(r_\tau(1+\varepsilon)^{1/2})$$

and hence

$$\begin{aligned}
\mu(F_\tau)(\log r_\tau)^{\beta'} C[\log^+ M_F((1+\varepsilon)^{1/2} r_\tau)]^m r_\tau^{-2m} \sigma_{[Y]}((1+\varepsilon)^{1/2} r_\tau) \\
\geqq \int_{F_\tau} \sigma_{[Y]}(w;r_\tau)\,d\mu(w)
\end{aligned}$$

and so $\mu(F_\tau)\leqq(\log r_\tau)^{-\beta'}$. Let $E_\tau=\bigcup_{j\geqq\tau} F_j$. Then $\mu(E_\tau)\leqq\sum_{j=\tau}^{\infty}(\log r_j)^{-\beta'}$ and for $\tau\geqq M(\varepsilon,\delta)$, $\mu(E_\tau)\leqq\delta$. For $a\notin E_\tau$ and $\tau\geqq M(\varepsilon,\delta)$

$$\sigma_{[Y]}(a;r_{\tau'})\leqq C(\log r_{\tau'})^{\beta'}[\log^+ M_F((1+\varepsilon)^{1/2} r_{\tau'})]^m r_{\tau'}^{-2m}\sigma_{[Y]}((1+\varepsilon)^{1/2} r_{\tau'})$$

and thus for $r\in[r_{\tau'-1}, r_{\tau'})$

$$\begin{aligned}
\sigma_{[Y]}(a;r) &\leqq C(\log r_{\tau'})^{\beta'}[\log^+ M_F((1+\varepsilon)^{1/2} r_{\tau'}]^m r_{\tau'}^{-2m}\sigma_{[Y]}((1+\varepsilon)^{1/2} r_{\tau'}) \\
&\leqq C'(\log r)^{\beta'}[\log^+ M_F((1+\varepsilon)r)]^m \sigma_{[Y]}((1+\varepsilon)r)\cdot r^{-2m}.
\end{aligned}$$

Thus $\mathscr{E}=\bigcap_{\tau=1}^{\infty} E_\tau$ and $\mu(\mathscr{E})=0$. □

§10. Upper and Lower Bounds for the Trace of an Analytic Variety on Complex Planes

Let $G_q(\mathbb{C}^n)$ be the Grassmannian of all complex linear subspaces of $\mathbb{C}^n$ of dimension q. Then $G_q(\mathbb{C}^n)$ can be given the structure of a compact complex manifold of dimension $q(n-q)$. We can describe the local coordinate neighborhoods U_I as follows: let $I=(i_1<\ldots<i_q)$, where $i_j \leqq n$, and let $l_p=(l_{p1},\ldots,l_{pn})$, $p=1,\ldots,q$ with $l_{pi_j}=\begin{cases}1 & \text{if } p=i_j\\ 0 & \text{if } p\neq i_j\end{cases}$ for $i_j\in I$ and $l_{pi_j}=c_{pj}$ for $i_j\notin I$ and identify with $c\in\mathbb{C}^{q(n-q)}$ the linear space spanned by l_p, $p=1,,\ldots,q$. To see that $G_q(\mathbb{C}^n)$ is a compact manifold, we embed it as a submanifold of a complex projective space. Let $\mathbb{C}^{s(q)}=\bigwedge_q \mathbb{C}^n$ be the linear subspace spanned by the exterior products of degree q. If $l_1,\ldots,l_q$ are elements in $\mathbb{C}^n$, we associate the subspace spanned by $l_1,\ldots,l_q$ with $\pi(l_1\wedge\ldots\wedge l_q)$, where $\pi(\mathbb{C}^{s(q)}-\{0\})\to\mathbb{P}(\mathbb{C}^{s(q)})$ is the projection of $\mathbb{C}^{s(q)}$ into its projective space. Note that if $l_1,\ldots,l_q$ and $l'_1,\ldots,l'_q$ span the same subspace, then $l_1\wedge\ldots\wedge l_q=C(l'_1\wedge\ldots\wedge l'_q)$, where C is the Jacobian of the transformation $(l_1,\ldots,l_q)\to(l'_1,\ldots,l'_q)$, and so $\pi(l_1\wedge\ldots\wedge l_q)$ is well defined and furthermore is holomorphic on U_I. (For a more complete discussion of $G_q(\mathbb{C}^n)$, see [H]).

Although the notion of the Lebesgue measure of a set in not well defined on a complex manifold, since it changes with a change in the choice of local coordinates, the notion of a Lebesgue measurable set and a set of measure zero are invariant with respect to a local change of variables and so preserve their sense on complex manifolds.

If $\varphi\in\mathscr{C}_0^\infty(\mathbb{C}^{n-1})$, $\varphi\geqq 0$ and $\int\varphi(w)d\tau(w)=C_\varphi>0$, we set

$$V^\varphi(z)=C_\varphi^{-1}\int\log\left|\sum_{i=1}^{n-1} z_i w_i+z_n\right|\varphi(w)d\tau(w).$$

Lemma 5.27. *The function $V^\varphi(z)$ is plurisubharmonic in $\mathbb{C}^n$ and $\mathscr{C}^\infty$ in $\mathbb{C}^n-\{0\}$.*

Proof. It is enough to prove that $V^\varphi(z)$ is locally in $\mathscr{C}^\infty$ for $z\in\mathbb{C}^n-\{0\}$, and what is more, it is sufficient to show that $\Delta V\in\mathscr{C}^\infty$ in $\mathbb{C}^n-\{0\}$, since if $z_0\in\mathbb{C}^n-\{0\}$ and $\alpha\in\mathscr{C}_0^\infty(\mathbb{C}^n-\{0\})$ is such that $\alpha\equiv 1$ in a neighborhood of z_0, then it follows from Green's Theorem that for $\tilde{C}_{2n-2}=[(2n-2)\,\omega_{2n}]^{-1}$

$$\alpha(z)V(z)=\tilde{C}_{2n-2}\int\frac{-1}{\|z-z'\|^{2n-2}}\,\Delta(\alpha(z')V(z'))d\tau(z')$$

and

$$\frac{\partial}{\partial z^\beta}(\alpha(z)V(z))=\tilde{C}_{2n-2}\int\frac{-1}{\|z-z'\|^{2n-2}}(-1)^{|\beta|}\frac{\partial}{\partial z'^\beta}(\Delta\alpha(z')V(z'))d\tau(z'),$$

which is continuous for every β.

Suppose $z_0\neq 0$. If $(z_0)_1=\ldots=(z_0)_{n-1}=0$ and $z_n\neq 0$, then $z_0\notin\operatorname{supp}\Delta V$ and hence $\Delta V(z')=0$ in a neighborhood of z_0. Suppose then that $(z_0)_k\neq 0$ for $k\neq n$. Set

$$s=\sum_{i=1}^{n-1} z_i w_i+z_n,$$

$$w_k=\left(s-z_n-\sum_{\substack{i=1\\ i\neq k}}^{n-1} z_i w_i\right)z_k^{-1},$$

and

$$\tilde{w}=(w_1,\ldots,w_{k-1},s,w_{k+1},\ldots,w_{n-1});$$

let $\xi_k(w,z)\to(\tilde{w},z)$ be the map described above and $J_k(\tilde{w},z)$ the Jacobian of ξ_k^{-1}. Since

$$\Delta_z\log\left|\sum_{i=1}^{n-1}z_iw_i+z_n\right|=2\pi\delta\left(\sum_{i=1}^{n-1}z_iw_i+z_n\right)=2\pi\delta(s),$$

where δ is the Dirac measure at zero and $\delta(s)$ represents the current of integration over the hyperplane $s=0$, we have

$$\begin{aligned}\Delta V(z)&=2\pi\int\delta(s)\varphi(\xi_k^{-1}(\tilde{w},z))\,|J_k(\tilde{w},z)|^2\,d\tau(\tilde{w})\\&=\int\delta(s)\tilde{\varphi}(\tilde{w},z)\,d\tau(\tilde{w}),\end{aligned}$$

where $\tilde{\varphi}(\tilde{w},z)=2\pi\varphi(\xi_k^{-1}(\tilde{w},z))\,|J_k(\tilde{w},z)|^2\in\mathscr{C}^\infty$ in a neighborhood of z_0, since

$$\frac{\partial}{\partial z^\beta}\Delta V(z)=\int\delta(s)\frac{\partial}{\partial z^\beta}\tilde{\varphi}(\tilde{w},z)\,d\tau(\tilde{w}).$$ □

Suppose now that $K\subset\mathbb{C}^{n-1}$ is a compact set of positive Lebesgue measure. Let χ_K be the characteristic function of K and let $\varphi_\nu\in\mathscr{C}_0^\infty(\mathbb{C}^{n-1})$ be a sequence of functions such that φ_ν decreases to χ_K.

Lemma 5.28. *There exist constants C_1, C_2 and ν_0 such that*

$$C_1+\log\|z\|\leqq V^{\varphi_\nu}(z)\leqq C_2+\log\|z\|\qquad\text{for }\nu\geqq\nu_0.$$

Proof. Since $V^{\varphi_\nu}(uz)=V^{\varphi_\nu}(z)+\log|u|$, it is enough to show that $C_1\leqq V^{\varphi_\nu}(z)\leqq C_2$ for $\|z\|=1$ and $\nu\geqq\nu_0$. Let A be such that $K\subset B(0,A)$. Then $V^{\varphi_\nu}(z)\leqq(A+1)+\log\|z\|$. Let $\alpha\in\mathscr{C}_0^\infty(B(0,3A))$ such that $0\leqq\alpha\leqq1$ and $\alpha\equiv1$ on $B(0,2A)$. Then if $\|z\|=1$ and ν_0 is so large that $\operatorname{supp}\alpha_\nu\subset B(0,2A)$ for $\nu\geqq\nu_0$, we have

$$\begin{aligned}V^{\varphi_\nu}(z)&=C_{\varphi_\nu}^{-1}\int\log\left|\frac{\sum_{i=1}^{n-1}z_iw_i+z_n}{(2A+1)}\right|\varphi_\nu(w)\,d\tau(w)+\log(2A+1)\\&\geqq C_{\varphi_\nu}^{-1}\int\log\left|\frac{\sum_{i=1}^{n-1}z_iw_i+z_n}{(2A+1)}\right|\alpha(w)\,d\tau(w)+\log(2A+1)\\&\geqq[m(K)]^{-1}\int\log\left|\frac{\sum_{i=1}^{n-1}z_iw_i+z_n}{(2A+1)}\right|\alpha(w)\,d\tau(w)+\log(2A+1)\\&\geqq C_1\qquad\text{by Lemma 5.27.}\end{aligned}$$ □

Lemma 5.29. *Let $Y\subset\mathbb{C}^n$ be an analytic variety of pure dimension $p\geqq1$. Then the set $\{L\in G_{n-1}(\mathbb{C})\colon L\cap Y\neq\emptyset$ and $\dim(L\cap Y)=p\}$ is of Lebesgue measure zero in $G_{n-1}(\mathbb{C})$.*

Proof. Let Y_j be the irreducible branches of Y, each of which is of dimension p. Let $w \in \mathbb{C}^n$. Then if $Y_j \cap \left\{z: \sum_{i=1}^{n} z_i w_i = 0\right\}$ is non-empty and is not of pure dimension $(p-1)$, we have $Y_j \subset \left\{z: \sum_{i=1}^{n} z_i w_i = 0\right\}$. The set $A_j = \{w \in \mathbb{C}^n: Y_j \subset \{z: \sum z_i w_i = 0\}\}$ is a linear subspace, and furthermore $A_j \neq \mathbb{C}^n$, since $Y \subset \bigcap_{w \in A} \left\{z: \sum_{i=1}^{n} z_i w_i = 0\right\}$ and $0 = \bigcap_{w \in \mathbb{C}^n} \left\{z: \sum_{i=1}^{n} z_i w_i = 0\right\}$. If π is the projection of $\mathbb{C}^n$ into $\mathbb{P}(\mathbb{C}^n)$, then $\pi(A_j)$ is of measure zero in $\mathbb{P}(\mathbb{C}^n)$, and if $Y \notin \bigcup_j \pi(A_j)$, $\dim(L \cap Y) = p-1$. □

Lemma 5.30. *Let $Y \subset \mathbb{C}^n$ be an analytic variety of pure dimension p and $\theta_{[Y]}$ the closed positive current of integration over the regular points $\tilde{Y}$ of Y. Then the set of $w \in \mathbb{C}^{n-1}$ such that the simple extension of*

$$(2\pi)^{-1} i \partial \bar{\partial} \log \left| \sum_{i=1}^{n-1} z_i w_i + z_n \right| \wedge \theta_{[Y]}$$

to $\mathbb{C}^n$ as a positive closed current of degree $(p-1)$ is not the current of integration on $Y \cap \left\{z: \sum_{i=1}^{n-1} z_i w_i + z_n = 0\right\}$ is of Lebesgue measure zero.

Proof. We begin by remarking that if $f \in \mathscr{H}(B(0,1))$, then f is constant on every connected component of the analytic variety

$$Z = \left\{z \in B(0,1): \frac{\partial f}{\partial z_1} = \ldots = \frac{\partial f}{\partial z_n} = 0\right\}.$$

Indeed, for every $z \in \tilde{Z}$, the set of regular points of Z, there exists a neighborhood U_z of z and a biholomorphic map, $\pi_z: U_z \to B(0,1) \subset \mathbb{C}^s$, where s depends on z. Set $\tilde{f}(u) = f(\pi_z^{-1}(u))$ in $B(0,1)$. Then $\dfrac{\partial \tilde{f}(u)}{\partial u_j} = \sum_{i=1}^{n} \dfrac{\partial f}{\partial z_i} \dfrac{\partial z_i}{\partial u_j} \equiv 0$ so $\tilde{f}$ is constant in $B(0,1)$, and hence f is constant in U_z and thus on every component of Z.

Let Y' be the analytic variety of singular points of Y, which is of dimension at most $(p-1)$. Set

$$A_1 = \left\{w \in \mathbb{C}^{n-1}: Y' \cap \left\{z: \sum_{i=1}^{n-1} z_i w_i + z_n = 0\right\} \neq \emptyset \quad \text{and} \quad \dim(Y') \cap \left\{z: \sum_{i=1}^{n-1} z_i w_i + z_n = 0\right\} > (p-2)\right\}$$

$$A_2 = \left\{w \in \mathbb{C}^{n-1}: Y \cap \left\{z: \sum_{i=1}^{n-1} z_i w_i + z_n = 0\right\} \neq \emptyset \quad \text{and} \quad \dim(Y) \cap \left\{z: \sum_{i=1}^{n-1} z_i w_i + z_n = 0\right\} > (p-1)\right\}.$$

It follows from Lemma 5.29 that A_1 and A_2 are of Lebesgue measure zero in $\mathbb{C}^{n-1}$.

Let $z \in \tilde{Y} - 0$. For simplicity, we assume $z_1 \neq 0$. Then there exists an open neighborhood U_z of z in $\tilde{Y}$ such that $\inf\{z_1' : z' \in U_z\} > 0$ and a biholomorphic map $\xi_z : U_z \to B(0, 1) \subset \mathbb{C}^p$. For w fixed, set

$$Y_w = \left\{ u \in B(0,1) : \sum_{i=1}^{n-1} z_i(u) w_i + z_n = 0 \right\}$$

$$\hat{Y}_w = \left\{ u \in B(0,1) : \frac{\partial}{\partial u_j} \left(\sum_{i=1}^{n-1} z_i(u) w_i + z_n \right) = 0,\ j = 1, \ldots, p \right\},$$

and $A = \{w \in \mathbb{C}^{n-1} : \overline{Y_w} \subset \overline{\hat{Y}_w}\}$. Then A is an F_σ-set (and hence measurable), since $F_v = \{w : \overline{Y_w \cap B(0, 1 - 1/v)} \subset \overline{\hat{Y}_w \cap B(0, 1 - 1/v)}\}$ is closed and $A = \bigcup_v F_v$. Let

$$f(u) = \frac{-\sum_{i=2}^{n-1} z_i(u) w_i - z_n(u)}{z_1(u)}$$

in $B(0, 1)$. Suppose for $w' = (w_2, \ldots, w_n)$ fixed, $w = (w_1, w') \in A$. Then for $u \in Y_w$

$$\frac{\partial f(u)}{\partial u_j} = z_1(u)^{-1} \left(-\sum_{i=2}^{n-1} \frac{\partial z_i w_i}{\partial u_j} - \frac{\partial z_n}{\partial u_j} \right)$$
$$+ z_1(u)^{-2} \left(\sum_{i=2}^{n-1} z_i(u) w_i + z_n(u) \right) \frac{\partial z_i}{\partial z_j} = 0; \qquad j = 1, \ldots, p.$$

But we have seen that $f \equiv$ constant on each connected component of $\left\{ u \in B(0,1) : \frac{\partial f}{\partial u_j} = 0,\ j = 1, \ldots, p \right\}$, and thus for every w', there are at most a countable number of $w_1 = f(u)$ such that $(w_1, w') \in A$, so A is of measure zero in $\mathbb{C}^{n-1}$. We now choose a countable dense set $z_i \in \tilde{Y} - 0$ and neighborhoods U_{z_i} and sets A_i as defined above. Then $A_3 = \bigcup_{i=1}^{\infty} A_i$ is of measure zero.

Suppose that $w \notin A_1 \cup A_2 \cup A_3$. Let $Z_w = Y \cap \left\{ z : \sum_{i=1}^{n-1} z_i w_i + z_n = 0 \right\}$ with $\tilde{Z}_w$ as regular points and Z_w' as singular points. Then, since $w \notin A_2$, $\dim(Z_w') \leq p - 2$. It follows from above discussion that since $w \notin A_3$,

$$\frac{1}{2\pi} i \partial \bar{\partial} \log \left| \sum_{i=1}^{n-1} z_i w_i + z_n \right| \wedge \theta_v$$

is the closed positive current of integration over Z_w in $\mathbb{C}^n - Y' - Z_w'$. Its simple extension $\tilde{t}$ is a closed positive current of degree $(p-1)$ in $\mathbb{C}^n - Y'$, since $\dim(Z_w) \leq p - 2$, and the simple extension of $\tilde{t}$ to $\mathbb{C}^n - Y'$ is again a closed positive current of degree $(p-1)$, since $\dim(Z_w \cap Y') \leq p - 2$ for $w \notin A_2$ (cf. Chapter 2).

Lemma 5.31. *Let $Y \subset \mathbb{C}^n$ be an analytic variety of dimension $p \geq 1$ and let $\theta_{[Y]}$ be the closed current of integration on Y. Let $K \subset \mathbb{C}^{n-1}$ be a compact set and*

$\varphi_\nu \in \mathscr{C}_0^\infty(\mathbb{C}^{n-1})$ a sequence of functions such that φ_ν decreases to χ_K, the characteristic function of K. Then for $r>1$, if V^{φ_ν} is defined as in Lemma 5.27,

$$\lim_{\nu\to\infty} \int_{B(0,r)-\overline{B(0,1)}} i\partial\bar{\partial} V^{\varphi_\nu}(z)\wedge\theta_{[Y]}\wedge\beta_{p-1}$$

$$=\frac{2\pi}{m(K)}\int_{\mathbb{C}^{n-1}}[\sigma_{[Y]}(w;r)-\bar{\sigma}_{[Y]}(w,1)]\chi_K(w)d\tau(w),$$

where θ_w is the current of integration over the set

$$Y\cap\left\{z\colon \sum_{i=1}^{n-1} z_i w_i+z_n=0\right\} \quad \text{and} \quad \sigma_{[Y]}(w;r)=\int_{B(0,r)}\theta_w\wedge\beta_{p-1},$$

$$\bar{\sigma}_{[Y]}(w,r)=\int_{\overline{B(0,r)}}\theta_w\wedge\beta_{p-1}.$$

Proof. Let Y' be the singular points of Y and let $\psi_\mu\in\mathscr{C}_0^\infty(B(0,r))$ be a sequence such that the ψ_μ increase to $\chi_{(B(0,r)-Y'-\bar{B}(0,1))}$. An integration by parts gives us

$$\int\psi_\mu i\partial\bar{\partial}V^{\varphi_\nu}\wedge\theta_{[Y]}\wedge\beta_{p-1}=\int V^{\varphi_\nu}i\partial\bar{\partial}\psi_\mu\wedge\theta_{[Y]}\wedge\beta_{p-1},$$

and it follows from Fubini's Theorem that

$$\int V^{\varphi_\nu}i\partial\bar{\partial}\psi_\mu\wedge\theta_{[Y]}\wedge\beta_{p-1}$$

$$=C_{\varphi_\nu}^{-1}\int_{\mathbb{C}^{n-1}}\left[\int\log\left|\sum_{i=1}^{n-1}z_iw_i+z_n\right|i\partial\bar{\partial}\psi_\nu\wedge\theta_{[Y]}\wedge\beta_{p-1}\right]\varphi_\nu(w)d\tau(w)$$

A second integration by parts gives us

$$C_{\varphi_\nu}^{-1}\int_{\mathbb{C}^{n-1}}\left[\int\psi_\mu i\partial\bar{\partial}\log\left|\sum_{i=1}^{n-1}z_iw_i+z_n\right|\wedge\theta_{[Y]}\wedge\beta_{p-1}\right]\varphi_\nu(w)d\tau(w)$$

$$=C_{\varphi_\nu}^{-1}\int_{\mathbb{C}^{n-1}}\left[\int\log\left|\sum_{i=1}^{n-1}z_iw_i+z_n\right|i\partial\bar{\partial}\psi_\mu\wedge\theta_{[Y]}\wedge\beta_{p-1}\right]\varphi_\nu(w)d\tau(w).$$

It follows from Lemma 5.30 and the Monotone Convergence Theorem that

$$\lim_{\mu\to\infty}\int\psi_\mu i\partial\bar{\partial}V^{\varphi_\nu}\wedge\theta_{[Y]}\wedge\beta_{p-1}$$

$$=C_{\varphi_\nu}^{-1}\int_{\mathbb{C}^{n-1}}\left[\lim_{\mu\to\infty}\int\psi_\mu i\partial\bar{\partial}\log\left|\sum_{i=1}^{n-1}z_iw_i+z_n\right|\wedge\theta_{[Y]}\wedge\beta_{p-1}\right]\varphi_\nu(w)d\tau(w).$$

or

$$\int_{B(0,r)-\overline{B(0,1)}} i\partial\bar{\partial}V^{\varphi_\nu}\wedge\theta_{[Y]}\wedge\beta_{p-1}$$

$$=2\pi C_{\varphi_\nu}^{-1}\int_{\mathbb{C}^{n-1}}[\sigma_{[Y]}(w;r)-\sigma_{[Y]}(w';1)]\varphi_\nu(w)d\tau(w).$$

Finally, it follows from the Lebesgue Dominated Convergence Theorem that

$$\lim_{\nu\to\infty}\int i\partial\bar{\partial}V^{\varphi_\nu}\wedge\theta_{[Y]}\wedge\beta_{p-1}$$

$$=2\pi m(K)^{-1}\int_{\mathbb{C}^{n-1}}[\sigma_{[Y]}(w;r)-\sigma_{[Y]}(w;1)]\chi_K(w)d\tau(w).$$ □

Lemma 5.32. *Let $Y \subset \mathbb{C}^n$ be an analytic variety of pure dimension $p \geqq 1$ and let $K \subset \mathbb{C}^{n-1}$ be a compact set and $\varphi_\nu \in \mathscr{C}_0^\infty(\mathbb{C}^{n-1})$ a sequence of functions such that φ_ν decreases to χ_K. Then if $k > 1$ is a constant, there exist constants $\gamma_1, \gamma_2, \gamma_3$ and γ_4 which depend only on K and k such that for $r > 1$*

$$r^{-2}\gamma_1 \sigma_{[Y]}(\gamma_2 r) \leqq \gamma_3 \sigma_{[Y]}(2) + \int_{B(0,r)-\bar{B}(0,1)} i\partial\bar{\partial} V^{\varphi_\nu} \wedge \theta_{[Y]} \wedge \beta_{p-1}$$

$$\leqq \gamma_3 \sigma_{[Y]}(2) + \left(\frac{\gamma_4}{k^2} + 1\right) \sigma_{[Y]}(kr) r^{-2}$$

and hence

$$r^{-2}\gamma_1 \sigma_{[Y]}(\gamma_2 r) \leqq \gamma_3 \sigma_{[Y]}(2) + \frac{2\pi}{m(K)} \int [\sigma_{[Y]}(w;r) - \sigma_{[Y]}(w;1)] \chi_K(w) d\tau(w)$$

$$\leqq \gamma_3 \sigma_{[Y]}(2) + \gamma_4' \sigma_{[Y]}(kr) r^{-2}.$$

Proof. Let $A_r^\nu = \{z \in \mathbb{C}^n \colon V^{\varphi_\nu}(z) \leqq \log r\}$ and let θ_μ be a sequence of positive closed currents of degree p with $\mathscr{C}^\infty$ coefficients which converges to $\theta_{[Y]}$ in $\mathscr{E}'$. Set $t_\nu(r) = \sup\limits_{bd A_r} \|z\|^2$.

Since by Lemma 5.27, $V^{\varphi_\nu} \in \mathscr{C}^\infty(\mathbb{C}^n - \{0\})$, it follows from Sard's Theorem that the set of r for which $bd A_r^\nu$ is not a $\mathscr{C}^\infty$ manifold is of measure zero in $\mathbb{R}$. For r in the complement of this exceptional set, we have by Stokes' Theorem

$$t_\nu(r) \int_{A_r^\nu} i\partial\bar{\partial} V^{\varphi_\nu} \wedge \theta_\mu \wedge \beta_{p-1} = t_\nu(r) \int_{bd A_r^\nu} i\bar{\partial} V^{\varphi_\nu} \wedge \theta_\mu \wedge \beta_{p-1}.$$

By Lemma 5.22, $i\bar{\partial} V^{\varphi_\nu} \wedge \theta_\mu \wedge \beta_{p-1}$ is a positive measure on $bd A_r^\nu$ and hence by Lemma 5.22 (ii)

$$t_\nu(r) \int_{bd A_r^\nu} i\bar{\partial} V^{\varphi_\nu} \wedge \theta_\mu \wedge \beta_{p-1} \geqq \int_{bd A_r^\nu} \|z\|^2 i\bar{\partial} V^{\varphi_\nu} \wedge \theta_\mu \wedge \beta_{p-1}$$

$$\geqq \int_{A_r^\nu} \|z\|^2 i\partial\bar{\partial} V^{\varphi_\nu} \wedge \theta_\mu \wedge \beta_{p-1}$$

$$+ \int_{A_r^\nu} (\log r - V^{\varphi_\nu}) \theta_\mu \wedge i\partial\bar{\partial} \|z\|^2 \wedge \beta_{p-1}.$$

By grouping these inequalities, we obtain for $s < 1$

$$t_\nu(r) \int_{A_r^\nu} i\partial\bar{\partial} V^{\varphi_\nu} \wedge \theta_\mu \wedge \beta_{p-1} \geqq c_s \int_{A_{sr}^\nu} \theta_\nu \wedge \beta_p.$$

By Lemma 5.28, for $\nu \geqq \nu_0$, $C_1 + \log\|z\| \leqq V^{\varphi_\nu}(z) \leqq C_2 + \log\|z\|$, and so $A_r^\nu \subset B(0, re^{C_2})$ and $B(0, sre^{-C_1}) \subset A_{sr}^\nu$. Thus

$$r^2 \int_{B(0,r)} i\partial\bar{\partial} V^{\varphi_\nu} \wedge \theta_\mu \wedge \beta_{p-1} \geqq \gamma_1 \int_{B(0,\gamma_2 r)} \theta_\mu \wedge \beta_p.$$

Let $\eta \in \mathscr{C}^\infty(B(0,2))$ such that $\eta \equiv 1$ on $B(0,1)$. Then

$$\int \eta i\partial\bar{\partial} V^{\varphi_\nu} \wedge \theta_\mu \wedge \beta_{p-1} = \int V^{\varphi_\nu} i\partial\bar{\partial}\eta \wedge \theta_\mu \wedge \beta_{p-1}$$

$$\leqq \gamma_3 \int_{B(0,2)} \theta_\mu \wedge \beta_p \quad \text{(cf. Chapter 2)},$$

since $|V^{\varphi_\nu}|$ is bounded on supp $i\partial\bar{\partial}\eta$ independantly of ν for $\nu \geqq \nu_0$. We now let $\mu \to \infty$ and obtain

$$\int_{\overline{B(0,r)-B(0,1)}} i\partial\bar{\partial} V^{\varphi_\nu} \wedge \theta_{[Y]} \wedge \beta_{p-1} - \gamma_3 \sigma_X(2) \geqq \gamma_1 \sigma_{[Y]}(\gamma_2 r) \cdot r^{-2}.$$

Since this is true for a dense set of r for which the function

$$\int_{\overline{B(0,r)-B(0,1)}} i\partial\bar{\partial} V^{\varphi_\nu} \wedge \theta_{[Y]} \wedge \beta_{p-1}$$

is continuous, we can choose a sequence r_τ increasing to r for which the inequality is valid. This establishes the left hand inequality for all r.

To establish the right hand inequality, we note that by subtracting a constant, we may assume that $\log \|z\| - C_3 \leqq V^{\varphi_\nu}(z) \leqq \log \|z\|$ for $\nu \geqq \nu_0$ and $C_3 \geqq 0$. We apply Lemma 5.22 (ii) to $V^{\varphi_\nu} - \log \|z\|$ and obtain

$$\begin{aligned}
&\int_{B(0,kr)} (k^2 r^2 - \|z\|^2) i\partial\bar{\partial} V^{\varphi_\nu} \wedge \theta_\mu \wedge \beta_{p-1} \\
&\quad - \int_{B(0,kr)} (k^2 r^2 - \|z\|^2) i\partial\bar{\partial} \log \|z\| \wedge \theta_\mu \wedge \beta_{p-1} \\
&\quad = \int_{bd\,B(0,kr)} (V^{\varphi_\nu} - \log \|z\|) \theta_\mu \wedge \beta_{p-1} \wedge i\bar{\partial} \|z\|^2 \\
&\quad - \int_{B(0,kr)} (V^{\varphi_\nu} - \log \|z\|) \theta_\mu \wedge \beta_p .
\end{aligned}$$

By Lemma 5.22 (i), the first term on the right hand side is negative, and so

$$\begin{aligned}
(k^2-1) r^2 &\int_{\overline{B(0,r)-B(0,1)}} i\partial\bar{\partial} V^{\varphi_\nu} \wedge \theta_\mu \wedge \beta_{p-1} \\
&\leqq C_3 \int_{B(0,kr)} \theta_\mu \wedge \beta_p + k^2 r^2 \int_{B(0,kr)} \theta_\mu \wedge i\partial\bar{\partial} \log \|z\| \wedge \beta_{p-1} \\
&\leqq C_4 \int_{B(0,kr)} \theta_\mu \wedge \beta_p \qquad \text{(cf. Chapter 2).}
\end{aligned}$$

We now let μ tend to infinity and obtain

$$\int_{\overline{B(0,r)-B(0,1)}} i\partial\bar{\partial} V^{\varphi_\nu} \wedge \theta_{[Y]} \wedge \beta_{p-1} \leqq \gamma_4 r^{-2} \int_{B(0,kr)} \theta_{[Y]} \wedge \beta_p .$$

Finally, we choose an increasing sequence of r for which $\sigma_{[Y]}(kr)$ is continuous and apply the above inequality. □

Theorem 5.33. *Let $Y \subset \mathbb{C}^n$ be an analytic set of pure dimension p. Then for $0 \leqq q \leqq p$, $\varepsilon > 0$, and $\beta > 1$*

$$\left\{ L \in G_{n-q}(\mathbb{C}) \colon \limsup_{r \to \infty} \frac{\sigma_{[L \cap Y]}(r)}{r^{-2q} (\log r)^{\beta q} \sigma_{[Y]}((1+\varepsilon) r)} \neq 0 \right\}$$

is of Lebesgue measure zero in $G_{n-q}(\mathbb{C})$.

Proof. We shall show by induction on q that

$$\left\{ L = (L_1, \ldots, L_q) \in [G_{n-1}(\mathbb{C})]^q \colon \limsup_{r \to \infty} \frac{\sigma_{[Y \cap L_1 \ldots \cap L_q]}(r)}{r^{-2q} (\log r)^{\beta q} \sigma_{[Y]}((1+\varepsilon)^{q/p} r)} \neq 0 \right\}$$

is of Lebesgue measure zero for the product measure on $[G_{n-1}(\mathbb{C})]^q$. Let $q=1$ and suppose that

$$\text{measure}\left(E=\left\{L\in G_{n-1}(\mathbb{C})\colon \limsup_{r\to\infty}\frac{\sigma_{[Y\cap L]}(r)}{r^{-2}(\log r)^{\beta}\sigma_{[Y]}((1+\varepsilon)^{1/p}r)}\neq 0\right\}>0\right).$$

Then we can find a compact set $K\subset E\subset U_0$ of strictly positive measure, where

$$U_0=\left\{L\in G_{n-1}(\mathbb{C})\colon L=\left\{z\colon \sum_{i=1}^{n-1} z_i w_i+z_n=0\right\},\ w\in\mathbb{C}^{n-1}\right\}.$$

Let $r_\tau=(1+\varepsilon)^{\tau/2p}$, $k=(1+\varepsilon)^{1/2p}$. By Lemma 5.32, we obtain

$$\int_{\mathbb{C}^{n-1}}[\sigma_{[Y]}(w;r)-\sigma_{[Y]}(w;1)]\,\chi_K(w)\,d\tau(w)\leqq C_{\varepsilon,K}\,\sigma_{[Y]}((1+\varepsilon)^{1/2p}r)\cdot r^{-2}.$$

By reasoning as in Theorem 5.26, we conclude that

$$\left\{w\in K\colon \limsup_{r\to\infty}\frac{\sigma_{[Y]}(w;r)-\sigma_{[Y]}(w;1)}{r^{-2}(\log r)^{\beta}\sigma_{[Y]}((1+\varepsilon)^{1/p}r)}\neq 0\right\}$$

is of Lebesgue measure zero, which is a contradiction. Thus measure $(E)=0$. We now assume the induction hypothesis for $t\leqq q-1$. Suppose now that $L\notin E$ as defined above. By the induction hypothesis, the set

$$E=\left\{\hat{L}(L_1,\dots,L_{q-1})\in[G_{n-1}(\mathbb{C})]^{q-1}\colon\right.$$

$$\left.\limsup_{r\to\infty}\frac{\sigma_{[Y\cap L_1\cap\dots\cap L_{q-1}]}(r)}{r^{-2(q-1)}(\log r)^{\beta(q-1)}\sigma_{[Y]}((1+\varepsilon)^{\frac{q-1}{2p}}r)}\neq 0\right\}$$

is of measure zero. This establishes the induction hypothesis for all q.

To terminate the proof, we show that if $\eta\colon[G_{n-1}(\mathbb{C})]^q$ is given by $\eta(L_1,\dots,L_q)=\bigcap_{i=1}^{q}L_i$, then for $E\subset G_{n-q}(\mathbb{C})$ such that the $2q(n-q)$ Lebesgue measure of E is positive, the product measure of $\eta^{-1}(E)$ is positive in $[G_{n-1}(\mathbb{C})]^q$. Let $\mathbb{C}^{s(q)}=\bigwedge^{q}\mathbb{C}^n$ as a linear space. For $I=(i_1<\dots<i_q)$, we let $\bar{I}=(j_1<\dots<j_{n-q}\colon j_i\notin I\}$ and $\operatorname{sign}(I,\bar{I})$ the sign of the permutation $(1,\dots,q)\to(I,\bar{I})$. If $e_1,\dots,e_n$ is the standard basis in $\mathbb{C}^n$, we set $e_I=e_{i_1}\wedge\dots\wedge e_{i_q}$, and we define the map $\xi_q\colon\mathbb{C}^{s(q)}\to\mathbb{C}^{s(n-q)}$ to be the linear map such that $\xi_q(e_I)=\operatorname{sign}(I,\bar{I})e_{\bar{I}}$.

Let $\pi_q\colon\mathbb{C}^{s(q)}-\{0\}\to\mathbb{P}(\mathbb{C}^{s(q)})$. Then the map

$$\tilde{\xi}_q=\pi_{n-q}\xi_q\pi_q^{-1}\colon G_q(\mathbb{C})\to G_{n-q}(\mathbb{C})$$

is holomorphic and $\tilde{\xi}_q(L)=L^{\perp}$, the space orthogonal to L. To see this, we calculate in local coordinates. Let U_I be the coordinate patch defined at the beginning of Section 9. For simplicity, we assume that $I=(1, \ldots, q)$. Then to

$$\begin{matrix}(1, \ldots, 0, c_{11}, \ldots, c_{1(n-q)}) \\ \vdots \\ (0, \ldots, 1, c_{q1}, \ldots, c_{q(n-q)})\end{matrix}$$

we associate

$$\begin{matrix}(-c_{11}, \ldots, -c_{q1}, 1, 0, \ldots, 0) \\ \vdots \\ (-c_{1(n-q)}, \ldots, -c_{q(n-q)}, 0, \ldots, 1)\end{matrix}$$

which is clearly a holomorphic homeomorphism of U_I onto $U_{\bar{I}}$. We define the holomorphic map ω_1: $[G_{n-1}(\mathbb{C})]^q \to \bigwedge_q \mathbb{C}^n$ by

$$\omega_1(L_1, L_2, \ldots, L_q) = \xi_1 \pi_{n-1}^{-1}(L_1) \wedge \ldots \wedge \xi_1 \pi_{n-1}^{-1}(L_q).$$

Then $\omega_1^{-1}(0)$ is a proper analytic subset of $[G_{n-1}(\mathbb{C})]^q$ and so of measure zero. Finally, we define the holomorphic map ω_2: $[G_{n-1}(\mathbb{C})]^q - \omega^{-1}(0) \to G_{n-q}(\mathbb{C})$ by

$$\omega_2\colon (L_1, \ldots \ldots, L_q) = \pi_{n-q} \xi_q [\xi_1 \pi_{n-1}^{-1}(L_1) \wedge \ldots \wedge \xi_1 \pi_{n-1}^{-1}(L_q)],$$

which is surjective. It follows from Lemma 5.24 that if the $2q(n-q)$ Lebesgue measure of $E \subset G_{n-q}(\mathbb{C})$ is positive, then the product measure of $\omega_2^{-1}(E)$ is positive. □

Theorem 5.34. *Let $Y \subset \mathbb{C}^n$ be an analytic variety of pure dimension p and let r_m be a sequence which increases to infinity. Then for $0 < q \leqq p$*

$$\left\{ L \in G_{n-q}(\mathbb{C})\colon \lim_{m\to\infty} \frac{\sigma_{[Y\cap L]}(k r_m)}{r_m^{-2q} \sigma_{[Y]}(r_m)} = 0 \text{ for all } k > 0 \right\}$$

is of Lebesgue measure zero in $G_{n-q}(\mathbb{C})$.

Proof. We begin by showing by induction on q that

$$\left\{ \tilde{L} = (L_1, \ldots, L_q) \in [G_{n-1}(\mathbb{C})]^q\colon \lim_{m\to\infty} \frac{\sigma_{[Y\cap L_1 \cap \ldots \cap L_q]}(k r_m)}{\sigma_{[Y]}(r_m) \cdot r_m^{-2}} = 0 \text{ for all } k > 0 \right\}$$

is of Lebesgue measure zero for the product measure. For $q=1$, we suppose that

$$\text{measure}\left(E = \left\{ L \in G_{n-1}(\mathbb{C})\colon \lim_{m\to\infty} \frac{\sigma_{[Y\cap L]}(k r_m)}{r_m^{-2} \sigma_{[Y]}(r_m)} = 0 \text{ for all } k > 0 \right\} > 0 \right).$$

Then we can find a compact set $K \subset E \cap U_0$, such that measure $(K) > 0$, where

$$U_0 = \left\{ L \in G_{n-1}(\mathbb{C})\colon L = \left\{ z\colon \sum_{i=1}^{n-1} z_i w_i + z_n = 0 \right\}, w \in \mathbb{C}^{n-1} \right\},$$

Let $f_m^{(t)}(L)=\frac{\sigma_{[Y\cap L]}(tr_m)}{r_m^{-2}\sigma_{[Y]}(r_m)}$, $t\in Z$. Then for every t, $\lim_{m\to\infty} f_m^{(t)}(L)=0$ for $L\in K$. It follows from Egorov's Theorem that we can find $K'\subset K$ compact with measure $(K')>0$ such that $\lim_{m\to\infty} f_m^{(t)}(L)=0$ uniformly on K' (for t fixed). By Lemma 5.32, we can find constants C_1, C_2 and C_3 which depend only on K' such that

$$r_m^{-2}C_1\sigma_{[Y]}(r_m)\leqq C_2\sigma_{[Y]}(2)+\int[\sigma_{[Y]}(w;\, C_3 r_m)-\sigma_{[Y]}(w;\, 1)]\chi_{K'}(w)d\tau(w),$$

which contradicts the fact that for $t\geqq C_3$, $\lim_{m\to\infty} f_m^{(t)}(L)=0$ uniformly on K'. Thus measure $(E)=0$. We now assume the induction hypothesis for $s\leqq q-1$. Suppose that $L\notin E$ as defined above. Then there exists $k_0>0$ and a subsequence $r_{m'}$ (which depend on L) such that $\lim_{m\to\infty}\frac{\sigma_{[Y\cap L]}(k_0 r_{m'})}{(r_{m'})^{-2}\sigma_{[Y]}(r'_m)}\neq 0$. By the induction hypothesis, the set

$$E_L=\left\{\hat{L}=(L_1,\dots,L_{q-1})\in[G_{n-1}(\mathbb{C})]^{q-1}:\right.$$
$$\left.\lim_{m'\to\infty}\frac{\sigma_{[L\cap Y\cap L_1\cap\dots\cap L_q]}(kk_0 r_{m'})}{r_{m'}^{-2(q-1)}\sigma_{[Y\cap L]}(k_0 r_{m'})}=0 \text{ for all } k>0\right\}$$

is of Lebesgue measure zero. The conclusion now follows in the same manner as that of Theorem 5.33. □

Of course, for the purpose of applications, one wants to choose the sequence r_m so that $\sigma_{[Y]}(r_m)$ has "maximal growth". As an example of an illustration, we present the following corollary:

Corollary 5.35. *Let $Y\subset\mathbb{C}^n$ be an analytic variety of pure dimension p such that $r^{-2p}\sigma_Y(r)$ is of finite order ρ. Then the set of $L\in G_{n-q}$, $0<p\leqq q$, such that $r^{-2(p-q)}\sigma_{[Y\cap L]}(r)$ is of order different from ρ is of Lebesgue measure zero in G_{n-q}. If $r^{-2p}\sigma_{[Y]}(r)$ is of normal type with respect to the proximate order $\rho(r)$, then the set $L\in G_{n-q}$ such that $r^{-2(p-q)}\sigma_{[Y\cap L]}(r)$ is of minimal type with respect to $\rho(r)$ is of Lebesgue measure zero in G_{n-q}.*

Proof. It follows from Theorem 5.33 that the set of $L\in G_{n-q}$ for which $r^{-2(p-q)}\sigma_{[Y\cap L]}(r)$ is of order bigger than ρ is of Lebesgue measure zero. Suppose that r_m is an increasing sequence such that $\sigma_{[Y]}(r_m)\geqq\lambda r_m^{\rho(r_m)+2p}$, $\lambda>0$. Then there exists a set $E\subset G_{n-q}$ of Lebesgue measure zero such that for $L\notin E$, we can find a subsequence $r_{m'}$ for which

$$\lim_{m'\to\infty}(r_{m'})^{-2(p-q)}\frac{\sigma_{[Y\cap L]}(r_{m'})}{(r_{m'})^{\rho(r_{m'})}}>0.$$

It then follows from Theorem 1.18 that

$$\limsup_{r\to\infty}\frac{\sigma_{[Y\cap L]}(r)\cdot r^{-2(p-q)}}{r^{\rho(r)}}>0. \qquad \square$$

Historical Notes

The representation with growth conditions of an analytic subvariety X of $\mathbb{C}^n$ of co-dimension greater than one poses certain delicate problems which are only partially resolved. First of all, X is not in general a complete intersection, and secondly even when it is, the Lelong-Poincaré equation $\theta_X = i\partial\bar{\partial}\log|F|$ (studied in Chapter 3) for X a subvariety of co-dimension one (which reduces the problem to a linear equation) is replaced by a nonlinear equation of complex Monge-Ampère type: $\theta_X = (i\partial\bar{\partial}\log\|F\|)^q$ for X a complete intersection of pure dimension q.

Two results which were published the same year (1972) have enriched the theory of holomorphic mappings. One is the counter-example of Cornalba and Shiffman [1], which shows that there is no control (even asymptotically) as a function of $\|F\|$ for the growth of the analytic subvariety $F^{-1}(a)$ of co-dimension two for F an entire holomorphic mapping of $\mathbb{C}^2$ into $\mathbb{C}^2$. An average estimate has been obtained by Carlson [2] and Gruman [8, 12]. The latter shows that the set of a for which no such estimate is possible is pluripolar.

Another result, due to Skoda [2, 3] was thus surprising. He showed that for X an analytic subvariety of $\mathbb{C}^n$ of pure dimension q, $0 \leqq q \leqq n-1$, one can express X as $X = F^{-1}(0)$ for F an entire holomorphic mapping $F: \mathbb{C}^n \to \mathbb{C}^{n+1}$ where $\|F\|$ has an asymptotic growth controlled by that of $\nu_X(r)$, the indicator of X. A comparaison of the results shows the "statistical" nature of the control (which is the point of view adopted by Stoll [8] in his study of the transcendental Bezout problem).

The study of the trace of an analytic subvariety X on linear subspaces (undertaken in § 9) was also motivated by the counter-example of Cornalba and Shiffman [1]. An upper bound outside exceptional sets was first given by Carlson [2] using the Crofton Formula and general properties of positive monotonic functions. Much more surprising was the fact, first shown by Gruman [8, 11, 12], that one could obtain a lower estimate of the trace of X on linear subspaces outside a small exceptional set. In fact, one can obtain estimates outside sets much smaller than those of Lebesgue mesure zero. For results in this direction, we refer the reader to the articles of Molzon, Shiffman and Sibony [1], Alexander [2], Gruman [8, 11, 12], and Molzon [4].

Chapter 6. Application of Entire Functions in Number Theory

As has already been noted, one of the basic motivations for studying entire functions of finite order is that all of the familiar transcendental functions fall into this category (and what is more, for meromorphic functions of finite order, one arrives at the familiar elliptic functions). Thus, by studying the algebraic or arithmetic properties of entire functions of finite order, one implicity studies the algebraic or arithmetic properties of the fundamental transcentental functions. This in turn has a wide variety of applications in transcendental number theory.

We give here an illustration where the methods of holomorphic functions of several complex variables were used to solve a problem which contains several classical problems of number theory.

The following method uses an idea which goes back to C.L. Siegel and consists of constructing an entire fonction with "many zeros" in the class of those whose values have given algebraic properties. We present here the solution of E. Bombieri in $\mathbb{C}^n$ given in connection with his joint paper with S. Lang. The solution relies heavily on the technics of positive closed currents and the resolution of the operator $\partial\bar{\partial}$.

§ 1. Preliminaries from Number Theory

A complex number α is said to be *algebraic of degree p* if there exists an irreducible polynomial $P(x)=a_0 x^p+\ldots+a_p$ with a_k a real integer and $a_0 \neq 0$ such that $P(\alpha)=0$. If we suppose that the a_k have no common divisor, then P is unique and is called the *minimal polynomial* for α. If $a_0=1$, we say that α is an *algebraic integer*. A complex number which is not algebraic is said to be *transcendental*. We denote by Q the field of rational numbers.

If K is a subfield of $\mathbb{C}$, we say that K is an *extension of finite type* if there exist $y_1, \ldots, y_m \in K$ such that $K=Q(y_1, \ldots, y_m)$. If K is a vector space of finite dimension over Q, we denote this dimension by $[K:Q]$. If K, an extension of finite type, is a field of algebraic numbers, then it follows from the Theorem of the Primitive Element that K is then *simple*, that is $K=Q(\alpha)$ for some algebraic α, and in addition we then have $Q(\alpha)=Q[\alpha]$, that is the field generated by adding α to the rationals is the same as the ring generated by adding α to the rationals. We then say that K is a *number field.*

Proposition 6.1. *Let α be an algebraic number. Then the following two properties are equivalent:*

i) *there exists a non-trivial unitary polynomial $Q(x) \in \mathbb{Z}[x]$ such that $Q(\alpha)=0$;*

ii) *there exists a non-trivial $\mathbb{Z}$-module M generated by a finite number of algebraic elements such that $\alpha M \subset M$.*

Proof. If p is the degree of Q in i), then $M=\mathbb{Z}[\alpha, \ldots, \alpha^p]$ is a non trivial $\mathbb{Z}$-module of finite dimension such that $\alpha M \subset M$, so i) implies ii). To see that ii) implies i), we let $v_1, \ldots, v_m$ be a basis for M. Then $\alpha v_i = \sum_{j=1}^{m} a_{ij} v_j$ with $a_{ij} \in \mathbb{Z}$. Let $A=[a_{ij}]_{i=1,\ldots,m}^{j=1,\ldots,m}$. Then $\det(A-\alpha I)$ annihilates the module M, so this determinant must be zero. Then $P(x)=\det(A-xI)$ is a polynomial with integral coefficients such that $P(\alpha)=0$. □

It follows from ii) above that the set of algebraic integers is a ring, since $(\alpha+\beta) MN = \alpha MN + \beta MN \subset MN$ and $\alpha\beta MN \subset MN$. It also follows from ii) that $\lambda\alpha$ is an algebraic integer for α an algebraic integer and $\lambda \in \mathbb{Z}$, since $\lambda\alpha M = \alpha(\lambda M) = \alpha M \subset M$. *For* α an algebraic number, we let D_α be the ideal in $\mathbb{Z}$ defined by

$$D_\alpha = \{\lambda \in \mathbb{Z} : \lambda\alpha \text{ is an algebraic integer}\}.$$

This ideal is non-zero, since if $Q(x)=\sum_{i=0}^{n} c_i x^i$ is the minimal polynomial of α, then $c_i \alpha \in D_\alpha$, for

$$Q(x) = x^n + c_{n-1} x^{n-1} + c_{n-2} c_n x^{n-2} + \ldots + c_0 c_n^{n-2} = \sum_{j=0}^{n} c_j c_n^{n-j-1} x^j$$

satisfies $Q(c_n \alpha)=0$, and hence (ii) of Proposition 6.1 is fulfilled. A positive element of D_α is called *a denominator* for α, and the positive generator $d(\alpha)$ of D_α is called *the denominator of* α.

Let α be an algebraic number and $P(x)$ its irreducible polynomial. Denote by $\alpha_1, \ldots, \alpha_m$ the complex roots of P (which are all different, since P is an irreducible polynomial in $Q[x]$). Thus $P(x)=\prod_{j=1}^{n}(x-\alpha_j)$. The α_j are said to be the conjugates of α, and we set $\overline{|\alpha|} = \max_{1 \leq j \leq m} |\alpha_j|$. We then define the *size* of α, $s(\alpha)$, by $s(\alpha) = \max(\log \overline{|\alpha|}, \log d(\alpha))$.

Proposition 6.2. *Let K be a number field. Then for $\alpha \in K$, $\alpha \neq 0$,*

$$-2[K:Q]\, s(\alpha) \leq -[K:Q] \log d(\alpha) - ([K:Q]-1) \log \overline{|\alpha|} \leq \log |\alpha|.$$

Proof. Let m be the degree of the irreducible polynomial for α, so that $m \leqq [K:Q]$. Notice that the irreducible polynomial for $d(\alpha)\cdot\alpha$ is

$$\tilde{P}(x) = x^m + a_{m-1} d(\alpha) x^{m-1} + a_{m-2} d(\alpha)^2 x^{m-2} + \ldots + a_0 d(\alpha)^m$$

(where $x^m + a_{m-1} x^{m-1} + a_{m-2} x^{m-2} + \ldots + a_0$ is the irreducible polynomial for α). Hence $\tilde{P}(x) = \prod_{j=1}^{m} (x - d(\alpha)\alpha_j)$. Since the coefficients of $\tilde{P}(x)$ are integral, we have $\prod_{j=1}^{m} d(\alpha)\cdot|\alpha_j| \geqq 1$, from which the conclusion follows. □

We shall need the following technical lemma from linear algebra.

Lemma 6.3. *Let A be a non-zero subgroup of $\mathbb{R}^n$. Assume that in any bounded region of $\mathbb{R}^n$ there exists only a finite number of elements of A. Let m be the maximal number of elements of A which are linearly independent over $\mathbb{R}$. Then we can select m elements of A which are linearly independent over $\mathbb{R}$ which form a $\mathbb{Z}$-basis for A.*

Proof. Let $\{w_1, \ldots, w_m\}$ be a maximal set of elements of A linearly independent over $\mathbb{R}$. We proceed by induction on $m \leqq n$. Suppose that $m = 1$. Let $\tilde{w} = \tilde{t} w_1$, where $|\tilde{t}|$ is the smallest possible non-zero value. Suppose that $w \in A$, and let t be such that $w = t w_1$. Choose $q \in \mathbb{Z}$ such that $q\tilde{t} \leqq t < (q+1)\tilde{t}$. Then $w - q\tilde{w} = (t - q\tilde{t}) w_1 \in A$ and $0 \leqq t - q\tilde{t} < (q+1)\tilde{t} - q\tilde{t} = \tilde{t}$, which contradicts the choice of $\tilde{t}$ unless $t = q\tilde{t}$, in which case $\tilde{w}$ is a $\mathbb{Z}$-basis for A.

Suppose now that $m > 1$. *Let* V be the vector space over $\mathbb{R}$ generated by $\{w_1, \ldots, w_m\}$ and let V_{m-1} be the space generated by $\{w_1, \ldots, w_{m-1}\}$. Set $A_{m-1} = A \cap V_{m-1}$. Then in any bounded region of V_{m-1}, there exists only a finite number of elements of A_{m-1}. By the induction hypothesis, we can find $\{w'_1, \ldots, w'_{m-1}\}$ which form a $\mathbb{Z}$-basis for A_{m-1}.

Let S be the set of elements of A which can be written in the form $\sum_{i=1}^{m} t_i w'_i$ with $0 < t_i \leqq 1$ if $i = 1, \ldots, m-1$, $w'_m = w_m$ and $0 \leqq t_m \leqq 1$. This set is certainly bounded and hence contains only a finite number of elements (including w_m). We select an element v_m in the set whose last coordinate t_m is the smallest possible non-zero value. We shall show that $\{w'_1, \ldots, w'_{m-1}, v_m\}$ is a $\mathbb{Z}$-basis for A.

We write $v_m = c_1 w'_1 + \ldots + c_m w_m$, $0 \leqq c_m \leqq 1$, $c_j \in \mathbb{R}$. Suppose that $v \in A$ and write $v = \sum_{j=1}^{m} x_j w'_j$, $x_j \in \mathbb{R}$. Let q_m be the integer such that $q_m c_m \leqq x_m < (q_m + 1) c_m$. Then the last coordinate of $v - q_m v_m$ with respect to $\{w'_1, \ldots, w'_m\}$ is equal to $x_m - q_m c_m$ and $0 \leqq x_m - q_m c_m < (q_m + 1) c_m - q_m c_m \leqq c_m \leqq 1$. Let q_i be integers such that $q_i \leqq x_i < q_i + 1$, $i = 1, \ldots, m-1$. Then $v - q_m v_m - q_1 w'_1 - \ldots - q_{m-1} w'_{m-1} \in S$. If its last coordinate is not zero, then it would be an element whose last coordinate is smaller than c_m, which would contradict the construction. Hence, its last coordinate is zero and it is in V_{m-1}. By the induction hypothesis, it can be written as a linear combination of $\{w'_1, \ldots, w'_{m-1}, v_m\}$ with integral coefficients. Furthermore, it is clear that $w'_1, \ldots, w_{m-1}, v_m$ are linearly independent over $\mathbb{R}$. □

Lemma 6.4. *Let K be a number field with $[K:Q]=n$. Then we can find a $\mathbb{Z}$-basis for I_K, the algebraic integers in K, of dimension n.*

Proof. By the Theorem of the Primitive Element, there exist exactly n embeddings $\sigma_1, \ldots, \sigma_n$ of K into $\mathbb{C}$. We map I_K into $\mathbb{C}^n$ by $\tau: \alpha \to (\sigma_1(\alpha), \ldots, \sigma_n(\alpha))$, which is an additive embedding and hence its image is an additive group. In any bounded region of $\mathbb{C}^n = \mathbb{R}^{2n}$, there are only a finite number of elements of $\tau(I_K)$, for in any bounded region, the $\sigma_j(\alpha)$ are bounded for $\alpha \in I_K$. Each such α is the root of a polynomial $(x-\sigma_1(\alpha)) \ldots (x-\sigma_n(\alpha))$ whose coefficients are integral, since on the one hand they are algebraic integers and on the other hand, being symmetric in the conjugates, they are rational. By Lemma 6.3, if m is the number of elements of I_K which are independent over $\mathbb{R}$, then we can find a $\mathbb{Z}$-basis for I_K of dimension m. Since $[K:Q]=n$, $m \leqq n$ because a basis $(w_1, \ldots, w_n)$ for K over Q is a basis for the Q-vector space generated by I_K (i.e. for $\alpha \in I_K$,

$$\alpha = \sum_{i=1}^{n} c_i w_i = \sum_{i=1}^{n} \frac{c_i}{d(w_i)} d(w_i) w_i, \qquad c_i \in Q,$$

and $d(w_i)w_i \in I_K$). On the other hand, $m \geqq n$ since a $\mathbb{Z}$-basis for I_K is a Q-basis for K. □

Lemma 6.5. *Let $y_j = \sum_{i=1}^{n} a_{ij} x_i$, $j=1, \ldots, m$, have integral coefficients $a_{ij} \in \mathbb{Z}$ and suppose $m<n$ and $|a_{ij}|<A$. Then there exists a non-trivial solution $x_1, \ldots, x_n$ in $\mathbb{Z}^n$ of $y_j=0$, $j=1, \ldots, m$, with $|x_i| < 1+(nA)^{\frac{m}{n-m}}$.*

Proof. As the x_i take on the $(2M+1)$ integral values between $-M$ and M, we obtain $(2M+1)^n$ points in $\mathbb{R}^m$ contained in the cube $-nAM \leqq y_j \leqq nAM$. Since there are exactly $(2nAM+1)^m$ different points with integral coordinates in this cube, it follows that there exist two different points $(x_1, \ldots, x_n)$ and $(x'_1, \ldots, x'_n)$ having the same image as soon as $(2nAM+1)^m < (2M+1)^n$. Then $x''_i = x_i - x'_i$ gives a non-trivial solution with $|x''_i| \leqq 2M$. We choose for $2M$ the even integer in the interval $(nA)^{\frac{m}{n-m}} - 1 \leqq 2M \leqq (nA)^{\frac{m}{n-m}} + 1$ so that

$$\begin{aligned}(2nAM+1)^m < (nA)^m (2M+1)^m &\leqq (2M+1)^{n-m}(2M+1)^m \\ &= (2M+1)^n.\end{aligned}$$

□

Lemma 6.6. *Let K be a number field with $[K:Q]=s$. Let $y_j = \sum_{i=1}^{n} a_{ij} x_j$, $j=1, \ldots, m$ with $n > ms$ and suppose that the a_{ij} are algebraic integers in K*

with $\overline{|a_{ij}|} < A$. *Then there exists* $x_i \in \mathbb{Z}$ *not all zero such that* $\sum_{i=1}^{n} a_{ij} x_i = 0$, $j = 1, \ldots, m$ *and* $|x_i| < 1 + (cnA)^{\frac{m}{n-ms}}$, *where* c *depends only on* K.

Proof. Let $w_1, \ldots, w_s$ be a basis for I_K over $\mathbb{Z}$, which exists by Lemma 6.4. Let $\sigma_1, \ldots, \sigma_s$ be the s embeddings of K into $\mathbb{C}$. For $a \in I_K$, let $\tau(a) = (\sigma_1(a), \ldots, \sigma_s(a))$, which embeds I_K as an additive subgroup of $\mathbb{C}^s = \mathbb{R}^{2s}$. Then $s_k = \tau(w_k)$, $k = 1, \ldots, s$, are linearly independent over the real numbers. Since $a = \sum_{k=1}^{s} \alpha^{(k)} w_k$, $\tau(a) = \sum_{k=1}^{s} \alpha^{(k)} s_k$ and so there exists a constant c depending only on K (or rather on the basis $(w_1, \ldots, w_s)$) such that

$$\overline{|a|} = \sup_{1 \leqq i \leqq s} |\sigma_i(a)| \geqq c^{-1} \sup_{1 \leqq i \leqq s} |\alpha^{(i)}|.$$

In particular, we can write $a_{ij} = \sum_{k=1}^{s} \alpha_{ij}^{(k)} w_k$ with $\alpha_{ij}^{(k)} \in \mathbb{Z}$ and $|\alpha_{ij}| < cA$. We now apply Lemma 6.5 to the ms equations

$$\sum_{i=1}^{n} \alpha_{ij}^{(k)} x_i = 0, \qquad j = 1, \ldots, m, \; k = 1, \ldots, s. \qquad \square$$

If $P(x_1, \ldots, x_m) = \sum_{|\alpha| \leqq d} c_\alpha x^\alpha$ is a polynomial with complex coefficients, we set $|P| = \max |c_\alpha|$. If the coefficients are algebraic numbers, we set $\|P\| = \max \overline{|c_\alpha|}$, and we define the *size* of P by $s(P) = \max(\deg P, \log \|P\|)$.

Let $P(x_1, \ldots, x_m) = \sum c_\alpha x^\alpha$ be a polynomial with complex coefficients c_α and $Q(x_1, \ldots, x_m) = \sum a_\alpha x^\alpha$ be a polynomial with real non-negative coefficients. We say that Q *dominates* P if $|c_\alpha| \leqq a_\alpha$ for all α, and we denote this by $P < Q$. It is immediate that if $P_1 < Q_1$ and $P_2 < Q_2$, then

i) $P_1 + P_2 < Q_1 + Q_2$

ii) $P_1 P_2 < Q_1 Q_2$

iii) $\dfrac{\partial P_1}{\partial x_j} < \dfrac{\partial Q_1}{\partial x_j}$ $j = 1, \ldots, m$.

Furthermore $P < |P|(1 + x_1 + \ldots + x_m)^{\deg P}$.

Lemma 6.7. *Let* K *be a number field. Let* $f_1, \ldots, f_m$ *be functions of* n *complex variables holomorphic in a neighborhood* U *of* $q \in \mathbb{C}^n$ *such that* $f_i(q) \in K$ *for all* i *and such that the ring* $K[f_1, \ldots, f_m]$ *is mapped into itself by the derivatives* $D_1 = \dfrac{\partial}{\partial z_1}, \ldots, D_n = \dfrac{\partial}{\partial z_n}$. *Then there exists a constant* C_0 *such that for every polynomial* $Q(x_1, \ldots, x_m)$ *with coefficients in* I_K *and of degree at most* d,

$$\|D_1^{k_1} \ldots D_n^{k_n}(Q(f_1, \ldots, f_m))(q)\| \leqq \|Q\| \, (d + |k|)^{|k|} \, C_0^{|k|+d}.$$

Furthermore, there is a denominator for $D_1^{k_1} \ldots D_n^{k_n}(Q(f_1, \ldots, f_m)(q))$ *bounded by* $C_0^{|k|+d}$.

Proof. There exist polynomials $P_{ij}(x_1, \ldots, x_m)$ with coefficients in K such that $D_j f_i = P_{ij}(f_1, \ldots, f_m)$. Let δ be the maximum of the degrees of P_{ij}, $i=1, \ldots, m$, $j=1, \ldots, n$. For any polynomial $P \in K[x_1, \ldots, x_m]$, we set

$$\hat{D}_j P(x_1, \ldots, x_m) = \sum_{i=1}^{m} \frac{\partial P}{\partial x_j}(x_1, \ldots, x_m) P_{ij}(x_1, \ldots, x_m).$$

Then $P < \|P\|(1+x_1+\ldots+x_m)^d$ and

$$P_{ij} < \|P_{ij}\|(1+x_1+\ldots+x_m)^{\delta} \quad \text{so} \quad \hat{D}_j P < \|P\| \, \tilde{C}_0 d(1+x_1+\ldots+x_m)^{d+\delta}.$$

By iteration, one then obtains a dominating polynomial

$$\hat{D}_1^{k_1} \ldots \hat{D}_n^{k_n} P < \|P\| \, \tilde{C}_0^{|k|}(d+|k|)^{|k|}(1+x_1+\ldots+x_m)^{d+|k|\delta}.$$

We now substitute the values $f_i(q)$ for x_i to obtain the first estimate.

To prove the result on the denominator, we proceed by induction on $|k|$. For $|k|=0$, the result is trivial. Thus, we assume the result for all $|k|$ up to $j_0 - 1$. Let α be the common denominator of $f_1(q), \ldots, f_m(q)$. Then $\alpha^{d+|k|\delta}$ is a common denominator for $\hat{D}_1^{k_1} \ldots \hat{D}_n^{k_n}(Q(f_1, \ldots, f_m)(q))$. □

If F is a field and A is a ring containing F, we say that the elements $\{x_1, \ldots, x_m\}$ are *algebraically dependent* over F if there is a polynomial P with coefficients in K such that $P(x_1, \ldots, x_n)=0$; otherwise, we say that $\{x_1, \ldots, x_m\}$ are *algebraically independent*. A subset E of A is algebraically independent if every finite subset of E is algebraically independent. If L is an extension of a field F, a subset B is a *transcendence basis* for L if B is a maximal set of algebraically independent elements of L over K. The *transcendence degree* of L is the number of elements in this basis (perhaps infinite).

§2. A Schwarz Lemma

Lemma 6.8. *Let $F(z)$ be holomorphic in a neighborhood of $\{z: \|z\| \leqq R\}$ in $\mathbb{C}^n$. For $0 \leqq r \leqq R$ set $M(r) = \sup_{\|z\|=r} \log |F(z)|$. Then*

$$M(r) \leqq M(R) - v(r) \log \frac{R^2+r^2}{(4n-2)Rr}$$

where $\tau = \frac{i}{\pi} \partial \bar{\partial} \log |F|$ and $v(r)$ is the indicator of τ.

Proof. By the Riesz representation of the $\mathbb{R}^{2n}$-subharmonic function $\log|F|$, for $z \in B(0, R)$, and $g(z, a)$ the Green function of the ball $B(0, R)$, we have

$$\log|F(z)| = H_R(z) - \omega_{2n}^{-1} \int d\sigma(a)\, g(a, z);$$

H_R is the harmonic function which takes on the values $\log|F|$ for $\|z\|=R$; $H_R(z)\leqq M(R)$, and $\sigma=\dfrac{1}{2\pi}\Delta\log|F|$ is the trace-measure of τ. Let $a'\in\mathbb{C}^n$ be on the line $0a$ such that $\|a\|\,\|a'\|=R^2$. Then for $\|a\|=t$ and $\|z\|=r$,

$$g(z,a)=\|a-z\|^{2-2n}-\left(\frac{\|a\|}{R}(\|z-a'\|)\right)^{2-2n}$$

from which we obtain

$$0\leqq g(z,a)\leqq g'(r,t)=(r+t)^{2-2n}-\left(R+\frac{r}{R}t\right)^{2-2n}$$

and

$$M(r)\leqq M(R)-\omega_{2n}^{-1}\int_0^R g'(r,t)\,d\sigma(t).$$

Since $g'(r,t)=0$ for $t=R$, we obtain after an integration by parts:

$$M(r)\leqq M(R)-\omega_{2n}^{-1}\int_0^R \sigma(t)\frac{\partial g'}{\partial t}(r,t)\,dt.$$

Thus if we replace $\sigma(t)\tau_{2n-2}t^{2n-2}$ by $v(0,t)$, we obtain

$$M(r)\leqq M(R)-\int_0^R \frac{v(t)}{t}\left[\left(\frac{t}{t+r}\right)^{2n-1}-\frac{r}{R}\left(\frac{Rt}{rt+R^2}\right)^{2n-1}\right]dt.$$

For the first integral, we obtain

$$A(r)=\int_0^R \frac{v_\tau(t)}{t}\left(\frac{t}{t+r}\right)^{2n-1}dt\geqq v_\tau(r)\int_r^R\left(\frac{t}{t+r}\right)^{2n-1}\frac{dt}{t}$$

$$\geqq v(r)\int_{1/2}^{\alpha}\frac{u^{2n-2}}{1-u}\,du$$

and $\frac{1}{2}<\alpha=\dfrac{R}{R+r}<1$. To calculate the integral we write

$$I_n=\left[-\log(1-u)-\left(u+\frac{u^2}{2}+\dots+\frac{u^{2n-2}}{2n-2}\right)\right]_{1/2}^{\alpha}$$

$$I_n\geqq\log\frac{R+r}{2r}+\sum_{k=1}^{2n-2}k^{-1}[2^{-k}-1].$$

For $n\geqq 3$, using $\int_{k-1/2}^{k+1/2}t^{-1}dt>\dfrac{1}{k}$, we obtain

$$I_n\geqq\sum_1^4 k^{-1}(2^{-k}-1)+\int_{9/2}^{2n-3/2}t^{-1}dt\geqq-\tfrac{17}{12}+\log 9-\log(4n-3)$$

or $A(r)\geqq\log\dfrac{R+r}{(4n-3)r}$.

The formula is proved to be true for $n=2$ by direct calculation.
For $B(r)$ the second integral, we obtain a bound

$$B(r) \leqq v(r)\left(\frac{R}{r}\right)^{2n-2} \int_{\lambda}^{\mu} \frac{u^{2n-2}}{1-u}\, du$$

$$0<\lambda=\frac{r^2}{R^2+r^2}<\mu=\frac{r}{R+r}<1$$

so that

$$B(r) \leqq v(r)\left(\frac{R}{r}\right)^{2n-2} \mu^{2n-2} \log \frac{1-\lambda}{1-\mu} = v(r)\left(\frac{R}{R+r}\right)^{2n-2} \log \frac{R(R+r)}{R^2+r^2}.$$

If we replace μ^{2n-2} by 1, we obtain for $n \geqq 2$

$$A(r)-B(r) \geqq v(r) \log \frac{R^2+r^2}{(4n-3)Rr} \geqq v(r) \log \frac{R^2+r^2}{(4n-2)Rr}.$$

To obtain a bound which is valid for $n=1$, it is sufficient to replace $4n-3$ by $4n-2$; moreover, we note that for the large values of n, a bound for $M(r)$ is obtained only for large values of $\frac{R}{r}$. □

§3. Statement and Proof of the Main Theorem

With these preliminaries out of the way, we are now in a position to undertake the proof of the main result. We will say that a meromorphic function is of finite order at most ρ is it is the quotient of two entire functions of finite order at most ρ.

Theorem 6.9. *Let K be a number field and let $f=(f_1, \dots, f_m)$ be meromorphic functions in $\mathbb{C}^n$ of finite order ρ. Assume furthermore that*

i) the transcendence degree of $K[f]$ is at least $(n+1)$ (that is, there are at least $(n+1)$ of the f_i's which are algebraically independent over K);

ii) the partial derivatives $\frac{\partial}{\partial z_i}$, $i=1, \dots, n$, map the ring $K[f]$ into itself.

Then the set of points $\xi \in \mathbb{C}^n$ where $f(\xi)$ is finite and in K^m is contained in an algebraic hypersurface of degree at most $n(n+1)\rho[K:Q]$.

Remark 1. Condition ii) can be relaxed so as to include the case where the field $K(f)$ is mapped into itself by $\frac{\partial}{\partial z_i}$, $i=1,, \dots, n$, for if $\frac{\partial f_j}{\partial z_i}=\frac{P_{ij}(f)}{Q(f)}$ with $P_{ij}[x]$, $Q[x] \in K[x]$, we set $f_{m+1}=Q(f)^{-1}$. Then $\tilde{f}=(f_1, \dots, f_{m+1})$ satisfies $\frac{\partial f_j}{\partial z_i} \in K[\tilde{f}]$, $i=1,, \dots, m+1$, and we can apply Theorem 6.9 with the additional restriction that $Q(f(\xi)) \neq 0$.

Remark 2. If $K \subset L \subset M$ are three fields, then $\dim_K M = \dim_K L + \dim_L M$. In particular, $K(f_1, f_2) = K(f_1)(f_2)$, etc. Thus, we can find f_j, $j = i_1, \ldots, i_{n+1}$ such that $f_{i_1}, \ldots, f_{i_{n+1}}$ are algebraically independent. We assume without loss of generality that these are $f_1, \ldots, f_{n+1}$.

Let S be a finite set of points ξ_i, $i = 1, \ldots, t$, in $\mathbb{C}^n$ for which $f(\xi)$ is finite and in K^m. Let $j = (j_1, \ldots, j_{m+1})$ be a multi-index and consider the function $F(z) = \sum_{0 \leqq j_\lambda \leqq J} a_j f_1^{j_1} \ldots f_{n+j}^{j_{n+1}}$. We use Lemma 6.6 to choose integers a_j in K such that $D^\lambda F(\xi) = 0$ for $\lambda = (\lambda_1, \ldots, \lambda_n)$, $\sup \lambda_i < L$, $\xi \in S$, with $\overline{|a_j|}$ not too large. In what follows, we will denote by $\{x\}$ the gratest integer smaller than or equal to x.

Lemma 6.10. *If* $J^{n+1} = [K:Q]\{t L^n \log L\}$, *then we can find integers* a_j *not all zero such that* $D^\lambda F(\xi) = 0$ *for all* λ *with* $\sup \lambda_j < L$ *and all* $\xi \in S$ *such that* $\log |a_j| < L + o(L)$, *where* o *depends only on* S *and* F.

Proof. By Lemma 6.7, $\|D^\lambda f_1^j \ldots f_{n+1}^{j_{n+1}}(\xi)\| < \{(n+1)J + L\}^L C_1^{(n+1)J+L}$ for $\xi \in S$, $\sup \lambda_j < L$, $\sup j_\alpha < J$. Furthermore, we can find a common denominator Δ for $D^\lambda f_1^{j_1} \ldots f_{n+1}^{j_{n+1}}(\xi)$ such that $\Delta \leqq C_1^{(n+1)J+L}$, where C_1 depends on f and S but not J and L. Solving for $D^\lambda F(\xi) = 0$ is equivalent to solving $t \binom{L+n}{n}$ equations $\sum a_j (\Delta D^\lambda f_1^{j_1} \ldots f_{n+1}^{j_{n+1}}(\xi)) = 0$ in J^{n+1} unknowns a_j. Since

$$\|\Delta D^\lambda f_1^{j_1} \ldots f_{n+1}^{j_{n+1}}(\xi)\| < ((n+1)J + L)^L C_1^{2(n+1)J+2L},$$

by Lemma 6.6, we can find integers a_j not all zero with

$$\log |a_j| < \frac{\tau}{(\eta - \tau)} (L \log (L+J) + J) + o(L)$$

with $\tau = t \binom{L+n}{n}$ and $\eta = J^{n+1}$. If we choose $J^{n+1} = [K:Q]\{t L^n \log L\}$ then $\frac{\tau}{(\eta-\tau)} \leqq (\log L)^{-1}$. □

For every L, we let $\gamma(L)$ be the integer such that $D^\sigma F(\xi) = 0$ for $\max \sigma_j < \gamma$, $\xi \in S$, but there exists $\xi' \in S$ and σ' with $\sigma'_j = \gamma$ such that $D^{\sigma'} F(\xi) \neq 0$. By the construction of Lemma 6.10, $L \leqq \gamma < +\infty$ (since $\gamma = +\infty$ would imply $F(z) \equiv 0$, contradicting the algebraic independence of the f_i). Let $g(z)$ be an entire function of order ρ such that $g f_j$ is entire for $j = 1, \ldots, n+1$ and of order at most ρ, and set $G_\gamma(z) = g(z)^{(n+1)J} F(z)$.

Lemma 6.11. *There exist* σ' *with* $\max \sigma'_j = \gamma$ *and* $\xi' \in S$ *such that*

$$\log |D^\sigma G_\gamma(\xi')| \geqq -([K:Q] - 1) \gamma \log \gamma + O(\gamma)$$

where $O(\gamma)$ *depends only on* f *and* S.

Proof. $D^{\sigma'} G_\gamma(\xi') = g(\xi')^{(n+1)J} D^{\sigma'} F(\xi')$ since $D^\sigma F(\xi')=0$ for $\max \sigma_j < s$. But $\xi = D^{\sigma'} F(\xi') \in K$ and is not zero for some σ' and some ξ', so

$$\begin{aligned}\log |D^{\sigma'} G_\gamma(\xi')| &= (n+1)J \log |g(\xi')| + \log |\xi| \\ &\geqq O(J) - ([K:Q]-1)s(\xi) + O(\log d(\xi))\end{aligned}$$

by Proposition 6.2. The result now follows from Lemmas 6.7 and 6.10. □

Proof of Theorem 6.9. Let $T_\gamma = \frac{i}{\pi\gamma} \log |G_\gamma(z)|$. Suppose that $r \geqq \|\xi'\| + n$. Then $\{\xi : |\xi_i| < 1, i = 1, \ldots, n\} \subset B(0,r)$ and by Cauchy's Inequality

$$|D^\sigma G_\gamma(\xi')| \leqq \gamma^\gamma \max_{\|z\|=r} |G_\gamma(z)|,$$

or equivalently

$$\log |D^{\sigma'} G_\gamma(\xi')| \leqq \gamma \log \gamma + \max_{\|z\|=r} \log |G_\gamma(z)|.$$

Since $v_{\gamma T_\gamma} = \gamma v_{T_\gamma}$, by Lemma 6.8, we have

$$\log |D^{\sigma'} G_\gamma(\xi')| \leqq \gamma \log \gamma + \max_{\|z\|=R} \log |G_\gamma(z)| - v_{T_\gamma}(0,r) \cdot \gamma \log \frac{R}{4nr}$$

for $R \geqq \|\xi'\| + n$.

Since g and each f_j are of order ρ, on $\|z\| = R$

$$\log |G_\gamma(z)| \leqq C J R^{\rho+\varepsilon} + n \log (J+L).$$

Let $R = \gamma^\alpha$ with $\alpha < \{(n+1)\rho\}^{-1}$. Then $R^{\rho+\varepsilon} < \gamma^{(n+1)^{-1}}$ for ε small enough. Thus

$$J \leqq t L^{\frac{n}{n+1}} (\log L)^{1/n+1} \leqq t (\log \gamma)^{\frac{1}{n+1}} \gamma^{\frac{n}{n+1}},$$

and so $\max\limits_{\|z\|=R} \log |G_\gamma(z)| = o(\gamma \log \gamma)$ for $R = \gamma^\alpha$, $\alpha < \{(n+1)\rho\}^{-1}$.

If r is fixed, $r > \|\xi'\| + n$ and γ is sufficiently large (depending on r and hence S), we obtain

$$\log |D^{\sigma'} G_\gamma(\xi')| \leqq \gamma \log \gamma + o(\gamma \log \gamma) - \gamma v_{T_\gamma}(0,r) [\gamma \log \gamma + C].$$

By Lemma 6.11, T_γ satisfies

i) $v_{T_\gamma}(0,r) \leqq [(n+1)\rho + o(1)][K:Q]$ for every r and γ sufficiently large depending on r;

ii) $v_{T_\gamma}(\xi', r) \geqq 1$ for $\xi' \in S$.

By i), the total mass of v_{T_s} is uniformly bounded in every compact subset of $\mathbb{C}^n$ as $\gamma \to \infty$, and so the sequence T_γ has a subsequence T_μ which converges weakly to a positive closed current T. For K a compact subset of $\mathbb{C}^n$, $\int_K d\sigma_T = \lim\limits_{\mu\to\infty} \int_K d\sigma_{T_\mu}$. In particular, this implies that $v_T(\xi') \geqq 1$ for $\xi' \in S$ and $v_T(a,r) \leqq (n+1)\rho[K:Q]$ for r large.

By translating if necessary, we may assume that $r^{-1} v_T(0,r)$ is integrable in a neighborhood of the origin. We then use the result of Chapter 3 and set

$$V(z) = \frac{(n-2)!}{2\pi^{n-1}} \int_{\xi \in \mathbb{C}^n} \left(\frac{1}{\|\xi\|^{2n-2}} - \frac{1}{\|\xi - z\|^{2n-2}} \right) d\sigma_T(\xi),$$

where $\sigma_T = T \wedge \frac{\beta^{n-1}}{(n-1)!}$, so that $V(z)$ is a plurisubharmonic function (cf. Theorems 3.17 and 3.26) with $V(z) \leqq (n+1)\rho[K:Q] \log \|z\|$ when $\|z\| \to \infty$ and $v_V(\xi) \geqq 1$ for $\xi \in S$. In particular, this implies that $\exp - 2nV(z)$ is non integrable in a neighborhood of $\xi \in S$. Thus, by Theorem 5.12, there exists a non-trivial entire function $F(z)$ such that

$$\int_{\mathbb{C}^n} \frac{|F(z)|^2 \exp - 2nV(z)}{(1+\|z\|^2)^{n+\varepsilon}} d\tau(z) < +\infty.$$

Since $\exp - 2nV(z)$ is non-integrable for $\xi \in S$ we obtain, $F(\xi) = 0$ for $\xi \in S$. For large $\|z\|$,

$$|F(z)|^2 (1+\|z\|^2)^{-n-\varepsilon} \exp - 2nV(z) \geqq |F(z)|^2 \|z\|^{-2n(n+1)\rho[K:Q]-2n-2\varepsilon}$$

for $\|z\| > R_\varepsilon$, so

$$\int_{\mathbb{C}^n} |F(z)|^2 (1+\|z\|^2)^{-n(n+1)\rho[K:Q]-n-\varepsilon} d\tau(z) < +\infty.$$

Thus, $F(z)$ is a polynomial of degree at most $n(n+1)\rho[K:Q]$, for if $F(z) = \sum C_\alpha z^\alpha$, since $z^\alpha, \bar{z}^\beta$ are orthogonal on the boundary of the ball of radius R for $\alpha \neq \beta$,

$$\int_{\mathbb{C}^n} |F(z)|^2 (1+\|z\|^2)^{-q} d\tau(z) = \sum_\alpha \int_0^\infty \frac{t_\alpha |C_\alpha|^2 R^{2|\alpha|} R^{2n-1}}{(1+R^2)^q} dR$$

for $t_\alpha = \int_{bd B(0,1)} |z^\alpha|^2 d\tau$, and this last sum is finite only if $C_\alpha = 0$ for $\alpha > q - n$. Thus, F is a polynomial P of bounded degree such that P vanishes on S. By multiplying by a constant, we may assume $|P| = 1$. We now choose an increasing sequence S_l such that $\bigcup_l S_l$ is dense in S and P_l the corresponding polynomials as constructed above. Then $\deg P_l \leqq n(n+1)\rho[K:Q]$. We can then find a subsequence which converges to a polynomial $\tilde{P} \not\equiv 0$ and $\tilde{P}(S) = 0$. □

Historical Notes

For $n=1$, Theorem 6.9 is known as the Schneider-Lang criterion, and one can deduce many famous results on transcendence from it: that e^α is transcendental for α algebraic (Hermite-Lindemann), that α^β is transcendental for α algebraic, β algebraic and not rational (Gelford-Schneider) (cf. the book by Waldschmidt [1]). Lang [1] proved an n dimensional version when the set S was a product of one dimensional sets. Since for $n=1$, card S is finite, Nagata conjectured that for $n \geqq 2$, the set S was contained in an algebraic hypersurface. This was proved by Bombieri [1] in 1970 using the properties of a closed positive current θ and the representation of the solution V of the equation $i\partial\bar{\partial}V=\theta$. The proof used the Schwarz Lemma for n variables given in a joint paper of Bombieri and Lang [1], which was written a short time before (cf. Lemma 6.8). The paper of Bombieri indicated implicitly the way to prove that the sets of density $v_t(z)\geqq c>0$ for t a positive closed current were analytic varieties (as obtained later by Siu [1] in 1974 using the L^2-estimates of Hörmander). For more details, see Lelong [12b]. The estimate for the degree of the algebraic hypersurface S was improved by Skoda [6] and Demailly [1]; Bertrand and Masser used the n dimensional result to obtain results on transcendence and algebraic independence. For a very thorough treatment of the above as well as many related topics, we refer the reader to the book by Waldschmidt [1].

Chapter 7. The Indicator of Growth Theorem

We have seen in Chapter 1 that the radial indicator $h_f^\star$ of an entire function $f(z)$ of normal type with respect to a proximate order $\rho(r)$ satisfies

i) $h_f^\star(z)=t^\rho h_f^\star(z)$ for $t\geqq 0$;

ii) $h_f^\star(z)$ is plurisubharmonic in $\mathbb{C}^n$.

We now show the converse for *strong* proximate orders; that is, suppose that $h(z)$ is a function which satisfies i) and ii) above. Then for every strong proximate order $\rho(r)$ there exists an entire function $f(z)$ such that $h_f^\star(z)=h(z)$.

Let k be an integer, $(1\leqq k\leqq n)$, and ψ a function plurisubharmonic in $\mathbb{C}^n$. Suppose that $\alpha\in\mathscr{C}_0^\infty(B(0,1))$, $\alpha\geqq 1$, $\int\alpha(z)\,d\tau(z)=1$, α a function of $\|z\|$. We define by induction a sequence $\{\psi_j^k\}$ of plurisubharmonic functions as follows:

i) $\psi_0^k=\ldots=\psi_k^k=\psi$

ii) $\psi_j^k=\alpha\star\psi_j'^k$ for $j>k$ with

$$\psi_j'^k=[\sup_{|\xi_j|<2}\psi_{j-1}^k(z_1,\ldots,z_{j-1},z_j+\xi_j,z_{j+1},\ldots,z_n)]^\star.$$

For $1\leqq j\leqq n$, we identify $\mathbb{C}^j$ with the subspace of $\mathbb{C}^n$ defined by $\{z\in\mathbb{C}^n$: $z_p=0$ for $p>j\}$ and $z^{(j)}$ will be the projection of z on $\mathbb{C}^j$, that is if $z=(z_1,\ldots,z_n)$, $z^{(j)}=(z_1,\ldots,z_j,0,\ldots,0)$. We will let $d\tau_j$ be the Lebesgue measure on $\mathbb{C}^j$.

We shall use this to prove the following result on the extension of entire functions with growth conditions.

Theorem 7.1. *Let* $\psi\in\mathrm{PSH}(\mathbb{C}^n)$ *and* f *holomorphic on* $\mathbb{C}^k$ $(k<n)$ *such that* $\int_{\mathbb{C}^k}|f|^2\exp-\psi\,d\tau_k<+\infty$. *Then* f *is the restriction of an entire function* $g(z)\in\mathscr{H}(\mathbb{C}^n)$ *such that*

$$\int_{\mathbb{C}^n}\frac{|g(z)|^2\exp-\psi_n^k(z)}{(1+\|z\|^2)^{3(n-k)}}\,d\tau_n(z)<+\infty.$$

Proof. We shall prove the result by induction on n by showing that if

$$f\in\mathscr{H}(\mathbb{C}^j)\quad\text{and}\quad\int_{\mathbb{C}^j}\frac{|f(z^{(j)})|^2\exp-\psi_k^k(z^{(j)})}{(1+\|z^{(j)}\|^2)^{3(j-k)}}\,d\tau_j(z^{(j)})=M_f<\infty$$

there exists $g \in \mathscr{H}(\mathbb{C}^{j+1})$ such that $g|_{\mathbb{C}^j} = f$ and

$$\int_{\mathbb{C}^{j+1}} \frac{|g(z^{(j+1)})|^2 \exp -\psi_{j+1}^k(z^{(j+1)})}{(1+\|z^{(j+1)}\|^2)^{3(j+1-k)}} d\tau_{j+1}(z) < C_j M_f.$$

Since ψ'^k_{j+1} is plurisubharmonic, we have $\psi^k_{j+1} \geqq \psi'^k_{j+1}$ and for

$$|z_{j+1}| < 1, \quad \psi'^k_{j+1}(z^{(j+1)}) \geqq \psi^k_j(z^{(j)}),$$

so

$$(7,1) \qquad \int_{\substack{\mathbb{C}^{j+1} \\ |z_{j+1}|<1}} \frac{|f(z^{(j)})|^2 \exp -\psi^k_{j+1}(z^{(j+1)})}{(1+\|z^{(j+1)}\|^2)^{3(j-k)}} d\tau_{j+1}(z^{(j+1)})$$

$$\leqq \int_{\mathbb{C}^j} \frac{|f(z^{(j)})|^2 \exp -\psi^k_j(z^{(j)}) d\tau_j(z^{(j)})}{(1+\|z^{(j)}\|^2)^{3(j-k)}}.$$

Let $\omega \in \mathscr{C}_0^\infty(|z|<1)$ such that $\omega(z)=1$ if $|z| \leqq 1/2$, $\omega(z)=0$ if $|z| \geqq 1$, and let $\tilde{C} = \sup \left|\frac{\partial \omega}{\partial \bar{z}}\right|$. We shall find $g(z^{(j+1)})$ so that

$$(7,2) \qquad g(z^{(j+1)}) = \omega(z_{j+1}) f(z^{(j)}) - z_{j+1} \chi(z^{(j+1)})$$

and such that $\bar{\partial} g = 0$ as a distribution. In fact, if β is the $(0,1)$ form $\beta = \frac{f(z^{(j)})}{z_{j+1}} \bar{\partial} \omega$ then β has $\mathscr{C}^\infty$ coefficients, $\bar{\partial}\beta = 0$, and $|\beta| \leqq 2\tilde{C}|f|$ so

$$\int_{\mathbb{C}^{j+1}} \frac{|\beta|^2 \exp -\psi^k_{j+1}(z^{(j+1)})}{(1+\|z^{(j+1)}\|^2)^{3(j-k)}} d\tau_{j+1}(z^{(j+1)}) \leqq 4\tilde{C}^2 M_f.$$

By Appendix III, we can find χ such that $\bar{\partial}\chi = \beta$ and

$$\int_{\mathbb{C}^{j+1}} \frac{|\chi|^2 \exp -\psi^k_{j+1}(z^{(j+1)})}{(1+\|z^{(j+1)}\|^2)^{3(j+2/3-k)}} d\tau_{j+1} \leqq 4\tilde{C}^2 M_f.$$

Then g defined by (7,2) satisfies $\bar{\partial} g = 0$ as a distribution. If $\alpha_\varepsilon(z) = \frac{1}{\varepsilon^{2(j+1)}} \alpha(z/\varepsilon)$ and $g_\varepsilon(z) = g \star \alpha_\varepsilon(z)$, then $g_\varepsilon \in \mathscr{C}^\infty$ and $\bar{\partial} g_\varepsilon = 0$, so g_ε is in fact a holomorphic function. Since $g_\varepsilon = g_\varepsilon \star \alpha_{\varepsilon'} = g_{\varepsilon'} \star \alpha_\varepsilon = g_{\varepsilon'}$ by the Mean Value Property for harmonic functions, g_ε is independent of ε. Furthermore, $g_\varepsilon \to g$ when $\varepsilon \to 0$ in $L^2(B(0,r))$ for every r so $g_\varepsilon = g$ almost everywhere, and hence g is (equivalent to) a holomorphic function. Finally, from (7,1) and (7,2), since $\|z_{j+1}\|^2 \leqq 1 + \|z^{(j+1)}\|^2$, it follows that

$$\int_{\mathbb{C}^{j+1}} \frac{|g(z^{(j+1)})|^2 \exp -\psi^k_{j+1}(z^{(j+1)})}{(1+\|z^{(j+1)}\|^2)^{3(j+1-k)}} d\tau_{j+1}(z^{(j+1)}) \leqq (1+4\tilde{C}^2) M_f. \qquad \square$$

Let $\mathrm{PSH}_{\rho(t)}(\mathbb{C}^n)$ be the set of plurisubharmonic functions of order ρ and normal type with respect to the proximate order $\rho(t)$.

Proposition 7.2. *Let $\psi \in \mathrm{PSH}_{\rho(t)}(\mathbb{C}^n)$ and $z_0 \neq 0$. Then given $\varepsilon > 0$, there exists $R(\varepsilon, z_0)$ and $\delta(\varepsilon, z_0)$ such that for $\|z - z_0\| < \delta$ and $t > R$*

$$\psi(tz) \leq \left(\frac{h_\psi^\star(z_0)}{\|z_0\|^\rho} + \varepsilon\right) t^{\rho(t)}.$$

Proof. Let $z_0' = \frac{z_0}{\|z_0\|}$. Since $h_\psi^\star(z)$ is upper semi-continuous, there exists a neighborhood $U_{z_0'}$ of z_0' such that $h_\psi^\star(z) \leq h_\psi^\star(z_0') + \varepsilon/2$ for $z \in U_{z_0'}$. We apply Hartog's Lemma (Theorem 1.31) to the family $V_t(z) = t^{-\rho(t)}\psi(tz)$ on $U_{z_0'}$ (which we suppose compact by passing to a smaller set if necessary). □

Corollary 7.3. *Let $\psi \in \mathrm{PSH}_{\rho(t)}(\mathbb{C}^n)$ and let Λ be a continuous positively homogeneous function of degree ρ with $\Lambda \geq h_\psi^\star$. Then given $\varepsilon > 0$, there exists $r > 0$ such that $\psi(z) \leq (\Lambda(z) + \varepsilon\|z\|^\rho)\|z\|^{\rho(\|z\|) - \rho}$ for $\|z\| \geq r$.*

Proof. Since the sphere of radius 1 is compact, for $t > t_1$, we have by Corollary 1.32: $\psi(tz) \leq (\Lambda(tz) + \varepsilon t^\rho) t^{\rho(t) - \rho}$ for $\|z\| = 1$ and $t > r_1$. □

Theorem 7.4. *Let $\psi \in \mathrm{PSH}_{\rho(t)}(\mathbb{C}^n)$. Then the functions ψ_j^k introduced in Theorem 7.1 all have the same radial indicator function $h_\psi^\star$.*

Proof. There exists $a > 0$ such that $\psi \leq \psi_j^k \leq \psi'$ for $\psi'(z) = [\sup_{\|\xi\| \leq a} \psi(z + \xi)]^\star$. For $z_0 \neq 0$ and $\varepsilon > 0$, there exist δ and r such that

$$\psi(tz) \leq \left(\frac{h_\psi^\star(z_0)}{\|z_0\|^\rho} + \varepsilon\right) t^{\rho(t)} \quad \text{for } t \geq r$$

and $\|z - z_0\| \leq \delta$, so that

$$\psi'(tz_0) \leq \left(\frac{h_\psi^\star(z_0)}{\|z_0\|^\rho} + \varepsilon\right) t^{\rho(t)} \quad \text{for } t \geq r'.$$

But this implies that $h_{\psi'}(z) \leq h_\psi^\star(z)$ for all z and hence $h_{\psi'}^\star(z) \leq h_\psi^\star(z)$.

Theorem 7.5. *Let $\psi \in \mathrm{PSH}_{\rho(t)}(\mathbb{C}^n)$ and f an entire function such that*

$$\int_{\mathbb{C}^n} |f|^2 \exp - \psi \, d\tau < \infty.$$

Then f is of normal type with respect to $\rho(t)$ and $h_f^\star(z) \leq 1/2 h_\psi^\star(z)$.

Proof. By the Cauchy Integral Formula, we have

$$f(z)^2 = \frac{1}{(2\pi)^n} \int f(r_1 e^{i\theta_1} + z_1, \ldots, r_n e^{i\theta_n} + z_n)^2 \, d\theta_1 \ldots d\theta_n,$$

so

$$|f(z)|^2 \leq \frac{1}{(2\pi)^n} \int_{1 \leq |\xi_j| \leq 2} \frac{|f(z+\xi)|^2 \, d\tau(\xi)}{|\xi_1| \ldots |\xi_n|} \leq C_n \int_{1 \leq |\xi_j| \leq 2} |f(z+\xi)|^2 \, d\tau(\xi).$$

Thus

$$|f(z)|^2 \leq C_n [\sup_{|\xi_j|\leq 2} \exp \psi(z+\xi)] \int_{1\leq|\xi_j|\leq 2} |f(z+\xi)|^2$$
$$\cdot \exp -(\sup_{|\xi_j|\leq 2} \psi(z+\xi))\, d\tau$$
$$\leq C_n [\sup_{|\xi_j|\leq 2} \exp \psi(z+\xi)] \int_{1\leq|\xi_j|\leq 2} |f(z+\xi)|^2 \exp -\psi(z+\xi) d\tau$$
$$\leq C_n' [\sup_{|\xi_j|\leq 2} \exp \psi(z+\xi)] \left(\int_{1\leq|\xi_j|\leq 2} |f(z+\xi)^2| \exp -\psi(z+\xi) d\tau \right)$$

so (7,3) $\log|f(z)| \leq C_n'' + 1/2 \sup_{\|\xi\|\leq 2n} \psi(z+\xi)$ and hence $f(z)$ is of finite order with respect to $\rho(t)$ and $h_f^\star(z) \leq \frac{1}{2} h_\psi^\star(z)$ by Theorem 7.4. □

Theorem 7.6. *Let $\psi \in \mathrm{PSH}(\mathbb{C}^n)$ be positively homogeneous of order ρ. Then we can find a decreasing sequence of plurisubharmonic functions $\{\psi_q\}$ each positively homogeneous of order ρ and $\mathscr{C}^\infty$ on $\mathbb{C}^n - \{0\}$ such that $\lim_{q\to\infty} \psi_q(z) = \psi(z)$.*

Proof. Let $\alpha(z) \in \mathscr{C}_0^\infty(B(0,1))$, $0 \leq \alpha(z) \leq 1$, $\int \alpha(z) d\tau(z) = 1$ and α depending only on $\|z\|$. Let $\alpha_\varepsilon(z) = \frac{1}{\varepsilon^{2n}} \alpha(z/\varepsilon)$. Then $\psi_\varepsilon(z) = \int \psi(z')\alpha_\varepsilon(z-z') d\tau(z')$ is $\mathscr{C}^\infty$, plurisubharmonic, and decreases to $\psi(z)$, but it is not in general positively homogeneous of order ρ, so we must change the construction slightly.

Let $\tilde{\psi}_\varepsilon(z) = \|z\|^{-2n} \int \psi(z') \alpha_\varepsilon\left(\frac{z-z'}{\|z\|}\right) d\tau(z')$ for $\|z\| \neq 0$ or equivalently $\tilde{\psi}_\varepsilon(z) = \int \psi(z - \|z\| w) \alpha_\varepsilon(w) d\tau(w)$ for all z. Then $\tilde{\psi}_\varepsilon(z)$ is $\mathscr{C}^\infty$ on $\mathbb{C}^n - \{0\}$, is positively homogeneous of order ρ (since ψ is), and since $\tilde{\psi}_\varepsilon = \psi_\varepsilon$ for $\|z\| = 1$, $\tilde{\psi}_\varepsilon$ decreases to ψ when ε tends to zero. It remains to show that ψ_ε is plurisubharmonic in $\mathbb{C}^n$.

Since $\alpha_\varepsilon(w)$ depends only on $\|w\|$, there exists a positive continuous function $A(r)$ such that

$$\psi_\varepsilon(z) = \int_0^\varepsilon A(r) T_r(z) \quad \text{with} \quad T_r(z) = \frac{1}{\omega_{2n}} \int_{\|w\|=r} \psi(z - \|z\| w) d\omega_{2n}(w).$$

Thus, it is sufficient to prove $T_r(z)$ plurisubharmonic in $\mathbb{C}^n$.

Let Γ be the unitary group on $\mathbb{C}^n$, which is compact, and $d\gamma$ the normalized Haar measure on Γ. Let z_0 be a fixed point of $\mathbb{C}^n$ of norm r and $\psi(\gamma) = \psi(z - \|z\| \gamma(z_0))$. Then $T_r(z) = \int_\Gamma \psi(\gamma) d\gamma$. Furthermore, there exists $\eta \in \Gamma$ such that $r\eta(z) = \|z\| z_0$, so $\psi(\gamma) = \psi(z - r\gamma\eta(z))$, and if $\phi(\gamma) = \psi(z - r\gamma(z))$,

$$T_r(z) = \int_\Gamma \phi(\gamma\eta) d\gamma = \int_\Gamma \phi(\gamma) d\gamma = \int_\Gamma \psi(z - r\gamma(z)) d\gamma.$$

Since for every $\gamma \in \Gamma$, $z - r\gamma(z)$ is a holomorphic function of z, $\psi(z - r\gamma(z))$ is plurisubharmonic in z and hence so is $T_r(z)$. □

By a Lipschitz continuous function, we will mean $|\psi(z)-\psi(z')| \leqq C\|z-z'\|$ for $z, z' \in S^{2n-1}$.

Proposition 7.7. *Let* $\rho(r)$ *be a strong proximate order and* $\psi(z)$ *a Lipschitz continuous plurisubharmonic function positively homogeneous of order* ρ. *Then there exists a plurisubharmonic function* $\tilde{\psi}(z)$ *such that* $\psi(z)t^{\rho(\|z\|)-\rho} \leqq \tilde{\psi}(z)$ *and* $h_{\tilde{\psi}}(z) \leqq \psi(z)$ *(where the indicator is taken with respect to* $\rho(r)$*).*

Proof. Suppose that $\varepsilon(r)$ is a continuous decreasing function of r such that $\lim_{r\to\infty} \varepsilon(r)=0$. Then there exists an increasing convex function ξ such that

i) $\xi'(\log r) \geqq \varepsilon(r) r^{\rho(r)}$ and $\xi''(\log r) \geqq \varepsilon(r) r^{\rho(r)}$ for $r \geqq r_0$,

ii) $\lim\limits_{r\to\infty} \dfrac{\xi(\log r)}{r^{\rho(r)}}=0$.

We define $\xi(s)$ by $\xi(s)=\int_0^s \xi'(t)\,dt$, $\xi'(t)=C_\rho \int_0^t \varepsilon(e^r) e^{r\rho(e^r)}\,dr$ with $C_\rho=2\sup(\rho,1)$. Then $\xi'(t) \geqq \varepsilon(e^t) C_\rho \int_0^t e^{r\rho(e^r)}\,dr \geqq \varepsilon(e^t) e^{t\rho(e^t)}$ for t large enough, and since $\varepsilon(s) \leqq \frac{1}{n}$ for $s \geqq s_n$, we have

$$\xi'(t) \leqq C_n + \frac{C_\rho}{n} \int_{s_n}^t e^{r\rho(e^r)}\,dr \leqq C_n + \frac{1}{n} C'_\rho e^{t\rho(e^t)}.$$

Thus $\xi(t) \leqq C_n t + \frac{C''_\rho}{n} e^{t\rho(et)}$, which shows ii). By adding a multiple of $\log(1+r^2)$, we may assume that i) holds for all r. We note that if $a=(a_1, \ldots, a_n)$ is a complex vector, then

$$(7,4) \qquad \sum_{j,k} \frac{\partial^2 \xi(z)}{\partial z_j \partial \bar{z}_k} a_j \bar{a}_k = \xi'(\log r)\left[\frac{\|a\|^2}{r^2} - \frac{|\sum_j a_j r_j|^2}{r^4}\right] + \xi''(\log r) \frac{|\sum_j a_j r_j|^2}{r^4}$$
$$\geqq \tfrac{1}{2} \varepsilon(r) r^{\rho(r)-2} \|a\|^2.$$

Since $\psi(z)$ is Lipschitz continuous and positively homogeneous of order ρ, $\left|\dfrac{\partial \psi(z)}{\partial z_j}\right| \leqq C\|z\|^{\rho-1}$; that is, as a distribution $\dfrac{\partial \psi(z)}{\partial z_j}$ is equivalent to a function with the above bound. Let a be a complex vector. Then as a distribution, setting $\|z\|=r$, we have

$$\begin{aligned}
\sum_{j,k} \frac{\partial^2(\psi(z) r^{\rho(r)-\rho})}{\partial z_j \partial \bar{z}_k} a_j \bar{a}_k &= \sum_{j,k} r^{\rho(r)-\rho} \frac{\partial^2 \psi(z)}{\partial z_j \partial \bar{z}_k} a_j \bar{a}_k \\
&\quad + \sum_{j,k} \frac{\partial r^{\rho(r)-\rho}}{\partial z_j} \frac{\partial \psi(z)}{\partial \bar{z}_k} a_j \bar{a}_k \\
&\quad + \sum_{j,k} \frac{\partial r^{\rho(r)-\rho}}{\partial \bar{z}_k} \frac{\partial \psi(z)}{\partial z_j} a_j \bar{a}_k \\
&\quad + \psi(z) \sum_{j,k} \frac{\partial^2 r^{\rho(r)-\rho}}{\partial z_j \partial \bar{z}_k} a_j \bar{a}_k \\
&\geqq -\varepsilon(r) r^{\rho(r)-2} \|a\|^2
\end{aligned}$$

for some $\varepsilon(r)$ such that $\lim_{r\to\infty}\varepsilon(r)=0$. Thus we can find $\xi(\log r)$ plurisubharmonic with $\lim_{r\to\infty}\frac{\xi(\log r)}{r^{\rho(r)}}=0$, such that $\tilde{\psi}(z)=\xi(\log r)+\psi(z)r^{\rho(r)}$ is plurisubharmonic. □

Theorem 7.8. *Let ψ be a subharmonic function in $\mathbb{C}$ positively homogeneous of order ρ. Then for $0\leqq\theta\leqq 2\pi$, there exists an entire function $f(z)$ of order ρ (depending perhaps on θ) such that*

$$\limsup_{t\to\infty}\frac{\log|f(te^{i\theta})|}{t^{\rho}}=\psi(e^{i\theta}) \quad \textit{and} \quad |f(z)|\leqq C\exp\tilde{\psi}(z)$$

where

$$\tilde{\psi}(z)=[\sup_{|\xi|\leqq 3}\psi(z+\xi)]^{\star}+C_1\log(1+|z|^2)+C_2(\log(1+|z|))^2.$$

Proof. Let $h(z)=\prod_{j=1}^{\infty}(1-2^{-j}z)$, which defines an entire function.

Suppose that for some j, $1/4\leqq|z-2^j|\leqq\frac{1}{2}$, so that

$$3\cdot 2^{j-2}\leqq 2^j-\tfrac{1}{4}\leqq|z|\leqq 2^j+\tfrac{1}{2}\leqq 3\cdot 2^{j-1}$$

and

$$|h(z)|\geqq\tfrac{1}{4}\cdot 2^{-j}\prod_{k\neq j}|1-2^{-k}z|\geqq\frac{2^{-j}}{4}\prod_{k>j}|1-\tfrac{3}{2}2^{j-k}|\prod_{k<j}|1-\tfrac{3}{4}2^{j-k}|$$

$$\geqq\tfrac{1}{4}2^{-j}C\prod_{k<j}|1-\tfrac{3}{4}2^{j-k}|\geqq C'$$

independent of j.

Let $\varphi\in\mathscr{C}_0^{\infty}(B(0,1/2))$ such that $0\leqq\varphi\leqq 1$ and $\varphi\equiv 1$ for $|z|\leqq\frac{1}{4}$, and set $g(z)=\sum_{j=1}^{\infty}\varphi(z-2^je^{i\theta})\exp\psi(2^je^{i\theta})$ (for every z, there is at most one summand). We shall write $f(z)=g(z)-h(ze^{-i\theta})v(z)$ where $v(z)$ is chosen so that $f(z)$ is holomorphic. We must have $\bar{\partial}f=0$ or $\bar{\partial}g=h\bar{\partial}v$. Since $\bar{\partial}g=0$ for $|z-2^je^{i\theta}|\leqq\frac{1}{4}$ if $p=\frac{\bar{\partial}g}{h}$, then $|p|\leqq C\left|\frac{\partial g}{\partial\bar{z}_j}\right|$ and

$$\int_{\mathbb{C}}|p|^2\exp-2\hat{\psi}(z)d\tau(z)<+\infty$$

where

$$\hat{\psi}(z)=[\sup_{|\xi|\leqq 1/2}\psi(z+\xi)]^{\star}+\log(1+|z|^2).$$

By Appendix III, there exists v such that $\bar{\partial}v=p$ and

$$\int_{\mathbb{C}}|v|^2\exp-(2\hat{\psi}(z)+\log(1+|z|^2))d\tau(z)<+\infty.$$

Then $g-hv$ defines a holomorphic function.

Suppose that $2^j \leqq |z| \leqq 2^{j+1}$. Then, since $\log(1+x) \leqq x$ for $x \geqq 0$,

$$\begin{aligned}\log|h(z)| &\leqq \sum_{k=1}^{\infty} \log\left(1+\frac{|z|}{2^k}\right) \leqq \sum_{k=1}^{j+1} \log(1+2^{j+1-k}) + \sum_{k=1}^{\infty} \frac{1}{2^k} \\ &\leqq \sum_{k=1}^{j+1} (j+2-k)\log 2 + 1 = \frac{(j+1)(j+2)}{2}\log 2 + 1 \\ &\leqq C_1(\log(|z|+1))^2 + 1.\end{aligned}$$

Let $\tilde{\psi}(z) = \hat{\psi}(z) + \log(1+|z|^2) + C_1(\log(1+|z|))^2$, which is subharmonic. Then $h_f^\star(e^{i\theta}) = \psi(e^{i\theta})$ and by Theorem 7.5,

$$|f| \leqq C\exp([\sup_{|\xi|\leqq 3} \psi(z+\xi)]^\star + 2\log(1+|z|^2) + C_1'(\log(1+|z|^2))^2. \qquad \square$$

Theorem 7.9. *Let $\psi(z)$ be a plurisubharmonic function positively homogeneous of order ρ in $\mathbb{C}^n$. Then for $z_0 \in \mathbb{C}^n$, there exists an entire function $g(z)$ (depending perhaps on z_0) such that $h_g^\star(z) \leqq \psi(z)$ and $h_g^\star(z_0) = \psi(z_0)$ (where the indicator is with respect to r^ρ).*

Proof. By a rotation, we may assume that $z_0 = (z_1, 0, \ldots, 0)$. Let $\psi(u) = \psi(u, 0, \ldots, 0)$. Then by Theorem 7.8, we can find an entire function in $\mathbb{C}$ $f(u)$ such that $h_f^\star(z_0) = \psi(z_0)$ and

$$\begin{aligned}\log|f(u)| \leqq [&\sup_{|\xi|\leqq 3} \psi(u+\xi)]^\star + C_1\log(1+|u|^2)^2 \\ &+ C_2\log(1+|u|^2) + \log C.\end{aligned}$$

By Theorem 7.1, there exists an entire function $g(z)$ such that $g(u, 0, \ldots, 0) = f(u)$ and $\int |g|^2 \exp - \tilde{\tilde{\psi}}(z) d\tau(z) < +\infty$ where

$$\tilde{\tilde{\psi}}(z) = 2[\sup_{\|\xi\|\leqq a} \psi(z+\xi)]^\star + C_n[\log(1+\|z\|^2)]^2 + C_n'\log(1+\|z\|^2)$$

for some $a > 0$. Then $h_g^\star(z) \leqq h_{\tilde{\tilde{\psi}}}^\star(z)$ by Theorem 7.5 and $h_{\tilde{\tilde{\psi}}}^\star(z) = h_\psi^\star(z) = \psi(z)$ by Proposition 7.2. $\square$

Corollary 7.10. *Let $\psi(z)$ be a Lipschitz continuous plurisubharmonic function positively homogeneous of order ρ in $\mathbb{C}^n$ and $\rho(t)$ a strong proximate order. Then for $z_0 \in \mathbb{C}^n$, there exists an entire function $g(z)$ (depending perhaps on z_0) such that $h_g^\star(z) \leqq \psi(z)$ and $h_g^\star(z_0) = \psi(z_0)$ (where the indicator is with respect to $r^{\rho(r)}$).*

Proof. Let $\tilde{\psi}(z)$ be the plurisubharmonic majorant of $\psi(z)t^{\rho(t)-\rho}$ constructed in Proposition 7.7. Then, as in Theorem 7.8, we construct an entire function $f(uz_0)$ of the variable u such that $f(2^j z_0) = \tilde{\psi}(2^j z_0)$ and

$$|f(z)| \leqq [\sup_{|\xi|\leqq 3} \tilde{\psi}(z+\xi)]^\star + C_1\log(1+\|z\|^2) + C_2(\log(1+\|z\|))^2$$

$\left(\text{we choose } g(uz_0)=\sum_{j=1}^{\infty}\varphi(uz_0-2^j z_0)\exp\tilde{\psi}(2^j z_0)\right)$. We then use Theorem 7.1, to extend f to $\mathbb{C}^n$, as in the proof of Theorem 7.9. □

If φ is a continuous positively homogeneous function of degree ρ in $\mathbb{C}^n$, we let $B_\varphi^{\rho(t)}$ be the Banach space of entire functions f such that

$$\lim_{\|z\|\to\infty} |f(z)\exp-\varphi(z)\|z\|^{\rho(\|z\|)-\rho}|=0$$

with supremum norm. Let $E_\varphi^{\rho(t)}=\bigcap_q B^{\rho(t)}_{\varphi+\frac{1}{q}\|z\|^\rho}$, which is a Fréchet space. If $\psi\in \mathrm{PSH}_{\rho(t)}(\mathbb{C}^n)$, let $m(\psi)$ be the set of continuous plurisubharmonic functions φ positively homogeneous of order ρ such that $h_\psi^\star(z)\leqq\varphi(z)$. Theorem 7.6 shows that $h_\psi^\star(z)=\inf_{\varphi\in m(\psi)}\varphi$ and that $m(\psi)$ is an ordered filtered set with a countable basis. We set $E_\psi^{\rho(t)}=\bigcap_{\varphi\in m(\psi)} E_\varphi^{\rho(t)}$, which is also a Fréchet space.

Theorem 7.11. *For $\psi\in \mathrm{PSH}_{\rho(t)}(\mathbb{C}^n)$, let f be an entire function of normal type with respect to $\rho(t)$. Then $f\in E_\psi^{\rho(t)}$ if and only if $h_f^\star(\psi)\leqq h_\psi^\star(z)$.*

Proof. If $f\in E_\psi$, then $f\in E_\varphi$ for every $\varphi\in m(\psi)$ and hence

$$|f(z)\exp(-\varphi(z)-\frac{1}{q}\|z\|^\rho)\|z\|^{\rho(\|z\|)-\rho}|$$

is bounded for every q. Thus

$$\int |f(z)|^2\exp-(2\varphi(z)+\varepsilon\|z\|^\rho)\|z\|^{\rho(\|z\|)-\rho}d\tau(z)<+\infty$$

for every $\varepsilon>0$. By Theorem 7.5, $h_f^\star(z)\leqq h_\varphi^\star(z)+\frac{\varepsilon}{2}\|z\|^\rho$, so $h_f^\star(z)\leqq h_\psi^\star(z)$.

On the other hand, for $\varphi\in m(\psi)$, $\log|f(z)|\leqq(\phi(z)+\varepsilon\|z\|^\rho)\|z\|^{\rho(\|z\|)-\rho}$ for $\|z\|\geqq r_\varepsilon$ by Corollary 7.3, so $f\in E_\varphi$ for every φ and hence $f\in E_\psi$. □

Theorem 7.12. *Let ψ be a plurisubharmonic function positively homogeneous of order ρ. There exists an entire function $f(z)$ in $\mathbb{C}^n$ whose indicator function $h_f^\star(z)$ with respect to r^ρ is $\psi(z)$.*

Proof. Suppose that $\theta\in \mathrm{PSH}_\rho(\mathbb{C}^n)$. Then if $\theta\leqq\psi$, $E_\theta\subseteq E_\psi$, and the injection is continuous for the topologies defined on these two spaces. If for some point z_0, we have $h_\theta^\star(z_0)<h_\psi^\star(z_0)$, then by Theorems 7.9 and 7.11, there exists $f\in E_\psi$ such that $f\notin E_\theta$ and hence E_θ is of first (Baire) category in E_ψ (the continuous linear image of a complete metric linear space into a second complete metric linear space is either the entire space or of first category).

For a function $\gamma(z)$ defined on $\mathbb{C}^n$, we define a domain in $\mathbb{C}^{n+1}$: $D=\{(\xi,z)\colon \xi\in\mathbb{C},\ z\in\mathbb{C}^n,\ |\xi|<\exp-\gamma(z)\}$. Then D_F is open if and only if $\gamma(z)$ is upper semi-continuous.

Let $\{\omega_k\}$ be a countable basis for the open sets of $\mathbb{C}^{n+1}$ and $\{\omega_s\}$ those elements such that $\omega \cap \complement D_\psi \neq \phi$. Set $u_s = D_\psi \cup \omega_s$. Then if $D_\varphi \not\supseteq D_\psi$, $u_s \subseteq D_\varphi$ for some s (depending on φ). Let

$$m_s = \{\varphi\colon \varphi \in \mathrm{PSH}(\mathbb{C}^n),\ \varphi \text{ positively homogeneous of order } \rho, u_s \subset D_\varphi\}$$

and let $V_s = \text{interior } (\bigcap_{\varphi \in m_s} D_\varphi)$. Then if $\psi_s = [\sup_{\varphi \in m_s} \varphi]^\star$, ψ_s is plurisubharmonic and positively homogeneous of order ρ and $V_s = D_{\psi_s}$. Thus $\psi_s \leqq \psi$, $\psi_s \neq \psi$.

Suppose that $f \in E_\psi$ and $h_f^\star(z) \neq \psi$. Then $u_s \subset D_{h_f^\star}$ for some s, and hence $h_f^\star \in m_s$. Thus, $h_f^\star(z) \leqq \psi_s(z)$ and $f \in E_{\psi_s}^{\rho(t)}$ by Theorem 7.11. Since $\bigcup_s E_{\psi_s}$ is of first category in E_ψ, we see by the Baire Category Theorem that there exists $f \in E_\psi \cap \complement(\bigcup_s E_{\psi_s})$, and for this f, $h_f^\star(z) = \psi(z)$. □

Corollary 7.13. *Let $\psi(z)$ be a Lipschitz continuous plurisubharmonic function positively homogeneous of order ρ. Then there exists an entire function f of normal type with respect to the strong proximate order $\rho(r)$ such that $h_f^\star(z) = \psi(z)$.*

The proof is the same as that of Theorem 7.12 based on Corollary 7.10. We note that for $n=1$, any positively homogeneous subharmonic function is Lipschitz continuous [*D*].

It seems unreasonable to hope for such a strong result for proximate orders in general. Nontheless, for the circled indicator function, we have the following result.

Theorem 7.14. *Let $\psi(z)$ be a plurisubharmonic function complex homogeneous of order ρ. Then there exists an entire function f of normal type with respect to the proximate order $\rho(r)$ such that $h_{f,c}^\star(z) = \psi(z)$.*

Proof. By Theorem 1.23, there exist non-zero constants $\varphi(q)$ such that the type σ of the entire function $f(u)$ of the complex variable u with respect to the proximate order $\rho(r)$ is given by $(\sigma \rho e)^{1/\rho} = \limsup_{q \to \infty} \varphi(q)\, C_q^{1/q}$, where $f(u) = \sum_q C_q u^q$ is the Taylor series expansion at the origin.

Since $\psi(z)$ is plurisubharmonic in $\mathbb{C}^n$, $D = \{z\colon \psi(z) < 1\}$ is a domain of holomorphy (cf. [*A, B*]), and so there exists a function $\tilde{f}(z)$ such that $\tilde{f}(z)$ cannot be extended as a holomorphic function to a neighborhood of any boundary point of D. Let $\tilde{f}(z) = \sum_q \tilde{P}_q(z)$ be the Taylor series expansion of $\tilde{f}(z)$ in terms of homogeneous polynomials. Then the entire function

$$f(z) = \sum_q \left[\frac{(e\rho)^{1/\rho}}{\varphi(q)}\right]^q \tilde{P}_q(z)$$

satisfies $h_{f,c}^\star(z) = \psi(z)$. □

Historical Notes

The indicator Theorem was first proved by C.O. Kiselman [2] for $\rho = 1$ (using Oka's Theorem on domains of holomorphy) and independently for all ρ by A. Martineau [4, 5] along a line already suggested by the work of Kiselman. The method of the proof relies on the resolution of the $\bar{\partial}$-equation and the L^2-estimates of Hörmander (see Appendix III). For $n > 1$, the indicator $h_f^\star(z)$ is not necessarily continuous (for an example see P. Lelong [13]) and an approximation process (see Theorem 7.6) and delicate arguments were needed to circumvent the possibility that the indicator is irregular. For Corollary 7.13 and related results, see Agranovič [1].

Chapter 8. Analytic Functionals

Let M be a complex submanifold of $\mathbb{C}^n$ of complex dimension m. Let $\mathscr{H}(M)$ be the space of holomorphic functions on M equipped with the topology of uniform convergence on compact subsets of M, and let $\mathscr{H}(M)'$ be its dual space of continuous linear functionals. We shall call the elements of $\mathscr{H}(M)'$ the *analytic functionals* on M.

Obviously, $\mathscr{H}(M)$ is contained in $\mathscr{C}(M)$, the space of continuous functions on M equipped with the compact-open topology (which is a Fréchet space topology): suppose K_i an exhaustive sequence of compact sets in M, that is $K_i \subset K_{i+1}$, $\bigcup_{i=1}^{\infty} K_i = M$ and for each given compact $K \subset M$, there exists m such $K_i \supset K$ for $i > m$; then if K_i is an exhaustive sequence in M of compact subsets, the topology on $\mathscr{C}(M)$ is defined by the semi-norms

$$p_i(f) = \sup_{z \in K_i} |f(z)|$$

and it is independent of the exhaustive sequence K_i. With this topology, $\mathscr{H}(M)$ is a closed subspace of $\mathscr{C}(M)$. By the Hahn-Banach Theorem, each element $\mu \in \mathscr{H}(M)'$ is the restriction to $\mathscr{H}(M)$ of $\tilde{\mu} \in \mathscr{C}(M)'$. The elements of $\mathscr{C}(M)'$ are just the complex measures with compact support in M. Thus, each analytic functional on M can be represented by a measure μ (not unique!) with compact support in M. Since the choice is not unique, this gives rise to an equivalence class $\gamma_\mu = \{\mu' \in \mathscr{C}(M)' \colon \mu'(f) = \mu(f) \, \forall f \in \mathscr{H}(M)\}$, and we study γ_μ to find certain representatives with extremal properties.

Definition 8.1. A *carrier* of an analytic functional $\mu \in \mathscr{H}(M)'$ is a compact subset K of M such that for every open neighborhood ω of K, there exists a constant C_ω such that $|\mu(f)| \leqq C_\omega \sup_\omega |f(z)|$, $f \in \mathscr{H}(M)$.

One would like to have some intrinsic idea of the support of an analytic functional similar to the notion of the support of a measure or distribution. In contrast to the preceeding examples, no such smallest support exists in general for analytic functionals (see the examples below). Basically, this is because a holomorphic function which vanishes on an open set is identically zero (in contrast to a $\mathscr{C}^\infty$ function, which can vanish on an open set without being identically zero). However, sometimes we shall be able to find a smallest carrier in a class of carriers. This is the problem we shall study here. We shall first study the problem in $\mathbb{C}^n$ and then reduce the general case to the problem in $\mathbb{C}^n$.

§1. Convex Sets and the Fourier-Borel Transform

Let $K \subset \mathbb{C}^n$ be a convex compact subset and set $h_K(u) = \sup_{z \in K} \mathbb{R}e\langle u, z\rangle$, where $\langle u, z\rangle = \sum_{i=1}^{n} u_i z_i$. Then

i) $h_K(u)$ is positively homogeneous of order 1;

ii) $h_K(u)$ is subadditive, that is $h_K(u_1 + u_2) \leqq h_K(u_1) + h_K(u_2)$.

Conversely, suppose that $h(u)$ is a positively homogeneous subadditive function, and define K by

$$K = \{z \in \mathbb{C}^n : \mathbb{R}e\langle z, u\rangle \leqq h(u) \text{ for all } u \in \mathbb{C}^n\}.$$

Then K is convex and compact and $h(u) = h_K(u)$. A real valued function which satisfies i) and ii) will be called a *support* function, and if K is a compact convex subset of $\mathbb{C}^n$, $h_K(u)$ will be called the support function of K.

Definition 8.2. Let $\mu \in \mathscr{H}(\mathbb{C}^n)'$. Then we define $\mathscr{F}_\mu(u)$, the *Fourier-Borel transform* of μ by

$$\mathscr{F}_\mu(u) = \mu(\exp\langle u, z\rangle) \in \mathscr{H}(\mathbb{C}^n).$$

We note that the functions $\exp\langle u, z\rangle$ are dense in $\mathscr{H}(\mathbb{C}^n)$, for the polynomials are dense in $\mathscr{H}(\mathbb{C}^n)$ and $z_i = \lim_{\lambda \to 0} \lambda^{-1}[\exp(\lambda z_i) - 1]$ uniformly on any compact subset of $\mathbb{C}^n$. Thus μ is uniquely determined by its Fourier-Borel transform. Furthermore, if μ is carried by the compact convex set K, it follows from Definition 8.1 that

$$|\mathscr{F}_\mu(u)| \leqq C_\varepsilon \exp(h_k(u) + \varepsilon\|u\|) \quad \text{for every } \varepsilon > 0. \tag{8,1}$$

In the following sections, we shall prove the converse to this inequality: that is, if (8,1) holds, then μ is carried by the convex compact subset K.

§2. The Projective Indicator

If K is a convex compact subset of $\mathbb{C}^n$, $\mathscr{H}(K)$ will be the linear space of functions defined and holomorphic in an open neighborhood of K. If $\{\Omega_j\}$ is a family of open convex sets with $\Omega_{j+1} \Subset \Omega_j$ and $K = \bigcap_{j=1}^{\infty} \Omega_j$, then $\mathscr{H}(K) = \bigcup_{j=1}^{\infty} \mathscr{H}(\Omega_j)$, and we equip $\mathscr{H}(K)$ with the inductive limit topology induced by the topologies on $\mathscr{H}(\Omega_j)$. This is independent of the choice of the family $\{\Omega_j\}$. Since $\mathscr{H}(\mathbb{C}^n)$ is dense in $\mathscr{H}(\Omega_j)$ (cf. [*A*, *B*]), $\mathscr{H}(\mathbb{C}^n)$ is dense in $\mathscr{H}(K)$ and it follows from Definition 8.1 that $\mu \in \mathscr{H}(K)'$ if and only if $\mu \in \mathscr{H}(\mathbb{C}^n)'$ and μ is carried by K.

Let $\mathbb{P}(\mathbb{C}^{n+1})$ be the complex projective space of dimension n. We will use coordinates $\bar{\xi}=(\xi_0, \xi)$ with $\xi \in \mathbb{C}^n$ and $\xi_0 \in \mathbb{C}$. For K a compact convex set we let $\overset{\star}{K}$ be the open set of $\mathbb{P}(\mathbb{C}^{n+1})$ formed by the hyperplanes $\bar{\xi}$ such that $\bar{\xi} \cap K = \emptyset$ ($\bar{\xi} \cap K = \emptyset$ if and only if $\langle \xi, z\rangle + \xi_0$ is never zero for $z \in K$). If $\mu \in \mathscr{H}(K)'$, we consider the value of μ on the holomorphic function $\dfrac{\xi_0}{\langle z, \xi\rangle + \xi_0}$. Since this is homogeneous of degree zero, it depends only on $\bar{\xi}$.

Definition 8.3. For $\mu \in \mathscr{H}(K)'$, we define the *projective indicator* of μ by $\varphi_\mu(\bar{\xi}) = \mu\left(\dfrac{\xi_0}{\langle z, \xi\rangle + \xi_0}\right)$. This is defined on $\overset{\star}{K}$.

Theorem 8.4. *The map $\mu \to \varphi_\mu$ is a bijection of the space $\mathscr{H}(K)'$ onto the linear space $\mathscr{H}_0(K)$ of functions holomorphic in $\bar{\xi}$ on $\overset{\star}{K}$ and zero at infinity (for the points $\bar{\xi}$ with $\xi_0 = 0$).*

We prove the theorem in several steps.

Lemma 8.5. *φ_μ is holomorphic and zero at infinity, that is for $\xi_0 = 0$.*

Proof. By linearity, $\varphi_\mu(\bar{\xi}) = \xi_0 \mu\left(\dfrac{1}{\langle z, \xi\rangle + \xi_0}\right)$, so $\varphi_\mu(\bar{\xi}) = 0$ when $\xi_0 = 0$. Let $K = \bigcap_{j=1}^{\infty} \Omega_j$ where Ω_j is an open convex neighborhood of K. Then since $\mathscr{H}(K) = \bigcup_{j=1}^{\infty} \mathscr{H}(\Omega_j)$, if $\mu \in \mathscr{H}(K)'$ for every j, $\mu \in \mathscr{H}(\Omega_j)'$, By the Hahn-Banach Theorem, since $\mathscr{H}(\Omega_j)$ is a closed subspace of the space $\mathscr{C}(\Omega_j)$, the continuous functions on Ω_j with the topology of uniform convergence on compact subsets, there is a measure μ_j with compact support in Ω_j such that $\mu(f) = \int f \, d\mu_j$ for $f_j \in \mathscr{C}(\Omega_j)$. Suppose $\bar{\xi} \cap \bar{\Omega}_j = \emptyset$, which for every $\bar{\xi}$ is true for $j > N_{\xi}$. Then

$$\varphi_\mu(\bar{\xi}) = \mu\left(\frac{\xi_0}{\langle z, \xi\rangle + \xi_0}\right) = \xi_0 \int \frac{d\mu}{\langle z, \xi\rangle + \xi_0}$$

is holomorphic in $\xi_0, \ldots, \xi_n$, since $\bar{\xi} \cap \operatorname{supp} \mu_j = \emptyset$. □

Lemma 8.6. *Suppose that K is convex and compact. Let ψ be a holomorphic function in $\overset{\star}{K}$ which is zero at infinity. If $\bar{f} \in \mathscr{H}(K)$, let f be a representation for $\bar{f}$ in an open neighborhood ω of K and let $\hat{K} = \{z : \rho(z) \leqq 0\}$ for a $\mathscr{C}^2$*

strictly convex function ρ such that $\hat{K} \subset \omega$. We let

$$(8,1) \qquad T_\psi(\bar{f}) = \frac{-(-1)^{\frac{n(n+1)}{2}}}{(2\pi i)^n} \int_{bd\hat{K}} f(z) \frac{\partial^{n-1}}{\partial \xi_0^{n-1}} \left(\frac{1}{\xi_0} \psi(\bar{\xi}) \right) \tilde{\Omega}(z, \bar{\xi}(z))$$

where

$$\bar{\xi} = \left(-\sum_{j=1}^{n} z_j \frac{\partial \rho}{\partial z_j}, \frac{\partial \rho}{\partial z_1}, \ldots, \frac{\partial \rho}{\partial z_n} \right) \quad \text{and} \quad \tilde{\Omega} = \sum_{j=1}^{n} (-1)^{j+1} \bar{\xi}_j \bigwedge_{k \neq j} \bar{\partial} \xi_k \bigwedge_{l=1}^{n} dz_l.$$

The map thus defined depends only on $\bar{f}$ and not on the choice of f or $\hat{K}$ and the map $\bar{f} \to T_\psi(\bar{f})$ is a continuous linear functional on $\mathscr{H}(K)$.

Proof. We begin by noting that the expression has a meaning. Indeed, outside of $\xi_0 = 0$, $\frac{\psi(\xi)}{\xi_0}$ is of degree -1 in ξ_0 and so $\frac{\partial^{n-1}}{\partial \xi_0^{n-1}} \left(\frac{\psi(\xi)}{\xi_0} \right)$ is of degree $(-n)$ in ξ. Since

$$\tilde{\Omega}(z, \bar{\xi}(z)) = \sum (-1)^{j+1} \bar{\xi}_j(z) \bigwedge_{k \neq j} \bar{\partial} \xi_k(z) \bigwedge_{l=1}^{n} dz_l$$

is of degree $+n$ in $\bar{\xi}$, $\frac{\partial^{n-1}}{\partial \xi_0^{n-1}} \left(\frac{\psi(\bar{\xi})}{\xi_0} \right) \tilde{\Omega}(z, \xi(z))$ is of degree zero. In a neighborhood of infinity ($\xi_0 = 0$), we can choose local coordinates in which, for instance, $\xi_j = 1$ and hence $\frac{\psi(\bar{\xi})}{\xi_0}$ is holomorphic since ψ is zero at infinity. Thus $\frac{\partial^{n-1}}{\partial \xi_0^{n-1}} \left(\frac{\psi(\bar{\xi})}{\xi_0} \right)$ is everywhere defined and holomorphic and the integration is well defined.

We show now that the value is independent of the choice of $\hat{K}$. It suffices to show that for $\hat{K}_2 \Subset \hat{K}_1 \Subset \omega$ the value remains unchanged, for if $\hat{K}$ and $\hat{K}'$ are two compact subsets of ω with nonempty intersection, we need only choose $K_2 \subset \hat{K} \cap \hat{K}'$ and if the integral has the same value for $\hat{K}$ and $\hat{K}_2$ and for $\hat{K}'$ and $\hat{K}_2$, then it has the same value for $\hat{K}$ and $\hat{K}'$. Thus, we verify

$$A = \int_{bd\hat{K}_1} f(z) \frac{\partial^{n-1}}{\partial \xi_0^{n-1}} \left(\frac{\psi(\bar{\xi})}{\xi_0} \right) \tilde{\Omega}(z, \bar{\xi}(z)) - \int_{bd\hat{K}_2} f(z) \frac{\partial^{n-1}}{\partial \xi_0^{n-1}} \left(\frac{\psi(\bar{\xi})}{\xi_0} \right) \tilde{\Omega}(z, \bar{\xi}(z)) = 0.$$

Suppose that ρ_j defines $\hat{K}_j$, that is $\hat{K}_j = [z, \rho_j(z) \leqq 0, \rho_j$ strictly convex$]$.

Let $X = \mathbb{C}^n \times \mathbb{P}(\mathbb{C}^n)$ and let Σ_j be the manifold $(z, \bar{\xi}^{(j)}(z))$, where $\xi_k^{(j)} = \frac{\partial \rho_j}{\partial z_k}$, $\xi_0^{(j)} = \sum_{k=1}^{n} -z_k \frac{\partial \rho_j}{\partial z_k}$, which we identify in a natural way with $bd\hat{K}_j$. Choose a point $z_0 \in \hat{K}_2$ (to remain fixed). For $z \in \hat{K}_1 - \hat{K}_2$, we let $m_j(z)$ be the point on $bd\hat{K}_j$ where the half line from z_0 to infinity passing through z intersects

bdK_j. If $t\in[0,1]$ set

$$\bar{\xi}^{(t)}=\frac{t\,\bar{\xi}^{(1)}}{\langle m_2,\bar{\xi}^{(1)}\rangle}-\frac{(1-t)\,\bar{\xi}^{(2)}}{\langle m_1,\bar{\xi}^{(2)}\rangle}.$$

Since for $z\in\hat{K}_1-\hat{K}_2$, we can write z in a unique way as $z=t\,m_1(z)+(1-t)\,m_2(z)$, we have

$$\langle\bar{\xi}^{(t)}(z),z\rangle=\frac{t\langle\bar{\xi}^{(1)}(z),z\rangle}{\langle m_2,\bar{\xi}^{(1)}(z)\rangle}-\frac{(1-t)\langle\bar{\xi}^{(2)}(z),z\rangle}{\langle m_1,\bar{\xi}^{(2)}(z)\rangle}$$

since $\langle m_j,\bar{\xi}^{(j)}(z)\rangle=0$.

We show that $Y_t=\{z'\colon\ \langle z',\xi^{(t)}(z)\rangle=0\}$ does not intersect K for $z\in\hat{K}_1-\hat{K}_2$ so that $\psi(\xi^{(t)}(z))$ is well defined. Let

$$\hat{Y}_t=\left\{z'\colon t\frac{\mathbb{R}e\langle z',\bar{\xi}^1(z)\rangle}{\mathbb{R}e\langle m_2,\bar{\xi}^1(z)\rangle}-(1-t)\frac{\mathbb{R}e\langle z',\bar{\xi}^{(2)}(z)\rangle}{\mathbb{R}e\langle m_1,\bar{\xi}^{(2)}(z)\rangle}=0\right\}.$$

Then $Y_t\subset\hat{Y}_t$ (we note that since we can multiply $\bar{\xi}^{(j)}(z)$ by any complex number, we may assume without loss of generality that $\langle m_2,\bar{\xi}^{(1)}(z)\rangle$ and $\langle m_1,\bar{\xi}^{(2)}(z)\rangle$ are both real, so the inclusion is trivial). For $\tilde{z}\in K$, $\mathbb{R}e\langle z,\bar{\xi}^{(1)}\rangle<0$ and since $m_2\in\hat{K}_2\subset\hat{K}_1$, $\mathbb{R}e\langle m_2,\bar{\xi}^{(1)}(z)\rangle<0$ so $t\dfrac{\mathbb{R}e\langle\tilde{z},\bar{\xi}^{(1)}(z)\rangle}{\mathbb{R}e\langle m_2,\bar{\xi}^{(1)}(z)\rangle}>0$. On the other hand $\mathbb{R}e\langle\tilde{z},\bar{\xi}^{(2)}(z)\rangle<0$ but

$$\mathbb{R}e\langle m_1,\bar{\xi}^{(2)}(z)\rangle>0,\quad\text{so}\quad\frac{-(1-t)\mathbb{R}e\langle\tilde{z},\bar{\xi}^{(2)}(z)\rangle}{\mathbb{R}e\langle m_1,\bar{\xi}^{(2)}(z)\rangle}>0.$$

The manifold $\Sigma_{12}=(z,\bar{\xi}^t(z))$ has boundary $\Sigma_1-\Sigma_2$. We apply Stokes' Theorem on this manifold:

$$\begin{aligned}A&=\int_{\Sigma_{12}}d\left(f(z)\frac{\partial^{n-1}}{\partial\xi_0^{n-1}}\left(\frac{\psi(\xi)}{\xi_0}\right)\tilde{\Omega}(z,\bar{\xi})\right)\\&=\int_{\Sigma_{12}}\bar{\partial}_z f(z)\frac{\partial^{n-1}}{\partial\xi_0^{n-1}}\left(\frac{\psi(\bar{\xi})}{\xi_0}\right)\tilde{\Omega}(z,\bar{\xi})+\int_{\Sigma_{12}}f(z)d_{\bar{\xi}}\frac{\partial^{n-1}}{\partial\xi_0^{n-1}}\left(\frac{\psi(\bar{\xi})}{\xi_0}\right)\tilde{\Omega}(z,\bar{\xi}).\end{aligned}$$

The first term is zero since $f(z)$ is holomorphic.

On Σ_{12}, $\langle z,\bar{\xi}\rangle=-1$ and $\sum_{k=1}^{n}(dz_k\bar{\xi}_k+z_k d\bar{\xi}_k)=0$ so, since

$$d_{\bar{\xi}}\left[\left(\frac{\partial^{n-1}}{\partial\xi_0^{n-1}}\left(\frac{\psi(\bar{\xi})}{\xi_0}\right)\right)\tilde{\Omega}(z,\bar{\xi})\right]=F(\bar{\xi})\bigwedge_{j=1}^{n}d\bar{\xi}_j\bigwedge_{j=1}^{n}dz_j,$$

this last factor is also zero and the Lemma is proved. □

Proof of Theorem 8.4. Let $C_{(\alpha)}=\mu(z^\alpha)$ for $\mu\in\mathscr{H}(K')$ and α a multi-index of positive numbers. Then there exists $C(K)$ such that for $\sup_j|\xi_j\xi_0^{-1}|<C(K)$ the series converges uniformly on K and if $(\xi z)=(\xi_1 z_1,\dots,\xi_n z_n)$:

$$\frac{\xi_0}{\xi_0+\xi_1 z_1+\dots+\xi_n z_n}=\sum_\alpha(-1)^{|\alpha|}\frac{|\alpha|!}{\alpha!}\left(\frac{\xi z}{\xi_0}\right)^\alpha.$$

Thus $\varphi_\mu(\bar{\xi})=\sum_{(\alpha)}(-1)^{|\alpha|}\frac{|\alpha|!}{\alpha!}\frac{\xi^\alpha}{\xi_0^{|\alpha|}}C_{(\alpha)}$ is a neighborhood of the origin in $\mathbb{P}(\mathbb{C}^{n+1})$. Set $\xi_0=1$ so that $\varphi_\mu(u)=\sum_{(\alpha)}(-1)^{|\alpha|}C_{(\alpha)}\frac{|\alpha|!}{\alpha!}u^\alpha$.

Let $\psi\in \overset{\star}{\mathrm{P}}_0(K)$. In order to calculate the $C_{(\alpha)}$, we consider the closed ball $\bar{B}_R$ with center at the origin and radius $R>R_0$ so that $K\subset\mathring{B}_R$. For all $\bar{\xi}\in\bar{B}_R$, the Taylor series at the origin of ψ converges uniformly on $\bar{B}_R$. Let $\psi(u)=\sum_{(\alpha)}a_{(\alpha)}u^\alpha$. Then

$$(8,2)\qquad (-1)^{n-1}\frac{\partial^{n-1}}{\partial\xi_0^{n-1}}\left(\frac{\psi(\xi)}{\xi_0}\right)=\sum_{(\alpha)}(|\alpha|+1)\dots(|\alpha|+n-1)\frac{a_{(\alpha)}\xi^\alpha}{\xi_0^{\alpha+n}}.$$

We shall show that for $\mu_\psi=T_\psi$ as defined by Lemma 8.6, $\varphi_{\mu_\psi}=\psi$.

The hyperplane tangent to the sphere of radius R at the point $z=(z_1,\dots,z_n)$ has the equation $-R^2+\sum_{j=1}^n(z'_j-z_j)\bar{z}_j=0$ with projective coordinates $\xi_0=-R^2$, $\xi_j=\bar{z}_j$. Thus

$$C_{(\beta)}=\frac{(-1)^{\frac{n(n+1)}{2}+n}}{(2\pi i)^n}\int_{\partial B_R}z^\beta\left(\sum_\alpha(|\alpha|+1)\dots(|\alpha|+n-1)a_{(\alpha)}\frac{\bar{z}^\alpha}{(-R^2)^{|\alpha|+n}}\right)$$
$$\times\left(\sum_{j=1}^n(-1)^{j+1}\bar{z}_j\bigwedge_{k\neq j}d\bar{z}_k\bigwedge_{k=1}^n dz_k\right).$$

Since the series converges uniformly on B_R, we can evaluate term by term, so we calculate

$$\lambda_{(\alpha),(\beta)}=\int_{bdB_R}\frac{(|\alpha|+1)\dots(|\alpha|+n-1)}{(-R^2)^{|\alpha|+n}}z^\beta\bar{z}^\alpha\left(\sum_{j=1}^n(-1)^{j+1}\bar{z}_j\bigwedge_{k\neq j}d\bar{z}_k\bigwedge_{k=1}^n dz_k\right).$$

We first apply Stokes' Theorem to obtain

$$\lambda_{(\alpha),(\beta)}=\int_{B_R}\frac{(|\alpha|+1)\dots(|\alpha|+n-1)}{(-R^2)^{(|\alpha|+n)}}\bar{\partial}\left(z^\beta\bar{z}^\alpha\sum_{j=1}^n(-1)^{j+1}\bar{z}_j\bigwedge_{k\neq j}d\bar{z}_k\bigwedge_{k=1}^n dz_k\right)$$
$$=(-1)^{\frac{n(n+1)}{2}}\int_{B_R}\frac{(|\alpha|+1)\dots(|\alpha|+n)}{(-R^2)^{(|\alpha|+n)}}z^\beta\bar{z}^\alpha\bigwedge_{k=1}^n(dz_k\wedge d\bar{z}_k)$$

so that $\lambda_{(\alpha),(\beta)}=0$ for $(\alpha)\neq(\beta)$ (we consider the integral on circles in every complex line and note that $\int_0^{2\pi}e^{in\theta}\cdot e^{-im\theta}d\theta=0$ for $m\neq n$) and

$$\lambda_{(\beta),(\beta)}=(-1)^{\frac{n(n+1)}{2}}\frac{(2i)^n(|\beta|+1)\dots(|\beta|+n)\pi^n}{(-1)^{|\beta|+n}}\frac{\beta!}{(|\beta|+n)!}$$

so

$$(8,3)\qquad C_{(\beta)}=(-1)^{|\beta|}\frac{\beta!}{|\beta|!}a_{(\beta)}.$$

Hence $\varphi_{\mu_\psi}(u)=\sum_\alpha a_{(\alpha)}u^\alpha$ in a neighborhood of $u=0$ so $\varphi_{\mu_\psi}=\psi$. □

§3. The Projective Laplace Transform

Let K be a compact convex set in $\mathbb{C}^n$ and $h_K(z)$ the support function of K. Suppose that $f(z)$ is an entire function of exponential type such that $h_f^\star(z) \leqq h_K(z)$ (here indicators are calculated with respect to $\rho=1$). Consider the integral

$$\Gamma_{\xi,\lambda}(\xi) = \xi_0 \int_0^\infty f(-\xi\lambda t) \exp -(\xi_0 \lambda t)(\lambda dt)$$

for $\xi \in \mathbb{C}^n$ fixed, which converges absolutely for $A(\xi,\lambda) = \{\xi_0 \colon \mathbb{R}e(\xi_0\lambda) > h^\star(-\xi\lambda)\}$ and defines a holomorphic function of ξ_0 in this set. In fact, this value is independent of λ, for if λ_1 and λ_2 are two distinct values such that $A = A(\xi,\lambda_1) \cap A(\xi,\lambda_2) \neq \emptyset$ then $\Gamma_1(\xi_0) = \Gamma_{\xi,\lambda_1}(\xi_0)$ and $\Gamma_2(\xi_0) = \Gamma_{\xi,\lambda_2}(\xi_0)$ are both holomorphic in A and if $|\xi_0|$ is small enough, it follows from Cauchy's Theorem that $\Gamma_1(\xi_0) - \Gamma_2(\xi_0) = 0$, since we are integrating around a closed curve (two half lines emanating from the origin). Thus, by the uniqueness of analytic continuation, $\Gamma_1(\xi_0) = \Gamma_2(\xi_0)$ in A, and the value $\Gamma_{\xi,\lambda}(\xi_0)$ depends only upon $(\xi_0, \xi) = \bar{\xi}$, the projection of ξ into $P(\mathbb{C}^{n+1})$. We set

$$\mathfrak{L}_f(\bar{\xi}) = \xi_0 \int_0^\infty f(-\xi\lambda t) \exp -(\xi_0 \lambda t) \lambda dt$$

for any value of λ for which the integral converges absolutely; $\mathfrak{L}_f(\bar{\xi})$ is called the *projective Laplace transform of f.*

Definition 8.7. The natural domain of convergence for $\mathfrak{L}_f(\bar{\xi})$ is the interior in $\mathbb{P}(\mathbb{C}^{n+1})$ of the set of points for which the integral converges absolutely.

Lemma 8.8. *Let $f(z)$ be an entire function of exponential type and K a compact convex set and suppose that $h_f^\star(z) \leqq h_K(z)$. Then the natural domain of convergence for $\mathfrak{L}_f(\bar{\xi})$ contains $\overset{\star}{K}$ and $\mathfrak{L}_f(\bar{\xi})$ is holomorphic in $\overset{\star}{K}$ and zero at infinity.*

Proof. Suppose $\bar{\xi} \neq \infty$, $\bar{\xi} \in \overset{\star}{K}$. By choosing local coordinates, we can let $\bar{\xi} = \{z \colon \langle z, u \rangle = 1\}$ be the equation of the plane determined by $\bar{\xi}$. We consider the map of $\mathbb{C}^n$ into $\mathbb{C}$ defined by $f(z) = \langle z, u \rangle$. Then $f(K) = G$ is a convex compact set in $\mathbb{C}$ which does not contain the point 1, since $\bar{\xi} \in \overset{\star}{K}$. Let $h_G(v)$ be the support function of G and let v be a value for which $h_G(v) < \mathbb{R}ev$ (which exists, since if $\mathbb{R}ev \leqq h_G(v)$ for all v then $1 \in G$ which is a contradiction). Then

$$\int_0^\infty f(uvt) \exp(-vt) v dt$$

is absolutely convergent, since

$$|f(uvt)| \leqq C_\varepsilon \exp(h_K(uvt) + \varepsilon \|uv\|),$$

$\varepsilon>0$, and $h_K(uvt)=h_G(vt)$. Hence $|f(uvt)\exp -vt|\leqq \exp -\eta|t|$ for some $\eta>0$, which implies that the integral converges absolutely.

Suppose $|f(z)|\leqq C_\varepsilon \exp(h_K(z)+\varepsilon\|z\|)$ for $\varepsilon>0$. By the Cauchy Integral Formula, if $\zeta_j=(0, \ldots, \zeta, 0, \ldots, 0)$ (ζ in the jth place), then

$$\frac{\partial f}{\partial z_j}=\frac{1}{2\pi i}\int_{|\zeta|=1}\frac{f(u_1, \ldots, u_j+\zeta, \ldots, u_n)}{\zeta}\,d\zeta$$

so

$$\begin{aligned}\left|\frac{\partial f}{\partial z_j}\right| &\leqq \sup_{|\zeta|=1}|f(u_1, \ldots, u_j+\zeta, \ldots, u_n)|\\ &\leqq C_\varepsilon \exp(\sup_{|\zeta|\leqq 1} h_K(\zeta_j))(\exp\varepsilon)\exp(h_K(z)+\varepsilon\|z\|)\end{aligned}$$

since $h_K(z+\zeta_j)\leqq h_K(z)+h_K(\zeta_j)$. Thus we also have the absolute convergence of each of the integrals $\int \frac{\partial f}{\partial z_j}(-uvt)\exp-(vt)(v\,dt)$, and we can differentiate under the integral sign, so $\mathfrak{L}_f(\xi)$ is holomorphic. □

Theorem 8.9. *Let $f(u)$ be an entire function of exponential type. We suppose for some compact convex set K and for every $\varepsilon>0$, $|f(u)|\leqq C_\varepsilon \exp(h_K(u)+\varepsilon\|u\|)$. Then f is the Fourier-Borel transform of an element $\mu\in\mathscr{H}(K)'$ and if $\mathfrak{L}_f(\bar\xi)$ is the projective Laplace transform of f then*

$$f(u)=\frac{-(-1)^{\frac{n(n+1)}{2}}}{(2\pi i)^n}\int_{bd\tilde K}\exp\langle z,u\rangle\frac{\partial^{n-1}}{\partial\xi_0^{n-1}}\left(\frac{\mathfrak{L}_f(\bar\xi)}{\xi_0}\right)\tilde\Omega(z,\bar\xi(z))$$

where $\tilde K$ is any bounded strictly convex neighborhood of K with $\mathscr{C}^2$ boundary.

Proof. It follows from Lemma 8.6 that $\mathfrak{L}_f(\bar\xi)$ determines a continuous linear functional on $\mathscr{H}(K)$ which we denote by μ_f. Then from (8,1) we have

$$f_{\mu_f}(u)=\mu_f(\exp\langle z,u\rangle)=\frac{-(-1)^{\frac{n(n+1)}{2}}}{(2\pi i)^n}\int_{bd\tilde K}\exp\langle z,u\rangle\frac{\partial^{n-1}}{\partial\xi_0^{n-1}}\left(\frac{\mathfrak{L}_f(\bar\xi)}{\xi_0}\right)\tilde\Omega(z,\bar\xi(z)).$$

Suppose $f(u)=\sum_{|\alpha|=0}^{\infty}\frac{a_{(\alpha)}u^\alpha}{\alpha!}$ so that

$$f(-u\lambda t)=\sum_m\left(\sum_{|\alpha|=m}\frac{a_{(\alpha)}u^\alpha}{\alpha!}(-1)^m(\lambda t)^m\right).$$

Thus, if $\|u\|$ is small enough so that $|f(-u\lambda t)|\leqq C\exp A|\lambda|t$ for $A<1$, we have

$$\mathfrak{L}_f(u)=\int_0^\infty\sum_m\left(\sum_{|\alpha|=m}\frac{a_{(\alpha)}u^\alpha}{\alpha!}\right)(-1)^m\lambda^m t^m e^{-t\lambda}\,dt.$$

Letting $\lambda=1$, we obtain

$$\mathfrak{L}_f(u)=\sum_\alpha(-1)^{|\alpha|}a_{(\alpha)}\frac{|\alpha|!}{\alpha!}u^\alpha$$

$$\mathfrak{L}_f(\bar{\xi})=\xi_0\sum_\alpha(-1)^{|\alpha|}a_{(\alpha)}\frac{|\alpha|!}{\alpha!}\left(\frac{\xi}{\xi_0}\right)^\alpha.$$

For fixed u_0, we have $\exp\langle u_0,z\rangle=\sum_{|\alpha|=1}^{\infty}\frac{u_0^\alpha z^\alpha}{\alpha!}$, the convergence being uniform on compact sets. Thus by (8,2) and (8,3)

$$\begin{aligned} f_{\mu_f}(u_0)&=\frac{-(-1)^{\frac{n(n+1)}{2}}}{(2\pi i)^n}\sum_{|\alpha|=1}^{\infty}\frac{u_0^\alpha}{\alpha!}\int_{bdB_R}z^\alpha\frac{\partial^{n-1}}{\partial\xi_0^{n-1}}\left(\frac{\mathfrak{L}_f(\bar{\xi})}{\xi_0}\right)\tilde{\Omega}(z,\bar{\xi}(t))\\ &=\frac{(-1)^{\frac{n(n+1)}{2}+n}}{(2\pi i)^n}u_0^\alpha\sum_{(\beta)}(-1)^{|\beta|}a_{(\beta)}\frac{|\beta|!}{\beta!}\int_{bdB_R}z^\alpha\bar{z}^\beta\tilde{\Omega}(z,\bar{\xi}(z))\\ &=\frac{u_0^\alpha}{\alpha!}a_{(\alpha)}\qquad\text{so } f_{\mu_f}(u)=f(u).\end{aligned}$$

§4. The Case of M a Complex Submanifold of $\mathbb{C}^n$

We begin by noting that a Stein manifold is biholomorphically equivalent to a complex submanifold of $\mathbb{C}^n$ (cf. [A, B]), so by transposition the following discussion applies to Stein manifolds.

Definition 8.10. Let K be a compact subset of M. We define the *supporting function* H_K of K by

$$H_K(\varphi)=\sup_{z\in K}\mathbb{R}e\,\varphi(z)\qquad\text{for } \varphi\in\mathscr{H}(M).$$

(The supremum over the empty set is defined as $-\infty$.)

Then H_K is positively homogeneous of order 1 and convex, that is

$$H_K(t\varphi)=tH_K(\varphi),\ t>0;\qquad H_K(\varphi+\psi)\leqq H_K(\varphi)+H_K(\psi).$$

The restriction of $H_K(\varphi)$ to the linear functions $\mathfrak{L}=\{\langle z,\xi\rangle;\ \xi\in\mathbb{C}^n\}$ is just the usual supporting function for a compact set.

Definition 8.11. Let $\mathscr{F}$ be an arbitrary subset of $\mathscr{H}(M)$. We define the $\mathscr{F}$-hull, $\hat{K}_{\mathscr{F}}$, of a compact set $K\subset M$ by

$$\hat{K}_{\mathscr{F}}=\{z\in M\colon \mathbb{R}e\,\varphi(z)\leqq H_K(\varphi)\ \forall\varphi\in\mathscr{F}\}.$$

We shall say that K is $\mathscr{F}$-convex if $\hat{K}_{\mathscr{F}}=K$ and M will be $\mathscr{F}$-convex if $\hat{K}_{\mathscr{F}}$ is compact whenever K is compact.

$\hat{K}_{\mathscr{F}}$ is the largest subset of M such that $H_{\hat{K}_{\mathscr{F}}}|_{\mathscr{F}} = H_K|_{\mathscr{F}}$. We note the following properties:

i) if $\mathscr{F} \subset \mathscr{G} \in \mathscr{H}(\mathbb{C}^n)$ and $K_1 \subset K_2 \Subset M$, then $\hat{K}_{1,\mathscr{G}} \subset \hat{K}_{2,\mathscr{F}}$;

ii) $\widehat{(\hat{K}_{\mathscr{G}})}_{\mathscr{F}} = \hat{K}_{\mathscr{F}} = \widehat{(\hat{K}_{\mathscr{F}})}_{\mathscr{G}}$ if $\mathscr{F} \subset \mathscr{G}$, so $\hat{K}_{\mathscr{F}}$ is $\mathscr{G}$-convex;

iii) $\widehat{(\bigcap K_i)}_{\mathscr{F}} \subset \bigcap (\hat{K}_i)_{\mathscr{F}}$, so the intersection of a family of $\mathscr{F}$-convex sets is $\mathscr{F}$-convex.

If $\mathscr{F} = \mathfrak{L}$, the family of linear functions, then $\hat{K}_{\mathfrak{L}}$ is just the convex hull of K intersected with M; if $\mathscr{F} = \mathbb{P}$, the polynomials, then $\hat{K}_{\mathbb{P}}$ is just the polynomially-convex hull of K intersected with M; if $\mathscr{F} = \mathscr{H}(M)$, then $\hat{K}_{\mathscr{H}(M)}$ is just the holomorphically-convex hull of K. To see this, it suffices to consider the family $e^{i\theta}\varphi$, $\varphi \in \mathscr{F}$, $0 \leqq \theta \leqq 2\pi$.

Definition 8.12. A compact set K is called an *$\mathscr{F}$-support* of $\mu \in \mathscr{H}'(M)$ if K is an $\mathscr{F}$-convex carrier of μ and for every carrier $L \subset K$, $\hat{L}_{\mathscr{F}} = \hat{K}_{\mathscr{F}}$.

By Zorn's Lemma, μ has an $\mathscr{F}$-support if and only if μ is carried by some $\mathscr{F}$-convex set. In general, an $\mathscr{F}$-support is not unique. For $n=1$, consider the linear functional $\mu(f) = \int_0^1 f(z)\,dz$. This has a unique convex support, namely the set $\{z: \operatorname{Im} z = 0, 0 \leqq \mathbb{R}e\, z \leqq 1\}$, but it does not have a unique polynomially-convex support, since any simple arc connecting 0 and 1 is polynomially convex (cf. [*B*, Theorem 1.3.1.]). Later on, we shall see that there does not always exists a unique convex support for an analytic functional.

§ 5. The Generalized Laplace Transform and Indicator Function

Definition 8.13. Let $\mu \in \mathscr{H}(M)'$. The generalized *Laplace transform* $\hat{\mu}$ of μ is defined by $\hat{\mu}(\varphi) = \mu(e^{\varphi})$, $\varphi \in \mathscr{H}(M)$ and the *indicator* of μ is defined by $p(\varphi) = \limsup_{t \to \infty} \dfrac{\log|\hat{\mu}(t\varphi)|}{t}$.

The restriction of $\hat{\mu}$ to $\mathfrak{L}$ is the Fourier-Borel transform of μ (when we identify $\mathfrak{L}$ with $\mathbb{C}^n$ by duality) and is an entire function of exponential type (order 1 and finite type); the restriction of p to $\mathfrak{L}$ is just the radial indicator of the Fourier-Borel transform of μ.

Suppose that μ is carried by K. Then for any open neighborhood L of K, there exists a constant C_L such that

$$|\hat{\mu}(t\varphi)| = |\mu(t e^{\varphi})| \leqq C_L \sup_L \exp \mathbb{R}e(t\varphi),$$

and so for $t>0$,

$$\frac{1}{t}\log|\hat{\mu}(t\varphi)|\leqq\frac{1}{t}\log C_L+H_L(\varphi).$$

Hence $p(\varphi)\leqq H_L(\varphi)$. Since this holds for all L, we have in fact $p(\varphi)\leqq H_K(\varphi)$.

Let E be a complex linear topological space and Ω an open subset of E. A function p defined in Ω with values in $[-\infty, +\infty)$ is plurisubharmonic ($p\in \mathrm{PSH}(\Omega)$) if $p\not\equiv-\infty$ is upper semi-continuous (i.e. $\{\theta\in\Omega: p(\theta)<C\}$ is open for topology of E for every real C) and if for every compact disc in Ω defined by $z=\theta_1+\lambda\theta_2\subset\Omega$ for $|\lambda|\leqq r$

$$p(\theta_1)\leqq\frac{1}{2\pi}\int_0^{2\pi} p(\theta_1+re^{i\varphi}\theta_2)\,d\varphi.$$

Another way to describe this situation is to say that $p\not\equiv-\infty$ is upper semicontinuous for the topology on E and its restriction to the intersection of Ω with any finite dimensional subspace M is plurisubharmonic or the constant $-\infty$ on any component of $\Omega\cap M$.

If $\mu\in\mathscr{H}(M)'$, then $\hat{\mu}$ is analytic in $\mathscr{H}(M)$, since it is continuous and $\hat{\mu}(\phi+\lambda\psi)=\mu(e^{\phi+\lambda\psi})$ is a holomorphic function of $\lambda\in\mathbb{C}$ for all $\varphi,\psi\in\mathscr{H}(M)$. Hence $\log|\hat{\mu}|\in\mathrm{PSH}(\mathscr{H}(M))$. We shall need the following result:

Theorem 8.14. *Let E be a complex linear topological space, separated and having a countable basis for the neighborhoods of the origin. Let Ω be an open subset of E and $\{p_i\}_{i\in I}$ a family of plurisubharmonic functions in Ω indexed by I, either the integers or the real numbers, with their natural order. Suppose that $\{p_i\}_{i\in I}$ is uniformly bounded above on every compact set of Ω. Then $p^\star(\theta)=\limsup_{\theta'\to\theta} p(\theta')$ is plurisubharmonic in Ω, where $p=\limsup_{i\in I} p_i$ ($p^\star$ is the upper regularization of p).*

Proof. Let θ_1 and θ_2 be fixed elements of E and r such that $\varphi+\lambda\psi\in\Omega$ for $|\lambda|\leqq r$. Let $\hat{\theta}$ be any element in a neighborhood U of the origin for which $\theta_1+\lambda\theta_2+U\subset\Omega$ for $|\lambda|\leqq r$. Since $\{p_i\}$ is bounded above on $\theta_1+\lambda\theta_2+\hat{\theta}$ for $|\lambda|\leqq r$ (this is a compact set), by Fatou's Lemma applied to the functions $\hat{p}(\lambda)=p_i(\theta_1+\lambda\theta_2+\hat{\theta})$,

$$\begin{aligned} p(\theta_1+\hat{\theta})&\leqq\limsup_{i\in I}\frac{1}{2\pi}\int_0^{2\pi} p_i(\theta_1+\hat{\theta}+re^{i\varphi}\theta_2)\,d\varphi\\ &\leqq\frac{1}{2\pi}\int_{0\star}^{2\pi} p(\theta_1+\hat{\theta}+re^{i\varphi}\theta_2)\,d\varphi\\ &\leqq\frac{1}{2\pi}\int_0^{2\pi} p^\star(\theta_1+\hat{\theta}re^{i\varphi}\theta_2)\,d\varphi\end{aligned}$$

(where $\int_{0\star}^{2\pi}$ denotes the lower Lebesgue integral over the interval $[0,2\pi]$). By our assumption on the topology of E, there exists a sequence $\{\theta_j\}_{j=1}^{\infty}$ which tends to 0 in E such that $\limsup_{j\to\infty} p(\theta_1+\hat{\theta}_j)=p^\star(\theta_1)$. Furthermore, $p^\star$ is bounded above on $\{\theta_1+\lambda\theta_2: |\lambda|\leqq r\}$, since the p_i are uniformly bounded above on $\bigcup_j \{\theta_1+\hat{\theta}_j+\lambda\theta_2: |\lambda|\leqq r\}\cup\{\theta_1+\lambda\theta_2: |\lambda|\leqq r\}$, which is compact. Thus, applying Fatou's Lemma to this sequence, we obtain

$$p^\star(\theta_1)=\limsup_{j\to\infty} p(\theta_1+\hat{\theta}_j)\leqq\frac{1}{2\pi}\int_0^{2\pi}\limsup_{j\to\infty} p^\star(\theta_1+\hat{\theta}_j+r\,e^{i\varphi}\,\theta_2)\,d\varphi$$

$$\leqq\frac{1}{2\pi}\int_0^{2\pi} p^\star(\theta_1+re^{i\varphi}\theta_2)\,d\varphi. \qquad \square$$

As far as our applications are concerned, we shall be intersected in the case where $\mathscr{F}$ is a linear subspace of $\mathscr{H}(M)$. Then the upper regularization of the restriction of p to $\mathscr{F}$, p the indicator function of an analytic functional μ, is plurisubharmonic and positively homogeneous of order 1 in $\mathscr{F}$. The upper regularization depends in general upon the subspace considered and even if $\mathscr{F}\subset\mathscr{G}$, we may have $p^\star_{\mathscr{F}}\nleqq p^\star_{\mathscr{G}}$.

§ 6. Support for Analytic Functionals

Theorem 8.15. *Let $\mu\in\mathscr{H}(\mathbb{C}^n)'$. Then for every $\xi\in\mathfrak{L}$, we have $p_{\mathfrak{L}}(\xi)=\inf_K(H_K(\xi); K$ carries $\mu)$ where $p_{\mathfrak{L}}$ is the upper regularization of the restriction of p to $\mathfrak{L}$.*

Proof. Since $p_{\mathfrak{L}}(\xi)\leqq H_K(\xi)$ for every convex subset K which carries μ, clearly $p_{\mathfrak{L}}(\xi)\leqq\inf_K H_K(\xi)$. To prove the converse, suppose that ξ_0 is fixed, $\|\xi_0\|=1$, and let α be any real number such that $\alpha>p_{\mathfrak{L}}(\xi_0)$. We show that $H_K(\xi_0)\leqq\alpha$ for some carrier K of μ. Set $q_s(\xi)=\alpha\,\mathbb{R}e\sum_j\xi_j\bar{\xi}_{0j}+s(\|\xi\|-\mathbb{R}e\sum_j\xi_j\xi_{0j})$ and set $N_s=\{\xi\in\mathfrak{L}: \|\xi\|=1, q_s(\xi)\leqq p_{\mathfrak{L}}(\xi)\}$, which is compact since $p_{\mathfrak{L}}(\xi)$ is upper semi-continuous and $\|\xi\|=1$ is compact. Then $\bigcap_{s>0} N_s=\{\emptyset\}$, for $q_s(\xi_0)=\alpha>p_{\mathfrak{L}}(\xi_0)$ and for $\xi\neq\xi_0$, $\|\xi\|=1$, $\lim_{s\to\infty} q_s(\xi)=+\infty$. By the Finite Intersection Property for compact sets, $N_{s_0}=\{\emptyset\}$ for some $s_0>0$ and so $q_{s_0}(\xi)\geqq p_{\mathfrak{L}}(\xi)$ for all ξ. Let K_{s_0} be the compact convex set with supporting function q_{s_0}. Then by Theorem 8.9, K_{s_0} carries μ and $H_{K_{s_0}}(\xi_0)=\alpha$ so $\inf_K H_K(\xi)\leqq p_{\mathfrak{L}}(\xi)$. $\square$

Corollary 8.16. *Let M be a complex submanifold of $\mathbb{C}^n$. If $\mu\in\mathscr{H}'(M)$, then μ has a unique $\mathfrak{L}$-support if and only if $p_{\mathfrak{L}}$ is convex.*

Proof. Suppose that $p_{\mathfrak{L}}$ is convex and let K_0 be the convex compact subset of M such that $H_{K_0}(\xi)=p_{\mathfrak{L}}(\xi)$. By Theorem 8.9, a convex compact subset K carries μ if and only if $K_0\subseteq K$, so K_0 is an $\mathfrak{L}$-support. $\square$

We shall use this specific case to study a more general situation. With some rather mild assumptions on the family $\mathscr{F}$, we shall be able to reduce the study of $\mathscr{F}$-convexity to that of linear convexity.

Let Ω be an open subset of M and $\alpha\colon \Omega \to \Omega' \subset \mathbb{C}^s$ a holomorphic map. We shall say that α is *regular* if its rank is everywhere equal to m-that is, the matrix $\frac{\partial \alpha_j}{\partial z_k}$, $k=1, \ldots, n$, $j=1, \ldots, s$ has rank m everywhere.

The map α is *proper* if $\alpha^{-1}(K)$ is compact on Ω when K is compact on Ω'. Finally if α is one-to-one, proper, and regular, we shall say that α is an *embedding*. We shall let $\alpha^\star\colon \mathscr{H}(\Omega') \to \mathscr{H}(\Omega)$ be defined by $\alpha^\star(\psi) = \psi \circ \alpha$ and we shall denote its transpose by α_t

$$\alpha_t\colon \mathscr{H}(\Omega)' \to \mathscr{H}(\Omega')', \qquad \alpha_t(\mu)(\psi) = \mu(\alpha^\star \psi) = \mu(\psi \circ \alpha).$$

Theorem 8.17. *Let $\mathscr{F}$ be a linear subspace of $\mathscr{H}(M)$ which contains elements $\alpha_1, \ldots, \alpha_s$ which embed M into $\mathbb{C}^s$. Then $\mu \in \mathscr{H}(M)'$ is carried by some $\mathscr{F}$-convex set K if and only if $p(\psi) \leqq H_K(\psi)$, $\psi \in \mathscr{F}$.*

Proof. If μ is carried by K, then the inequality is obvious, so we prove the converse. Let L be an open neighborhood of K and $L \Subset M$. For every point $z \in bdL$, there exists $\psi_z \in \mathscr{F}$ such that $|\psi(z)| > \sup_K |\psi|$, since $\mathscr{F}$ is a complex linear space, and by continuity, there exists a neighborhood N_z of z in which this inequality continues to hold. Since bdL is compact, there exists a finite number $N_i = N_{z_i}$, $i=1, \ldots, q$, which cover bdL. Then K is a compact subset of $A = \{z\colon |\psi_{z_i}(z)| \leqq |\psi_{z_i}(z_i)| = a_i,\ 1 \leqq i \leqq q\}$.

Suppose that $\mathbb{B}$ is some finite dimensional subset of $\exists$ spanned by $(\beta_1, \ldots, \beta_t)$. Then

$$\tilde{\alpha} = (\tilde{\alpha}_1, \ldots, \tilde{\alpha}_{s'}) = (\alpha_1, \ldots, \alpha_s, \psi_{z_1}, \ldots, \psi_{z_q}, \beta_1, \ldots, \beta_t), \qquad s' = s+q+t$$

is an embedding of M into $\mathbb{C}^{s'}$ (we assume without loss of generality that the coordinates $\tilde{\alpha}_j$ are linearly independent, for otherwise, we extract a maximal linearly independent subset). Then $M' = \tilde{\alpha}(M)$ is an m-dimensional manifold in $\mathbb{C}^{s'}$.

Let $b_j = \sup_A \tilde{\alpha}_j$ and let δ_j be so small that

$$K \subset \{z\colon |\tilde{\alpha}_j(z)| < b_j - 2\delta_j\}.$$

Let

$$D = \{w \in \mathbb{C}^{s'}\colon |w_j| < b_j\}$$

and

$$D' = \{w \in \mathbb{C}^{s'}\colon |w_j| < b_j - \delta_j\}.$$

We consider $\mu^\star = \alpha_t^\star(\mu)$, which is an analytic functional on $\mathbb{C}^{s'}$. If $p_\star$ is the indicator of $\mu^\star$ and $p_\star(\xi)$ is its restriction to the linear functions, then

$$p_\star(\xi) = p(\langle \xi, \alpha^\star \rangle) \leqq H_K(\langle \xi, \alpha^\star \rangle) \leqq \sum_j (b_j - 2\delta_j)|\xi_j|.$$

Thus by Theorem 8.9, $\mu^\star$ is carried by some compact subset $\hat{K}$ of D'.

Suppose $f \in \mathscr{H}(\mathbb{C}^{s'})$ such that $f|_{M'} \equiv 0$. Then $\mu^\star(f) = \mu(f \circ \tilde{\alpha}) = 0$. Now if $f \in \mathscr{H}(D')$ such that $f|_{M'} \equiv 0$, we can find a sequence $f_\nu \in \mathscr{H}(\mathbb{C}^{s'})$ with $f_\nu|_{M'} \equiv 0$ such that $f_\nu \to f$ uniformly on compact subsets of D'. Thus $\mu^\star(f) = 0$ in this case also, so that $\mu^\star$ extends to a linear functional on $\mathscr{H}(D')|_{M' \cap D'}$ (cf. [*B*, Theorem 7.2.7]). Furthermore, for every holomorphic function f defined on M', there exists a holomorphic function $\tilde{f}$ defined on $M'' = M' \cap D'$ such that $\tilde{f}|_{M''} = f$ (cf. [*A*, *B*]).

Thus, the map of Fréchet spaces $\omega\colon \mathscr{H}(D) \to \mathscr{H}(M'')$ defined by $\omega(F) = F|_{M''}$ is surjective, and by the Open Mapping Theorem, for $\hat{K}$, there exists a compact $K \subset M''$ such that $\sup_{\hat{K}} |\tilde{f}| \leqq B \sup_K |f|$ for some constant $B > 0$. Then for $\psi \in \mathscr{H}(M')$ and its extension $\tilde{\psi}$ (to $\mathscr{H}(D')$):

$$|\mu(\psi)| = |\mu^\star(\tilde{\psi})| \leqq C \sup_{\hat{K}} |\tilde{\psi}| \leqq CB \sup_K |\psi| \leqq CB \sup_{M''} |\psi|$$
$$\leqq CB \sup_L |\psi|.$$

Thus ψ is carried by K, since L was any compact neighborhood of K. □

Theorem 8.18. *Let $\mathscr{F}$ be a linear subspace of $\mathscr{H}(M)$ which contains elements $\alpha_1, \ldots, \alpha_s$ which embed M into $\mathbb{C}^s$. Then $\mu \in \mathscr{H}(M)'$ has a unique $\mathscr{F}$-support if and only if $p_{\mathscr{F}}$ is convex.*

We shall break up the proof of the theorem into two steps.

Proposition 8.19. *An analytic functional $\mu \in \mathscr{H}(M)'$ has a unique $\mathscr{F}$-support if and only if μ has a unique $\mathscr{F}$-support for every finite-dimensional subspace $\mathbb{B}$ of $\mathscr{F}$ such that $\alpha_j \in \mathbb{B}$, $j = 1, \ldots, s$.*

Proof. Suppose that μ has a unique $\mathscr{F}$-support K and let $\mathscr{F}'$ be any subset of $\mathscr{F}$ such that $K_{\mathscr{F}'}$ is compact. Then the $\mathscr{F}'$-hull of K is contained in every $\mathscr{F}'$-convex carrier of μ, since every $\mathscr{F}'$-convex set is $\mathscr{F}$-convex. Thus μ has a unique $\mathscr{F}'$-support.

Suppose that μ has two different $\mathscr{F}$-supports K_1 and K_2. Let L_1 and L_2 be compact neighborhoods of K_1 and K_2 respectively such that $K_1 - L_2 \neq \emptyset$ and $K_2 - L_1 \neq \emptyset$. In addition, we choose L_1 and L_2 sufficiently small so that μ is not carried by $L_1 \cap L_2$. Since μ is not carried by $K_1 \cap K_2$, the same is true for some compact neighborhood L of $K_1 \cap K_2$ and then we take L_1 and L_2 so that $L_1 \cap L_2 \subset L$. Now as in the proof of Theorem 8.17, we choose finite dimensional subspaces $\mathscr{F}_1$ and $\mathscr{F}_2$ such that $\hat{K}(j, \mathscr{F}_j) \subset L_j$, $j = 1, 2$.

Then if $\mathscr{F}' = \mathscr{F}_1 + \mathscr{F}_2$, $K(j, \mathscr{F}') \subset L_j$, so that μ has two $\mathscr{F}'$-convex carriers, $\hat{K}_{1,\mathscr{F}}$ and $\hat{K}_{2,\mathscr{F}}$, but μ is not carried by their intersection. Thus μ has no unique $\mathscr{F}'$-support. □

Proof of Theorem 8.18. Clearly $p_{\mathscr{F}}$ is convex if and only if its restriction to every finite dimensional subspace $\mathbb{B}$ is convex. Thus, by Proposition 8.19, it is sufficient to prove the theorem for every finite dimensional subspace $\mathbb{B}$ such that $\alpha_1, \ldots, \alpha_s$ belong to $\mathbb{B}$. Let $\alpha_1, \ldots, \alpha_s, \alpha_{s+1}, \ldots, \alpha_{s'}$, span $\mathbb{B}$ and define $\alpha = (\alpha_1, \ldots, \alpha_{s'})$ as an embedding of M into $\mathbb{C}^{s'}$. Let $\mu^\star = \alpha_t^\star(\mu)$. Then as in Theorem 8.17, $\mu^\star$ defines a linear functional on $\mathscr{H}(\alpha(M))$, and if K is an $\mathfrak{L}$-support for $\mu^\star$, $\alpha^{-1}(K \cap \alpha(M))$ is a $\mathbb{B}$-support for μ. Let $p_\star(\xi) = p_{\mathbb{B}}(\langle \xi, \alpha \rangle)$. Then by Corollary 8.16, $\mu^\star$ has a unique $\mathfrak{L}$-support if and only if $p_\star(\xi)$ is convex, so μ has a unique $\mathbb{B}$-support if and only if $p(\psi)|_{\mathbb{B}}$ is convex. □

§7. Unique Supports for Domains in $\mathbb{C}^n$

We have already seen an example of functionals which do not admit unique supports. For $n=1$, an analytic functional always has a unique convex support. This is because $p(\xi)$ is always convex for $n=1$ (cf. [D]). For $n>1$, however, in general there is no unique support even in the class of convex supports. Let μ be the analytic functional whose Fourier-Borel transform is $\cos(\xi_1 \xi_2)^{1/2}$ for $n=2$. Since the circled indicator of this function is $|\xi_1 \xi_2|^{1/2}$ and $\left(|t\xi_1|^{1/2} - \frac{1}{2}\left|\frac{\xi_2}{t}\right|^{1/2}\right)^2 \geqq 0$, $t|\xi_1| + \frac{1}{4t}|\xi_2| \geqq |\xi_1 \xi_2|^{1/2}$, μ is carried by the polydisc $K_t = \left\{(z_1, z_2)\colon |z_1| \leqq t,\ |z_2| \leqq \frac{1}{4t}\right\}$, by Theorem 8.9. Let $a_t = \left(t, -\frac{1}{4t}\right)$. Then $a_t \in K_s$ if and only if $s=t$. If $\hat{K}$ is any convex carrier contained in K_t, then $a_t \in \hat{K}$ since $2 = p\left(\frac{1}{t}, -4t\right) = \sup_{K_t}(\mathbb{R}e\langle a_t, 1\rangle)$ and $\mathbb{R}e\langle z, \xi_t\rangle < 2$ for $z \in K_t$, $z \neq a_t$ where $\xi_t = \left(\frac{1}{t}, -4t\right)$. Thus K_t is a convex support. The problem is that the convex supports have corners. We shall see below that this is a feature of convex supports which are not unique (as well as other types of non-uniques supports).

Proposition 8.20. *Let Ω be a domain of holomorphy in $\mathbb{C}^n$ and K a compact holomorphically convex subset of Ω. Then for every open neighborhood ω of K, there exists a constant C_ω such that if $f \in \mathscr{C}^\infty_{(0,1)}(\Omega)$ and $\bar\partial f = 0$, then given $\varepsilon > 0$, there exists $u \in \mathscr{C}^\infty(\Omega)$ with $\bar\partial u = f$* and $\sup_K u \leqq C_\omega \sup_\omega |f| + \varepsilon$.

Proof. Let ω_1 be an open neighborhood of K, $\omega_1 \Subset \Omega$ and ω_1 holomorphically convex in Ω. Then $-\log d_\Omega(z)$ is plurisubharmonic in Ω, and so we can find a plurisubharmonic function φ (depending only on ω_1, $\hat{C}$ and f) such that $\varphi \equiv 0$ on ω_1 and

$$\Big[\int_\Omega |f|^2 \exp - \varphi\, d\tau\Big]^{1/2} = \Big[\int_{\omega_1} |f|^2 \exp - \varphi\, d\tau + \int_{\Omega - \omega_1} |f|^2 \exp - \varphi\, d\tau\Big]^{1/2}$$
$$\leqq C_0(\omega_1) \sup_{\omega_1} |f| + \varepsilon / \hat{C}$$

where $\hat{C}$ is a constant to be fixed later (if $c=\sup_{\omega_1}(-\log d_\Omega(z))$, we choose $\gamma(t)$ to be a sufficiently rapidly increasing function of t such that $\gamma\equiv 0$ for $t\leqq c$ and set $\varphi(z)=\gamma(\sup(-\log d_\Omega(z),c))$. Then there exists a solution $u\in\mathscr{C}^\infty(\Omega)$ of the equation $\bar\partial u=f$ such that

$$\int_\Omega |u|^2[\exp-\varphi](1+\|z\|^2)^{-3n}d\tau\leqq\int_\Omega |f|^2\exp-\varphi\,d\tau \qquad \text{(cf. Appendix III).}$$

Let $\psi\in\mathscr{C}^\infty(\Omega)$ such that $\psi\equiv 1$ on K. Then if $C_n=\dfrac{(n-2)!}{4\pi^n}$, for $z\in K$,

$$\begin{aligned}
u(z)&=C_n\int_\Omega \frac{-1}{\|z-a\|^{2n-2}}\Delta(\psi u)d\tau=\frac{i}{2}C_n\int_\Omega\frac{-1}{\|z-a\|^{2n-2}}\partial\bar\partial(\psi u)\wedge\beta_{n-1}\\
&=\frac{i}{2}C_n\int_\Omega\partial\left(\frac{1}{\|z-a\|^{2n-2}}\right)\wedge\bar\partial(\psi u)\wedge\beta_{n-1}\\
&=\frac{i}{2}C_n\left[\int_\Omega u\partial\left(\frac{1}{\|z-a\|^{2n-2}}\right)\wedge\bar\partial\psi\wedge\beta_{n-1}\right.\\
&\qquad\left.+\int_\Omega\psi\partial\left(\frac{1}{\|z-a\|^{2n-2}}\right)\wedge f\wedge\beta_{n-1}\right].
\end{aligned}$$

Since for $z\in K$, $a\in\operatorname{supp}\bar\partial\psi$, $\|z-a\|>\delta>0$, it follows from the Schwarz Inequality that for $z\in K$,

$$|u(z)|\leqq\hat{C}(\omega_1)\Big[\int_{\omega_1-K}|u|^2\Big]^{1/2}+C'\sup_{\omega_1}|f|,$$

where $\hat{C}(\omega_1)$ depends only on ω_1

$$|u(z)|\leqq\hat{C}'(\omega_1)\Big[\int_\Omega|f|^2\exp-\varphi\,d\tau\Big]^{1/2}+C'\sup_{\omega_1}|f|,$$

where $\hat{C}'(\omega_1)$ depends only on ω_1, since $\varphi\equiv 0$ on ω_1. Thus

$$|u(z)|\leqq\hat{C}'(\omega_1)\left[C_0(\omega_1)\sup_{\omega_1}|f|+\frac{\varepsilon}{\hat{C}}\right]+C'\sup_{\omega_1}|f|,$$

so we choose $\hat{C}=\hat{C}'(\omega_1)$. □

Theorem 8.21. *Let K_0 and K_1 be compact sets in a domain of holomorphy $\Omega\subset\mathbb{C}^n$ and let L be the holomorphically convex hull of $K_0\cup K_1$. Suppose that K is such that $L-K$ is a disjoint union of two sets M_0 and M_1 closed in $L-K$ such that $K_j-K\subset M_j$, $j=0,1$. Then every analytic functional $\mu\in\mathscr{H}(\Omega)'$ carried by K_0 and K_1 is carried by K.*

Proof. Let ω be any open neighborhood of K. We begin by constructing a function $\psi\in\mathscr{C}_0^\infty(\Omega)$, such that

i) $0\leqq\psi\leqq 1$;

ii) $\psi=j$ on $\omega_j-\omega$ for some open neighborhoods ω_j of K_j;

iii) ψ is constant on every component of $U-\bar{\omega}$ for some open neighborhood U of the holomorphically convex hull of $\omega_0 \cup \omega_1$.

Let $m_j = M_j - \omega$. Then $L-\omega = m_0 \cup M_1$ and $m_0 \cap m_1 = \emptyset$. Furthermore, the m_j are closed in L, hence compact. Let

$$m_j^\delta = \{z \in \Omega : \inf_{w \in m_j} \|z-w\| < \delta\}.$$

Then for some $\varepsilon > 0$, the sets $m_j^{3\varepsilon}$ are disjoint and contained in Ω. Let $\alpha \in \mathscr{C}_0^\infty(B(0,\varepsilon))$ be such that $\int_{\mathbb{C}^n} \alpha(z)\, d\tau(z) = 1$ and set $\psi = \chi_{m_1} \star \alpha$ (χ_{m_1} is the characteristic function of m_1). Then $\psi = j$ in m_j^ε and $0 \leqq \psi \leqq 1$. Furthermore, $m_0^\varepsilon \cup m_1^\varepsilon \cup \omega$ is a neighborhood of L, so we can find two open neighborhoods U and V of L such that $V' \subset U \subset (m_0^\varepsilon \cup m_1^\varepsilon \cup \omega)$, where V' is the holomorphically convex hull of V. Set $\omega_j = (m_j^\varepsilon \cup \omega) \cap \bar{V}$ so that $\omega_0 \cup \omega_1 = V$; hence the holomorphically convex hull of $\omega_0 \cup \omega_1$ is contained in U. Since $U - \bar{\omega} \subset (m_0^\varepsilon \cup m_1^\varepsilon)$, ψ is constant on every component of this set.

By Theorem 8.20, we can find a constant C' such that for every $f \in \mathscr{H}(\Omega)$ and every $\varepsilon > 0$ there exists $u \in \mathscr{C}_0^\infty(\Omega)$ with $\bar{\partial} u = f \bar{\partial} \psi$ and

$$\sup_{\omega_0 \cup \omega_1} |u| \leqq C' \sup_U |f \bar{\partial} \psi| + \varepsilon \leqq C' \sup_\omega |f \bar{\partial} \psi| + \varepsilon,$$

since $\bar{\partial} \psi = 0$ on $U - \bar{\omega}$.

Now $\mu(f) = \mu(\psi f - u) + \mu((1-\psi) f + u)$, and since μ is carried by K_0 and K_1, we obtain

$$\begin{aligned} |\mu(f)| &\leqq C_0 \sup_{\omega_0} |\psi f - u| + C_1 \sup_{\omega_1} |(1-\psi) f + u| \\ &\leqq C_0 \sup_\omega |\psi| + C_0 \sup_\omega |u| + C_1 \sup_\omega |(1-\psi) f| + C_1 \sup_{\omega_1} |u| \end{aligned}$$

because $\psi = j$ in $\omega_j - \bar{\omega}$. Hence

$$|\mu(f)| \leqq (C_0 + C_1)(\sup_\omega |f| + C' \sup_\omega |f \bar{\partial} \psi| + \varepsilon)$$

and since ε was arbitrary, the proof is complete. □

Corollary 8.22. *Let Ω be a domain of holomorphy in $\mathbb{C}^n$ and K_0 and K_1 carriers of $\mu \in \mathscr{H}(\Omega)'$. Then μ is carried by $K = K_0 \cap (\overline{(L-K_0)} \cup K_1)$, where L is the holomorphically convex hull of $K_0 \cup K_1$. If $K_0 \cup K_1$ is holomorphically convex, then μ is carried by $K_0 \cap K_1$.*

Proof. Set

$$S = \overline{(L-K_0)} \cup K_1, \quad M_0 = K_0 - K = K_0 - S = L - S, \quad M_1 = (L-K) - M_0.$$

Then $M_0 \cap M_1 = \emptyset$, $M_0 \cup M_1 = L - K$ and M_0 is closed in $L-K$, for

$$\begin{aligned} (L-K) \cap M_0 &= L \cap \complement(K_0 \cap \complement M_0) \cap \bar{M}_0 \\ &= L \cap ((\complement K_0 \cap \bar{M}_0) \cup M_0) = L \cap M_0 = M_0. \end{aligned}$$

On the other hand, $M_0 = M_0 - K = (L-S) - K = (L-K) \cap \complement S$ is open since $\complement S$ is open in $L-K$. Finally $K_0 - K \subset M_0$ and $K_1 - K \subset \complement K_0 \subset \complement M_0$, so $K_1 - K \subset (L-K) \cap \complement M_0 = M_1$. We can then apply Theorem 8.21. □

Theorem 8.23. *Let Ω be a domain of holomorphy in $\mathbb{C}^n$ and $\mu \in \mathscr{H}(\Omega)'$. If K_0 is an $\mathscr{H}(\Omega)$-convex support of μ whose boundary is twice continuously differentiable, then K_0 is the unique $\mathscr{H}(\Omega)$-support of μ.*

Proof. We show that every convex carrier K_1 of μ contains K_0. To show this, it suffices to construct for every $\mathscr{H}(\Omega)$-convex compact set K_1 with $K_0 - K_1 \neq \emptyset$ two plurisubharmonic functions F and G continuous in Ω such that

i) $\sup_{K_1} F \leqq 0$, $\sup_{K_0} F > 0$;

ii) $\sup_{K_0 \cup K_1} G \leqq 0$, hence $\sup_{L} G \leqq 0$ (where L is the holomorphically convex hull of $K_0 \cup K_1$) and $z \notin K_0$, $G(z) \leqq 0$ implies $F(z) \leqq 0$. If $z \in L - K_0$ then $F(z) \leqq 0$ by (ii) and so by (i) $\sup_{(L-K_0) \subset K_1} F \leqq 0$. Thus $\sup_{K} F \leqq 0$, where $K = K_0 \cap L'$ and L' is the hull of $(L-K_0) \cup K_1$. Hence K is a holomorphically convex proper subset of K_0, since $F > 0$ somewhere in K_0. Then Corollary 8.22 shows that K carries μ, which is a contradiction. Hence K_0 is the unique holomorphically convex support of μ.

We now carry cut the construction of F and G. An essential ingredient in the proof will be the fact that the hull of a compact set $K \subset \Omega$ with respect to the holomorphic functions, the plurisubharmonic functions, and the continuous plurisubharmonic functions in Ω, is the same if Ω is a domain of holomorphy.

Since K_1 is supposed holomorphically convex, there exists a $\mathscr{C}^\infty$ plurisubharmonic fonction $\tilde{G}$ in Ω which satisfies $\sup_{K_1} \tilde{G} < 0$ and $\sup_{K_0} \tilde{G} \geqq 0$ (cf. [*B*, Theorem 2.6.11]). Then for δ sufficiently small $\hat{G} = \tilde{G} + \delta \|z\|^2$ also has these properties. Let $\hat{H} = \hat{G} - \sup_{K_0} G$ and let $a \in bd\, K_0$ such that $\hat{H}(a) = 0$.

We now choose b on the interior normal to $bd\, K_0$ at a, so that $z \neq a$, $\|z-b\| \leqq \|a-b\|$ implies $z \in \mathring{K}_0$ (it is at this point that we make use the differentiability conditions) and set $H_j(z) = \hat{H}(z) - (3-j)\, \varepsilon(\|z-b\|^2 - \|a-b\|^2)$, $j = 0, 1, 2$ where $0 < \varepsilon < \delta$ is so small that $\sup_{K_1} H_j < 0$. Then H_j is a $\mathscr{C}^\infty$ plurisubharmonic function. Furthermore $H_j(z) \leqq H_3(z) \leqq 0$ on $bd\, K_0$ with equality only at a. If $H_j(z) = 0$, $z \neq a$, then $z \in K_0$ and so by the maximum principle, $H_j \equiv 0$ in K_0, which is impossible since it is strictly plurisubharmonic.

We now construct a function $f \in \mathscr{C}^\infty(\Omega)$ such that $K_0 = \{z \in \Omega : f(z) \leqq 0\}$ and $f \geqq H_2$ in Ω (we construct the function locally and then piece together the local functions via a $\mathscr{C}^\infty$ partition of unity; of course, f is not plurisubharmonic). Then $f - H_1$ is convex in some neighborhood ω_1 of a since $f - H_1 \geqq H_2 - H_1$ and $f = H_1$ only at a. The matrix $\left(\partial^2 \dfrac{(H_2 - H_1)}{\partial x_j \partial x_k}\right)_{j,k=1}^{2n}$ is positive definite at a (where the x_j are the underlying real coordinates in $\mathbb{C}^n=\mathbb{R}^{2n}$). Thus, in ω_1, $f-H_1=\sup(A;\, A\leqq f-H_1 \text{ in } \omega_1)$ where the supremum is taken over all functions $A(z)=\mathbb{R}e\langle z,\theta\rangle+C$. We define a norm on these functions by $\|A\|=\sup\limits_{\|z\|\leqq 1}|A(z)|$, and for $\eta>0$ we let

$$G_\eta=H_1+A_0+\sup_A\{A;\, A_0+A\leqq f-H_1 \text{ in } \omega_1 \text{ and } \|A\|<\eta\},$$

where A_0 is defined by

$$H_2(z)-H_1(z)=H_1(z)-H_0(z)=A_0(z)+o(z-a)$$

(A_0 is the first order approximation in the Taylor series development of $(\|z-b\|^2-\|a-b\|^2)$). Then G_η is continuous and plurisubharmonic in Ω, and $G_\eta=f$ in some open neighborhood ω_η of a, since A_0 is the best affine approximation to $f-H_1$ at a. Furthermore, $G_\eta\leqq f$ in ω_1 by construction, and $H_1+A_0\leqq G_\eta$ in Ω, $H_0+A_0\leqq H_1$ in Ω and $H_1+A_0\leqq H_2$ in Ω.

Since G_η decreases to H_1+A_0 and $H_1+A_0\leqq H_2<0$ in $K_0-\omega_1$, by Dini's Theorem, for $\eta<\eta_0$, $G_\eta<0$ in $K_0-\omega_1$; hence $G_\eta\leqq 0$ in K_0, since $G_\eta\leqq f\leqq 0$ in $K_0\cap\omega_1$. Similarily for $\eta<\eta_1$, $G_\eta<0$ in K_1. Thus for $G=G_\eta$ and $\eta=\frac{1}{2}\inf(\eta_1,\eta_2)$, we have the first part of (ii).

Let ω_2 be the neighborhood of a where $G=f$. Then

$$\begin{aligned} q&=\sup(H_0(z)+A_0(z);\, z\notin K_0 \quad\text{and}\quad G(z)\leqq 0)\\ &\leqq\sup(H_0(z)+A_0(z);\, z\notin K_0\cup\omega_2 \quad\text{and}\quad H_1(z)+A_0(z)\leqq 0)\\ &\leqq\sup(-\varepsilon(-\varepsilon(\|z-b\|^2-\|a-b\|^2));\, z\notin K_0\cup\omega)<0 \end{aligned}$$

by the choice of b. Let $F(z)=H_0(z)+A_0(z)+C$ with

$$C=\inf(-q,\, -\sup_{K_1}(H_0+A_0))\geqq\inf(-q,\, -\sup_{K_1}H_2)>0.$$

Then (i) and (ii) are satisfied, which completes the proof. □

§8. Unique Convex Supports

The condition of differentiability imposed on the boundary in Theorem 8.23 is somewhat artificial, and we would like to replace it with some weaker condition. We do this below for the case of convex supports.

Definition 8.24. Let K_1 and K_2 be two compact convex subsets of $\mathbb{C}^n$. A convex set L *linearly separates* K_1 and K_2 if for every ξ a complex linear functional, $\xi(K_1)\cap\xi(K_2)\subset\xi(L)$.

In particular, for every λ, $0\leqq\lambda\leqq 1$, $\lambda K_1+(1-\lambda)K_2$ separates K_1 and K_2 linearly.

Lemma 8.25. *If an analytic functional $\mu \in \mathcal{H}(\mathbb{C}^n)'$ is carried by two convex compact sets K_1 and K_2, it is carried by every compact set which linearly separates K_1 and K_2.*

Proof. If φ_μ is the indicator of μ then φ_μ extends holomorphically to $\overset{\star}{\complement} K_1$ and $\overset{\star}{\complement} K_2$. Suppose that L linearly separates K_1 and K_2 and let $\hat{\xi}$ be a hyperplane of $\overset{\star}{\complement} L$ (which is not the element at infinity). Let $D(\hat{\xi}) = \{(1, \lambda\xi_1, \ldots, \lambda\xi_n), \lambda \in \mathbb{C}\}$ be the complex line in projective space determined by $\hat{\xi}$ and 0. For a point $(1, \lambda\xi)$, we expand φ_μ in its Taylor series expansion as a function of λ: $\varphi_\mu((1, \lambda\xi)) = \sum_{n=0}^{\infty} S_n(\lambda, \xi)(\lambda' - \lambda)^n$ which has radius of convergence $R(\lambda, \xi)$. Furthermore, if $R^\star(\lambda, \xi) = \liminf_{(\lambda', \xi') \to (\lambda, \xi)} R(\lambda', \xi')$ then $R(\lambda, \xi) = R^\star(\lambda, \xi)$ except perhaps on a set without interior. Thus if we can show that φ_μ extends to a holomorphic function of the complex variable λ on $\overset{\star}{\complement} L \cap D(\hat{\xi})$ for every $\hat{\xi}$, the series for φ_μ will converge in $\overset{\star}{\complement} L$, which will prove the lemma, since $\overset{\star}{\complement} L$ is open.

The hyperplane $\hat{\xi}$ is defined by an equation $1 + \langle z, \xi \rangle = 0$ or $\langle z, \xi \rangle = 0$. Let ξ_μ be the analytic functional on $\mathcal{H}(\mathbb{C})$ defined by $\xi_\mu(g) = \mu(g(\langle z, \xi \rangle))$. The hyperplane (η_0, η) where $\eta = \lambda\eta_0\xi$, $\lambda \in \mathbb{C}$, $\eta_0 \in \mathbb{C}$ is the set of points $\eta_0 + \lambda\eta_0\langle z, \xi \rangle = 0$ and is mapped by $\hat{\xi}$ onto the point $-\dfrac{1}{\lambda}$ if $\eta_0 \neq 0$ or 0 if $\eta_0 = 0$.

Now ξ_μ is carried by $\xi(K_1)$ and $\xi(K_2)$ since

$$\varphi_{\xi_\mu}(\lambda) = \xi_\mu\left(\frac{1}{1+\lambda\xi}\right) = \mu\left(\frac{\eta_0}{\eta_0 + \langle z, \lambda\eta_0\xi \rangle}\right) = \varphi_\mu((1, \lambda\xi))$$

if $\eta_0 \neq 0$ or $\varphi_{\xi_\mu}(0) = \xi_\mu(0) = 0 = \varphi_\mu(0)$. Then by the unicity of the convex support for $n = 1$,

$$\overset{\star}{\complement}(\xi(K_1) \cap \xi(K_2)) = \overset{\star}{\complement}(\xi(K_1)) \cup \overset{\star}{\complement}(\xi(K_2))$$

and φ_{ξ_μ} is carried by the intersection $\xi(K_1) \cap \xi(K_2)$. Thus φ_μ extends holomorphically to $\overset{\star}{\complement} L$ and μ is carried by L. □

Corollary 8.26. *If $\mu \neq 0$ and K_1 and K_2 are two convex carriers of μ, then $K_1 \cap K_2 \neq \emptyset$.*

Proof. Suppose $K_1 \cap K_2 = \emptyset$. If φ_μ is the indicator of μ then, φ_μ extends to $\overset{\star}{\complement} K_1$ and $\overset{\star}{\complement} K_2$, so it is defined and holomorphic everywhere hence it is constant, thus zero since it is zero at infinity. □

An open set $\Omega \subset \mathbb{P}(\mathbb{C}^{n+1})$ is starlike with respect to the origin if for every $\xi \in \Omega$, $t\xi \in \Omega$ for $0 \leqq t \leqq 1$. If Ω_α is a family of starlike domains with respect to the origin, then $\bigcup_{\alpha \in A} \Omega_\alpha$ is also, and it is simply connected. Thus there exists a largest set which is starlike with respect to the origin in which φ_μ is holomorphic for $\mu \in \mathscr{H}(\mathbb{C}^n)'$. We will denote this set by $\overset{\star}{\Omega}_\mu$ and call it the *starlike domain* of φ_μ. If K is a convex carrier of μ containing the origin, then $\overset{\star}{\complement} K$ is starlike with respect to origin in $\mathbb{P}(\mathbb{C}^{n+1})$ and hence $\overset{\star}{\complement} K \subset \overset{\star}{\Omega}_\mu$. *Thus a convex set L containing the origin will be a convex support of μ (relative to the family of all convex sets containing the origin) if and only if $\overset{\star}{\complement} L$ is maximal among the open sets $\overset{\star}{\complement} M$ where M runs through the family of convex compact sets which contain the origin and are contained in $\overset{\star}{\Omega}_\mu$.*

Definition 8.27. A complex hyperplane will be said to be a *supporting hyperplane* at $x_0 \in bd\,K$, K a compact set, if it is contained in a real supporting hyperplane at x_0.

Lemma 8.28. *Let K be a compact convex set. Then for $x \in K$, the convex hull of $(K \cap \complement B(x, \varepsilon))$ does not contain x for every $\varepsilon > 0$ if and only if x is extremal.*

Proof. Suppose that K is contained in some p dimensional subspace of $\mathbb{R}^{2n} = \mathbb{C}^n$. If there exists $\varepsilon_0 > 0$ such that the convex hull of $K \cap \complement(B(x, \xi_0))$ contains x, there exist $(p+1)$ points of K at a distance at least ε_0 from x such that x is in the convex hull of these points. Thus the point x is not extremal in this simplex and thus not extremal in K. On the other hand, if x is not extremal, x is contained in the interior of some line segment in K, say $x = tx_0 + (1-t)x_1$ $0 < t < 1$, and their exists ε_0 such that the convex hull of $(tx_0, x_1) \cap \complement B(x, \varepsilon)$ contains x for $\varepsilon < \varepsilon_0$. □

Lemma 8.29. *Let K be a convex support of μ which contains the origin. For every extremal point $x_0 \in K$, there exists a complex supporting hyperplane at x_0 which is a boundary point of $\overset{\star}{\Omega}_\mu$. If every complex supporting hyperplane at x_0 which contains the origin also contains K, then there exists a complex supporting hyperplane at x_0 which does not contain the origin and is a boundary point of $\overset{\star}{\Omega}_\mu$.*

Proof. If all the complex supporting hyperplanes at x_0 contain the origin, then they are by the very definition of $\overset{\star}{\Omega}_\mu$ boundary points of the domain (they correspond to the point at infinity and φ_μ is defined in a neighborhood of the points at infinity). Thus, we assume the contrary.

Let V be the smallest complex subspace of $\mathbb{C}^n$ containing K and let U be a complementary subspace such that $U \cap V = \{0\}$ and $V \times U = \mathbb{C}^n$. We assume without loss of generality that $V=\mathbb{C}^q$. Let $\overset{\star}{\complement}_V K$ be the complex hyperplanes in $\mathbb{P}(\mathbb{C}^{q+1})$ which do not meet K. Then $\overset{\star}{\complement} K=(\overset{\star}{\complement}_V K)\times\mathbb{C}^{n-q}$.

Suppose that φ_μ is holomorphic in $\overset{\star}{\complement}_V K\times\mathbb{C}^{n-q}$ and that it can be extended to a neighborhood ω of a boundary point (v_0', u') of $\overset{\star}{\complement} K$. Since $\mathbb{P}(\mathbb{C}^{q+1})$ is a complex manifold of dimension q, we can assume without loss of generality that $\omega=\Delta'\times\Delta''$ where Δ' is a polydisc in $\mathbb{C}^q$ and Δ'' is a polydisc in $\mathbb{C}^{n-q}$. Let $f(z', z'')=\sum_{(\alpha)} c_{(\alpha)}(z'')\, z'^{\alpha}$ be the Taylor series expansion of φ_μ in $z'\in\Delta'$ with coefficients in $\mathscr{H}(\Delta'')$. If $R(z'')$ is such that this series converges absolutely for $\sup_{1\leqq i\leqq n-q} |z_i'|\leqq R(z'')$ for z'' fixed, then

$$\psi(z'')=-\log R(z'')=\limsup_{|\alpha|\to\infty} |\alpha|^{-1}\log|c_{(\alpha)}(z')|,$$

and hence $R(z'')$ is either identically $+\infty$ in Δ'' or $\psi(z'')^\star$ is a plurisubharmonic function. But $\psi(z'')^*$ is identically $-\infty$ on an open subset of Δ''. Thus, we conclude that $R(z'')$ is identically $+\infty$ in Δ'', so φ_μ can be extended to a neighborhood $\omega_V(v_0')\times\mathbb{C}^{n-q}$, where $\omega_V(v_0')$ is a convex neighborhood of v_0' in V.

The trace of a complex hyperplane on V is either all of V or a complex hyperplane of V. In particular, the trace on V of a complex hyperplane with support at x_0 which does not contain the origin is of the form (v_0', u') where v_0' is a boundary point of $\overset{\star}{\complement}_V K$, $u'\in\mathbb{C}^{n-q}$.

Let $\pi_V(x_0)$ be the set of these boundary points. Then $\pi_V(x_0)$ is compact and $\left(\bigcup_{v_0'\in\pi_V(x_0)} \omega_V(v_0')\times\mathbb{C}^{n-q}\right)\cup\overset{\star}{\complement} K$ contains a starlike set which is thus simply connected and of the form $(\bar{\omega}_V(x_0)\times U')\cup\overset{\star}{\complement} K$, where $\bar{\omega}_V(x_0)$ is a neighborhood of $\pi_V(x_0)$. Thus, φ_μ is holomorphic in $(\bar{\omega}_V(x_0)\times U')\cup\overset{\star}{\complement} K=\Omega$.

Let L_n be the convex hull of $\left(K\cap B\left(x_0, \frac{1}{n}\right)\right)$. Then for $n>n_0$, $\overset{\star}{\complement} L_n\subset\Omega$. This will be true if we can show that $\overset{\star}{\complement}_V L_n\subset\left(\overset{\star}{\complement}_V K\right)\cup\bar{\omega}_V(x_0)$ for $n>n_0$. If this last statement does not hold, since $\overset{\star}{\complement}_V K\subsetneqq\overset{\star}{\complement} L_n$, for every n there exists a complex hyperplane $\overset{\star}{\xi}_n$ of V such that

i) $\overset{\star}{\xi}_n\cap L_n=\emptyset$,

ii) $\overset{\star}{\xi}_n\cap K\neq\emptyset$,

iii) $\overset{\star}{\xi}_n\notin\bar{\omega}_V(x_0)$.

If $\overset{\star}{\xi}_0$ is a limit point of $\overset{\star}{\xi}_n$ then from ii) $\overset{\star}{\xi}_0\cap K\neq\emptyset$ and $\overset{\star}{\xi}_0$ is a supporting plane at x_0 by i), which contradicts iii). Thus, L_{n_0} is a carrier of μ which contains the origin. This is the final contradiction which proves the lemma. □

Corollary 8.30. *Suppose that K is strictly convex and that for each point of $bd K$ there is only one complex tangent plane. Then if K is the support of an analytic functional μ, it is the unique convex support and $\overset{\star}{\complement} K$ is the domain of definition of φ_μ.*

Proof. By translating K, we may assume without loss of generality that K contains the origin in its interior (a strictly convex set always has non-empty interior). Then φ_μ is holomorphic in $\overset{\star}{\complement} K$ and cannot be extended to a neighborhood of any boundary point by Lemma 8.29. Thus, if L is a convex carrier $\overset{\star}{\complement} L \subset \overset{\star}{\complement} K$ so $K \subset L$. □

We say that a compact set K is linearly convex if its complement is a union of complex hyperplanes or equivalently if $K = \overset{\star}{\complement}\overset{\star}{\complement} K$. Let $\mathbb{P}$ be the family of linearly convex sets. A $\mathbb{P}$-support will be said to be linearly convex. If μ is carried by K, then φ_μ extends to $\overset{\star}{\complement} K$ and if the hypotheses of Corollary 8.30 are fulfilled, K is the unique linearly convex support.

Suppose that K is a convex set and V the smallest linear subspace of $\mathbb{C}^n$ containing V. We shall say that K has the property (u) if

$$(u)\begin{cases}\text{for every extremal point } x \text{ of } bd K, \text{ there exists at most}\\ \text{one complex hyperplane which has support } x_0.\end{cases}$$

Lemma 8.31. *If K is a convex carrier of μ and μ is carried by a complex subspace V, then μ is carried by $K \cap V$.*

Proof. We first show that $K \cup \tilde{V}$ is holomorphically convex if $\tilde{V}$ is a convex subset of V. Let V be defined by $f_1, \ldots, f_{n-q}$ (i.e. $V = \{z\colon f_1 = \ldots = f_{n-q} = 0\}$). Suppose $K \cap V \neq K$. For $x \notin K \cup \tilde{V}$ there exists φ_x such that $\varphi_x(x) = 1$ and $\varphi(y) = 0$ for $y \in V$.

Let $M = \sup\limits_K |\varphi_x(x)|$. Since K is convex, there exists ψ_x such that $\psi_x(x) = 1$ and $\sup\limits_{z\in K} |\psi_x(z)| < \dfrac{1}{M}$, so $\psi_x \varphi_x(x) = 1$ and $\sup\limits_{K \cup V} |\psi_x \varphi_x(z)| < 1$. Thus $K \cup V$ is holomorphically convex.

Let ω be a neighborhood of $K \cap V$. Then there exist open holomorphically convex neighborhoods of K, V and $K \cap V$ which we denote respectively by ω_1, ω_2 and ω_3 such that $(\omega_1 \cap \omega_2) \subset \omega$ and $\omega_3 \subset (\omega_1 \cup \omega_2)$. Then μ is carried by $\omega_2 \cap \omega_3$ and $\omega_1 \cap \omega_3$, and since $\omega_3 = (\omega_2 \cap \omega_3) \cup (\omega_1 \cap \omega_3)$ is holomorphically convex, by Corollary 8.22, μ is carried by $(\omega_1 \cap \omega_2) \subset \omega$. □

Theorem 8.32. *Let K be a compact convex set which satisfies condition (u). Then if K is the support of an analytic functional μ, it is the unique support.*

Proof. If K is a point then the condition (u) is satisfied and furthermore K is the unique support of μ, since by Lemma 8.25 every convex support contains K. Suppose now that K is not a point. By Corollary 8.26, if K and L are two convex supports, then $K\cap L\neq\emptyset$ so we can suppose $0\in K\cap L$. Since L is a support, there exists a point of K which is not in L.

Suppose that $K\subset V$ for V some linear subspace of $\mathbb{C}^n$. We first show that L lies in V. Since μ is carried by a compact subset of V and by L, it follows from Lemma 8.31 that μ is carried by $V\cap L$. Since L is a support, we must have $L=V\cap L$. Suppose that l is a complex hyperplane supporting $x_0\in K$ such that K is not contained in l.

Let $\tilde{l}$ be the trace of a real supporting hyperplane at x_0 such that $l\subset\tilde{l}$ and let $\tilde{l}$ be defined by the equation $\mathbb{R}e\langle z,\eta^\star\rangle=1$ chosen so that $\sup\limits_{z\in L}\operatorname{Re}\langle z,\eta^\star\rangle=a<1$. Then $\sup\limits_{z\in\lambda K+(1-\lambda)L}\operatorname{Re}\langle z,\eta^\star\rangle\leqq\lambda+(1-\lambda)\;a<1$. This means that l does not meet $\lambda K+(1-\lambda)L$ for $0\leqq\lambda\leqq 1$. Since the trace of $\lambda K+(1-\lambda)L$ on V is $\xi_V^\star$, l does not intersect $\xi_V^\star$ either.

Let $\overset{\star}{\xi}$ be a complex supporting hyperplane at x_0 such that $\overset{\star}{\xi}\cap V=\overset{\star}{\xi}_V$. By Lemma 8.30, μ is carried by $\lambda K+(1-\lambda)L$ and hence φ_μ extends analytically to $\overset{\star}{\xi}$. Thus by Lemma 8.30, φ_μ does not extend to one of these hyperplanes, which is a contradiction. □

Historical Notes

The use of the Laplace transform for one complex variable is due to E. Borel. Theorem 8.9 for $n=1$ is due to Polya [1] and the projective Laplace transform is due to Martineau [8], whom we follow closely here. Theorem 8.4 was also proved by Aizenberg [1]. Another proof of Theorem 8.9 can be found in [*B*].

The original work on carriers of analytic functionals in several variables goes back to the thesis of Martineau [1, 2]. We have followed closely the work of Kiselman [1] and a later work of Martineau [6]. The original work of Martineau [1, 2, 3] includes many complementary results and in particular explores many notions close to that of carrier – i.e. pseudocarrier, weak carrier, etc.

Chapter 9. Convolution Operators on Linear Spaces of Entire Functions

Suppose that $f(z)$ is an entire function and μ is a measure in $\mathbb{C}^n$ with compact support. We define the convolution operator $\hat{\mu}(f)$ by

$$\hat{\mu}(f)=g=f\star\mu=\int_{\mathbb{C}_n} f(z+w)\,d\mu(w) \tag{9,1}$$

It is a simple consequence of Cauchy's Theorem in the polydisc, for instance, that this includes all finite order differential operators with constant coefficients, and if we choose $\mu=\sum_{\nu=1}^{s}\lambda_\nu\delta(z^{(\nu)})$, δ the Dirac measure, then we obtain the finite difference operator $\hat{\mu}(f)=\sum_{\nu=1}^{s}\lambda_\nu f(z-z^{(\nu)})$. It is easy to see that if f satisfies certain growth conditions, then $\hat{\mu}(f)$ will grow asymptotically like f – for instance we have $h^\star_{\hat{\mu}(f)}(z)\leqq h^\star_f(z)$, where the indicator is with respect to a proximate order for which f has normal type. Eventually, we shall even consider measures which do not have compact support, but then we shall have to impose some conditions of decay at infinity on the measure μ.

The problem that will interest us is the following: given a convolution operator $\hat{\mu}$, when can we find an entire function f which is a solution of the equation $\hat{\mu}(f)=g$ such that the growth of f is close to that of g? As we have already seen, conditions on the growth of an entire function are intimately related to the distribution of the values it takes. We shall formulate the growth conditions which interest us in terms of weight functions which will turn the problem into one concerning linear operators between certain complex topological vector spaces.

§ 1. Linear Topological Spaces of Entire Functions

Suppose that $w(z)$ is a continuous real-valued function in $\mathbb{C}^n$. We define the linear spaces

$$B_w=\{f\in\mathscr{H}(\mathbb{C}^n)\colon \sup_{\mathbb{C}^n}|f(z)\exp(-w(z))|<+\infty\} \tag{9,2}$$

$$(\text{resp. } B^\star_w=\{f\in\mathscr{H}(\mathbb{C}^n)\colon \lim_{\|z\|\to\infty}|f(z)\exp(-w(z))|=0\}).$$

These become Banach spaces when we equip them with the norm

$$\|f\| = \sup_{\mathbb{C}^n} |f(z)\exp(-w(z))| \tag{9,3}$$

and the topology is finer than that of uniform convergence on compact sets. Suppose now that $\{w_m(z)\}$ is a sequence of real-valued functions such that $w_{m+1}(z) \leqq w_m(z)$. Then $B^\star_{w_{m+1}} \subset B^\star_{w_m}$ and $\bigcap_{m=1}^{\infty} B^\star_{w_m} = E$ is a Fréchet space when we give it the projective limit topology.

Lemma 9.1. *An element of the dual space of $B^\star_w$ can be represented by a complex measure μ in $\mathbb{C}^n$ such that $\int_{\mathbb{C}^n} \exp w(z)\, d|\mu|(z) < +\infty$. An element of the dual space of $\bigcap_{m=1}^{\infty} B^\star_{w_m}$ can be represented by a complex measure μ in $\mathbb{C}^n$ such that $\int_{\mathbb{C}^n} \exp w_m(z)\, d|\mu|(z) < +\infty$ for some m.*

Proof. If $(B^\star_{w_m})'$ is the dual space of continuous linear functionals on B_w, then the space of continuous linear functionals on $\bigcap_{m=1}^{\infty} B^\star_{w_m}$ is just $\bigcup_{m=1}^{\infty} (B^\star_{w_m})'$, so the second statement follows from the first.

Let $\tilde{B}_w = \{g$ continuous in $\mathbb{C}^n$: $\lim_{\|z\| \to \infty} |g(z)\exp(-w(z))| = 0\}$. When we give $\tilde{B}_w$ the sup norm as in (9,3), it becomes a Banach space, and B_w is a closed subspace. If C_0 is the Banach space of all continuous functions in $\mathbb{C}^n$ which tend to zero at infinity, then there exists an isometric isomorphism of B_w onto C_0 given by $g \to g \exp w(z)$. The dual of C_0 is just the set of bounded measures on $\mathbb{C}^n$, so the dual of $\tilde{B}_w$ is just the set of complex measures μ for which $\int_{\mathbb{C}^n} \exp w(z)\, d|\mu|(z) < +\infty$. It follows from the Hahn-Banach Theorem that every element of $B^\star_w$ extends to $\tilde{B}_w$. □

A complex semi-norm $p(z)$ is a semi-norm for which $p(\lambda z) = |\lambda| p(z)$ for $\lambda \in \mathbb{C}$. If $\rho(r)$ is a proximate order and $p(z)$ is a complex semi-norm on $\mathbb{C}^n$, we let $E_p^{\rho(r)}$ be the Fréchet space we obtain by letting

$$w_m(z) = \left\{ p(z) + \frac{1}{m} \|z\| \right\}^{\rho(\|z\|)}$$

and we let E^0 be the space we get by letting $w_m(z) = \|z\|^{1/m}$; $E_p^{\rho(r)}$ is just the space of entire functions f whose indicators with respect to $\rho(r)$ are less than or equal to $p(z)^\rho$ and E^0 is the space of entire functions of order zero. We note that by Theorem 1.18, $\lim_{t \to \infty} \frac{t^{\rho(st)}}{t^{\rho(t)}} = 1$, $0 < s < +\infty$, so if instead of $w_m(z) = \left\{ p(z) + \frac{1}{m} \|z\| \right\}^{\rho(\|z\|)}$ we define

$$w'_m(z) = \left\{ p(z) + \frac{1}{m} \|z\| \right\}^{\rho\left(p(z) + \frac{1}{m}\|z\|\right)},$$

we define an equivalent metric on $E_p^{\rho(r)}$.

Lemma 9.2. *Suppose that for $f \in E_p^{\rho(r)}$ (resp. E^0), $f(z) = \sum_{q=0}^{\infty} P_q(z)$ is the Taylor series expansion of f in homogeneous polynomials of degree q. Then $A_\nu(z) = \sum_{q=0}^{\nu} P_q(z)$ converges to f for the topology of $E_p^{\rho(r)}$ (resp. E_0).*

Proof. We let $p_m(z) = p(z) + \frac{1}{m}\|z\|$, $M_m = \sup_{\mathbb{C}^n} |f \exp - p_m^{\rho(p_m)}|$ and $g(z, \lambda) = f(\lambda z)$, $\lambda \in \mathbb{C}$. By the Cauchy Integral Formula, we have

$$P_q(z) = \frac{1}{2\pi i} \int_{|\lambda|=1} \frac{g(z,\lambda)}{\lambda^{q+1}} d\lambda$$

and so

$$\sup_{p_m(z)=R} \frac{|P_q(z)|}{p_m(z)^q} \leqq M_m \frac{\exp R^{\rho(R)}}{R^q}.$$

Let $\mu_q = \min \frac{\exp R^{\rho(R)}}{R^q}$ and r_q be the solution of $q = \rho r^{\rho(r)}$, so that $\mu_q \leqq \frac{e^{q/\rho}}{r_q^q}$. Let $r = \varphi(t)$ be the inverse function of $t = r^{\rho(r)}$. Since $\lim_{t \to \infty} \frac{\varphi(kt)}{\varphi(t)} = k^{1/\rho}$ by (1) of Theorem 1.23, given $\eta > 0$, there exists q_η such that for $q \geqq q_\eta$

$$\frac{1}{M_m} \sup_z \frac{|P_q(z)|}{p_m(z)^q} \leqq \mu_q \leqq (1+\eta)^q \left(\frac{e\rho}{\varphi(q)^\rho}\right)^{q/\rho}. \tag{9,4}$$

Furthermore, there exists $\delta_m > 0$ such that $\frac{p_k}{p_m} > (1+\delta_m)$ for $k < m$. We set

$$\tau_{qk} = \sup_z |P_q(z) \exp - (p_k(z))^{\rho(p_k(\|z\|))}|$$

so that

$$\begin{aligned} \tau_{qk} &\leqq \sup_z \frac{|P_q(z)|}{p_m(z)^q} \sup_z \frac{p_m(z)^q}{p_k(z)^q} \sup_z \frac{\exp - (p_k(z))^{\rho(p_k(z))}}{p_k(z)^q} \\ &\leqq M_m \mu_q \frac{1}{(1+\delta_m)^q} \frac{1}{\mu_q} \leqq \frac{M_m}{(1+\delta_m)^q}, \end{aligned}$$

and this serie converges. Thus $A_\nu \to f$ for the topology on $E_p^{\rho(r)}$. The proof for E^0 is identical. □

If $f \in E_p^{\rho(r)}$, we expand f at the origin in homogeneous polynomials, $f(z) = \sum_{q=0}^{\infty} P_q(z)$. Let $A_q^{(\rho)} = \left(\frac{\varphi(q)^\rho}{e\rho}\right)^{q/\rho}$ where $r = \varphi(t)$ is the inverse function of $t = r^{\rho(r)}$ and set $f_t(z) = \sum_{q=0}^{\infty} A_q^{(\rho)} P_q(z)$. If we apply Theorem 1.23 to the function $f_t(\lambda z)$ as a function of one complex variable, we see that the power series for f_t converges for $|\lambda| < p(z)$. Thus $f_t \in \mathscr{H}(D)$ for $D = \{z \colon p(z) < 1\}$, and in fact as we have seen in Theorem 7.14, this mapping is onto $\mathscr{H}(D)$. If we give $\mathscr{H}(D)$

the topology of uniform convergence on compact subsets, it follows from (9,4) that the mapping $f \to f_t$ is a topological isomorphism of $E_p^{\rho(r)}$ onto $\mathscr{H}(D)$. When the proximate order will be clear from the context, we will not note the dependence of A on $\rho(r)$.

If μ is a continuous linear functional on $E_p^{\rho(r)}$, we define the continuous linear functional μ_t on $\mathscr{H}(D)$ by $(f_t, \mu_t) = (f, \mu)$. This is an isomorphism between the dual spaces. Let K_m be the convex compact set $K_m = \{z: p_m(z) \leqq 1\}$ and let $p'_m = \sup_{z \in K_m} \mathbb{R}e\langle z, u\rangle$, which is also a complex norm which satisfies $p'_{m+1}(u) > p'_m(u)$. Let $\tilde{\mu}_t(u) = \mu_t(\exp\langle z, u\rangle)$ be the Fourier-Borel transform of μ_t. Since μ_t is carried by some K_m, we see by Theorem 8.9 that

$$\tilde{\mu}_t(u) \leqq C_m \exp[p'_m(u)] \quad \text{for } m \geqq m_{\mu_t}.$$

Since the Taylor series of a function f in $\mathscr{H}(D)$ converges to f for the topology introduced on $\mathscr{H}(D)$,

$$\mu_t(\exp\langle z, u\rangle) = \mu_t\left(\sum_{q=0}^{\infty} \sum_{|\alpha|=q} \frac{z^\alpha u^\alpha}{\alpha!}\right) = \sum_q \left(\sum_{|\alpha|=q} \mu_t(z^\alpha) \frac{u^\alpha}{\alpha!}\right) = \sum_{q=0}^{\infty} P_q^{\mu_t}(u).$$

If we now apply Theorem 1.9 to the function $\tilde{\mu}_t(u)$, which in the complex line λu is of order 1 and type at most $p'_m(u)$, we see that

$$\limsup_{q \to \infty} \left\{\frac{q}{e} |P_q^{\mu_t}(u)|^{1/q}\right\} \leqq p'_m(u) \quad \text{for } m \geqq m_{\mu_t}.$$

From the relation $\mu_t(z^\alpha) = \dfrac{1}{A_{|\alpha|}} \mu(z^\alpha)$ we see that $\mu \in (E_p^{\rho(r)})'$ (resp. $(E^0)'$) if and only if

$$\text{(9,5)} \qquad \limsup_{q + \infty} \left\{\frac{q}{e} \left|\frac{1}{A_q} \sum_{|\alpha|=q} \mu(z^\alpha) \frac{u^\alpha}{\alpha!}\right|^{1/q}\right\} \leqq p'_m(u), \qquad m \geqq m_\mu.$$

For $\mu \in (E_p^{\rho(r)})'$ (resp. $(E^0)'$), we define its Fourier-Borel transform to be the formal power series

$$\tilde{\mu}(u) = \mu(\exp\langle z, u\rangle) = \sum_q \sum_{|\alpha|=q} \mu(z^\alpha) \frac{u^\alpha}{\alpha!} = \sum_q \mathrm{P}_q^\mu(u).$$

Of course, in general this will not converge, since if $\rho < 1$, $\exp\langle z, u\rangle \notin E_p^{\rho(r)}$; however, (9.5) gives us a method of controlling the growth of the coefficients of this formal series.

If $\rho > 1$ and $\rho(r)$ is an associated proximate order, we assume that in addition for all $r \geqq 0$

i) $\rho(r) > 1$

ii) $\dfrac{d}{dr}(r^{\rho(r)-1}) > 0.$

Since both of these properties hold for r sufficiently large, there is no loss of generality. In this case, the equation $r = t^{\rho(t)-1}$ has a unique solution for all $r \geqq 0$.

Definition 9.3. Let $\rho^\star(r)=\dfrac{\rho(t)}{\rho(t)-1}$ where t is the unique solution of the equation $r=t^{\rho(t)-1}$. We define $\rho^\star(r)$ to be proximate order conjugate to $\rho(r)$.

Proposition 9.4. *For $\rho>1$, the conjugate proximate order is indeed a proximate order.*

Proof. We first note that $\lim\limits_{r\to\infty}\rho^\star(r)=\dfrac{\rho}{\rho-1}$ exists, so (i) of Definition 1.15 is verified. Furthermore

$$\begin{aligned}\frac{d}{dr}\rho^\star(r)&=\frac{d}{dt}\frac{\rho(t)}{\rho(t)-1}\cdot\left(\frac{dr}{dt}\right)^{-1}\\&=-\rho'(t)(\rho(t)-1)^{-2}t^{2-\rho(t)}[t\rho'(t)\log t+\rho(t)-1]^{-1},\end{aligned}$$

so

$$\lim_{r\to\infty} r\log r\frac{d}{dr}\rho^\star(r)=\lim_{t\to\infty}\frac{-t(\log t)\rho'(t)}{(\rho(t)-1)^2}=0$$

by (ii) of Definition 1.15. Thus, the same property is verified for $\rho^\star(r)$. □

For $\rho>1$, set $A_\rho=\dfrac{(\rho-1)^{\frac{\rho-1}{\rho}}}{\rho}$ and $F^{\rho^\star(r)}_{A_\rho'}=\bigcup\limits_m E^{\rho^\star(r)}_{A_\rho\, m'}$.

Theorem 9.5. *The mapping $\mu\to\tilde\mu(u)$ is a one-to-one linear mapping of $(E^{\rho(r)}_p)'$, (resp. $E^0)'$ onto*

i) $F^{\rho^\star(r)}_{A_\rho p'}$ *for* $\rho>1$

ii) *the set $Q^{\rho(r)}_{p'}$ of formal power series at the origin which satisfy* (9,5) *for some m for $\rho<1$ (resp. the set Q_0 of formal power series at the origin which satisfy* (9,5) *for some $\rho>0$* for $(E^0)'$).

Proof. Of course (ii) is just a restatement of facts already observed, so we must only verify (i).

Since $A_q^{1/q}=\dfrac{\varphi(q)}{(e\rho)^{1/\rho}}$ where $\varphi(q)=r_q$ with $q=r_q^{\rho(r_q)}$, we have

$$\frac{q}{e}\frac{1}{A_q^{1/q}}=\frac{q}{r_q}e^{1/\rho-1}\rho^{1/\rho}=\frac{A_\rho^{-1}r_q^{\rho(r_q)-1}}{(e\rho^\star)^{1/\rho^\star}}.$$

Let $r_q'=r_q^{\rho(r_q)-1}$. Then

$$\begin{aligned}\rho^\star(r_q')\log r_q'&=\rho^\star(r_q^{\rho(r_q)-1})\log(r_q^{\rho(r_q)-1})\\&=\rho(r_q)[\rho(r_q)-1]^{-1}\log(r_q^{\rho(r_q)-1})\\&=\rho(r_q)\log r_q=\log q,\end{aligned}$$

and if $r=\varphi'(t)$ is the inverse function of $t=r^{\rho^\star(r)}$, we have

$$\frac{q}{e}\frac{1}{A_q^{1/q}}=A_\rho^{-1}\frac{\varphi'(q)}{(e\rho^\star)^{1/\rho^\star}},$$

so the mapping is into. Since the calculations are all reversible, the mapping is also onto. □

Suppose that $\mu \in (E_p^{\rho(r)})'$. Then for any other element v, we define the convolution of v and μ, $\mu \star v = \tau$ on $E_p^{\rho(r)}$ as $(f(z), \mu \star v) = (\mu_w f(z+w), v)$. This is equivalent to the convolution of the two measures associated with μ and v, and hence $v \star \mu = \mu \star v$. Of course, it is not clear that $(f, \mu \star v)$ is defined for all elements of $E_p^{\rho(r)}$, but it is well defined on the polynomials, which are dense. An easy calculation shows that $\tilde{\tau}(u) = \tilde{\mu}(u)\tilde{v}(u)$ in terms of formal series.

Lemma 9.6. *Suppose that μ is any element in $(E_p^{\rho(r)})'$ (resp. $(E^0)'$) for $\rho < 1$ or an element of $(E_p^{\rho(r)})'$ such that $\int_{\mathbb{C}^n} \exp m \|z\|^{\rho(\|z\|)} d|\mu|(z) \leqq A_m$ for $\rho > 1$ and every positive integer m. Then the map $\hat{\mu}(E_p^{\rho(r)})$ given by $\hat{\mu}(f) = f \star \mu = \int_{\mathbb{C}^n} f(z+w)\, d\mu(w)$ is a continuous linear map of $E_p^{\rho(r)}$ into itself.*

Proof. Let m be chosen so that $\int_{\mathbb{C}^n} \exp p_{m-1}(z)\, d|\mu|(z) < +\infty$. Set $\alpha(z) = p_m(z)$, $\beta(w) = p_m(w)$, $\gamma(z, w) = \alpha(z) + \beta(w)$. Then there exists $M_m > 0$ such that $|f(z)| \leqq M_m \exp p_m(z)^{\rho(p_m(z))}$. We first treat the case $\rho > 1$. Choose $\delta_m > 0$ such that $p_{m-1}(z)[p_m(z)]^{-1} > (1+\delta_m)$. Then

$$\begin{aligned} M_m^{-1} \Big|\int_{\mathbb{C}^n} f(z+w)\, d\mu(w)\Big| &\leqq \Big|\int_{\beta < \delta_m \alpha} \exp \gamma^{\rho(\gamma)} d\mu(w)\Big| \\ &\quad + \Big|\exp \alpha^{\rho(\alpha)} \int_{\beta \geqq \delta_m \alpha} \exp\{\gamma^{\rho(\gamma)} - \alpha^{\rho(\alpha)}\}\, d\mu(w)\Big| \\ &= I_1 + I_2. \end{aligned}$$

Then $I_1 \leqq \exp p_{m-1}(z)^{\rho(p_{m-1}(z))} \int_{\mathbb{C}^n} d|\mu|(w)$. Since by the Mean Value Theorem,

$$\frac{r^{\rho(r)} - r'^{\rho(r')}}{r - r'} = [t\, \rho'(t) \log t + \rho(t)]\, t^{\rho(t)-1}$$

for some $t \in [r', r]$, we see that for $\rho > 1$ and $\beta \geqq \delta_m \alpha$, since $t^{\rho(t)-1}$ is increasing,

$$\gamma^{\rho(\gamma)} - \alpha^{\rho(\alpha)} \leqq \beta\, C \gamma^{\rho(\gamma)} \leqq \beta\, C[(1+\delta_m)\beta]^{\rho((1+\delta_m)\beta)}$$

and $I_2 \leqq C' \exp p_m(z)^{\rho(p_m(z))}$ by the hypothesis on μ.

Suppose now that $\rho < 1$. We can assume that $r^{\rho(r)-1}$ is decreasing and $\frac{d}{dr} r^{\rho(r)} < r^{\rho(r)-1}$, since this holds in any case for r large. Let $0 \leqq a \leqq b$. If we set $r = a+b$, $r' = b$ in the above formula, we obtain for some $t \geqq b$

$$(a+b)^{\rho(a+b)} - b^{\rho(b)} = a \cdot t^{\rho(t)-1} \leqq a \cdot b^{\rho(b)-1} \leqq a^{\rho(a)},$$

so $(a+b)^{\rho(a+b)} \leqq a^{\rho(a)} + b^{\rho(b)}$, from which it follows that

$$M_m^{-1} \Big|\int_{\mathbb{C}^n} f(z+w)\, d\mu(w)\Big| \leqq \exp p_m^{\rho(p_m(z))} \int \exp p_m(w)^{\rho(p_m(w))} d|\mu|(w). \quad \square$$

Definition 9.7. A convolution operator $\hat{\mu}$ on the space $E_p^{\rho(r)}$ (resp. E_0) will be any measure in $(E_p^{\rho(r)})'$ for $\rho<1$ (resp. $(E_0)'$ for E^0) or any measure in $(E_p^{\rho(r)})'$ for which $\int_{\mathbb{C}^n} \exp m\|z\|^{\rho(\|z\|)} d|\mu|(z)=A_m<\infty$ for all $m>0$ for $\rho>1$.

If $\hat{\mu}$ is a convolution operator in any one of these spaces, we want to show that we can always find a solution $\tilde{f}$ of the equation $\hat{\mu}(x)=f$ with $\tilde{f}$ in the same space. This is equivalent to showing that the continuous linear map $\hat{\mu}$ maps $E_p^{\rho(r)}$ onto itself. To do this, we shall use the principle of duality.

Proposition 9.8 (see [G]). *If E and F are two Fréchet spaces and α a continuous linear map from E into F, then the following two properties are equivalent:*

i) *α is onto*

ii) *${}^t\alpha$: $E'\to F'$, the transpose map defined by $(f, {}^t\alpha(\mu))=(\alpha(f), \mu)$ is one-to-one with weakly closed image in E'. If in addition, ${}^t\alpha$ is onto, then α is one-to-one.*

If $\hat{\mu}$ is the convolution operator, then the transpose map is given by $\nu\to\mu\star\nu$. We shall actually prove that the image is closed in the equivalent space as determined by Theorem 9.5, but first we must equip these spaces with topologies. For $\rho<1$, we equip $Q_{p'}^{\rho(r)}$ (resp. Q_0) with the topology of convergence of each coefficient. This is equivalent to convergence on the polynomials in $(E_p^{\rho(r)})'$ and so is weaker than the topology induced by the weak topology. For $\rho>1$, we equip $F_{A_\rho p'}^{\rho^\star(r)}$ with the topology of convergence of the Taylor series coefficients at each point in $\mathbb{C}^n$. If $\tilde{\mu}(u)$ is the Fourier-Borel transform of μ, and $\sum C_\alpha(u-u_0)^\alpha$ is its Taylor series expansion in a neighborhood of u_0, then $C_\alpha=\mu\left(\frac{z^\alpha}{\alpha!}\exp\langle z, u_0\rangle\right)$, and so this topology is also weaker than the weak topology.

§2. Theorems of Division

Lemma 9.9. *Let $A_q(u)=\dfrac{B_{q+m}(u)}{C_m(u)}$ be a homogeneous polynomial of degree q which is the ratio of two homogeneous polynomials of degree $(q+m)$ and m respectively. Suppose that for some complex norm $p_0(u)$ that $|B_{q+m}(u)|\leqq C[p_0(u)]^{q+m}$. Then given $\delta>0$, there is a constant K_δ (depending only on C_m and δ) such that $|A_q(u)|\leqq CK_\delta[p_0(u)]^q(1+\delta)^{q+m}$.*

Proof. Let $\Omega=\{u\colon 1-\delta\leqq p_0(u)\leqq 1+\delta\}$. For every point $u\in\mathring{\Omega}$ we can find a polydisc (by making a non-linear change of variable if necessary) $\Delta(u; r^u)$

centered at u and lying in Ω such that $C'_m(u'_1, \ldots, u'_{n-1}, \xi_n) \neq 0$ for $|\xi_n - u_n| = r^u_n$ and $|u'_i - u_i| \leqq r^u_i$, $i = 1, \ldots, n-1$ (cf. [A]). Let $\Omega' = \{u: p_0(u) = 1\}$ and $\Delta'_u = \Delta\left(u; \frac{r^u}{2}\right)$. Since Ω' is compact, it can be covered by a finite number of the Δ'_{u^j}, $j = 1, \ldots, N$ and the function $\frac{1}{C_m(u)}$ is bounded, say by $\frac{K_\delta}{2}$ on the compact set

$$K = \bigcup_j \{u': u' \in \Delta_{u^j}, |u'_i - u_i| \leqq r^{u^j_i}, i = 1, \ldots, n-1, |u'_n - u_n| = r^{u^j_n}\}.$$

Suppose the function A_q takes on its maximum on Ω' at the point u^0. Then $u^0 \in \Delta'_{u^j}$ for some j. By Cauchy's Formula

$$|A_q(u^0)| = \left| \frac{1}{2\pi i} \int_{|\xi_n - u^j_n|} \frac{B_{q+m}(u^0_1, \ldots, u^0_{n-1}, \xi_n)\, d\xi_n}{C_m(u^0_1, \ldots, u^0_{n-1}, \xi_n)(\xi_n - u^0_n)} \right|$$
$$\leqq K_\delta C (1+\delta)^{q+m}. \qquad \square$$

Theorem 9.10 (Division Theorem for $\rho < 1$). *Let $H(u)$, $F(u) \in Q^{\rho(r)}_p$ for $\rho < 1$ (resp. Q_0) with $H(u) = F(u) G(u)$, where $G(u)$ is a formal power series at the origin. Then $G(u) \in Q^{\rho(r)}_p$ (resp. Q_0).*

Proof. Let $\varepsilon > 0$ be given and let

$$G(u) = \sum_{q=0}^{\infty} R_q(u), \qquad H(u) = \sum_{q=0}^{\infty} P_q(u), \quad \text{and} \quad F(u) = \sum_{q=0}^{\infty} T_q(u)$$

with s the smallest integer such that $T_s(u) \not\equiv 0$. We choose m_0 so large that (9,5) holds for both $H(u)$ *and* $F(u)$ for $m \geqq m_0$. Thus, there exist constants C_1 and C_2 such that for $m \geqq m_0 + 1$ (since $p_m(u) \geqq p_{m_0}(u) + \eta \|u\|$ for some $\eta > 0$)

$$|P_q(u)| \leqq C_1 [p'_m(u)]^q \left(\frac{\varphi(q)^\rho}{e\rho}\right)^{q/\rho} \left(\frac{e}{q}\right)^q$$

$$|T_q(u)| \leqq C_2 [p'_m(u)]^q \left(\frac{\varphi(q)^\rho}{e\rho}\right)^{q/\rho} \left(\frac{e}{q}\right)^q.$$

Since $P_{q+s}(u) = \sum_{l+k=q} R_l(u) T_{k+s}(u)$,

$$R_q(u) = T_s^{-1}(u) [P_{q+s}(u) - \sum_{\substack{l+k=q \\ l \neq q}} R_l(u) T_{k+s}(u)].$$

We now show by induction that there exist constants K_q with $K_{q-1} \leqq K_q$ and $K_q = K_{q-1}$ for $q \geqq \tilde{q}$ such that for $\delta > 0$

$$|R_q(u)| \leqq K_q [p'_m(u)]^q (1+\delta)^q q \left(\frac{\varphi(q)^\rho}{e\rho}\right)^{q/\rho} \left(\frac{e}{q+s}\right) \left(\frac{\varphi(q)}{q}\right)^{s+1}.$$

For $q = 0$, by Lemma 9.9, we have

$$|R_q(u)| \leqq C_2 K_\delta [p'_m(u)]^q (1+\delta)^q \left(\frac{\varphi(q+s)^\rho}{e\rho}\right)^{\frac{q+s}{\rho}} \left(\frac{e}{q+s}\right)^{q+s}$$

and if $\zeta(r)=r^{1-\rho(r)}$, we have

$$\left[\frac{\varphi(q+s)}{q+s}\right]^{q+s}=\zeta(r_{q+s})^{q+s}=\left[\frac{\varphi(q)}{q}\right]^{q+s}\left[\frac{\zeta(r_{q+s})}{\zeta(r_q)}\right]^{q+s}$$
$$\leqq(1+\delta)^{q+s}\left[\frac{\varphi(q)}{q}\right]^{q+s+1}$$

for q sufficiently large. We now assume the conclusion for $q\leqq q_0-1$. Then by Lemma 9.9

$$|R_{q_0}(u)|\leqq|T_s(u)|^{-1}\Big[|P_{q_0+s}(u)|+\sum_{\substack{l+k=q_0\\ l\neq q_0}}|R_l(u)T_{k+s}(u)|\Big]$$
$$\leqq K_\delta(1+\delta)^s[p_m'(u)]^{q_0}(1+\delta)^{q_0}\left(\frac{\varphi(q_0+s)^\rho}{e\rho}\right)^{\frac{q_0+s}{\rho}}\left(\frac{e}{q_0+s}\right)^{q_0+s}$$
$$\times\left\{C_1+\sum_{\substack{l+k=q_0\\ l\neq q_0}}K_{q-1}C_2\,l\left[\frac{\varphi(l)^l\,\varphi(k+s)^{k+s}}{\varphi(k+l+s)^{k+l+s}}\right]\frac{(k+l+s)^{k+l+s}}{l^l(k+s)^{k+s}}\left(\frac{\varphi(l)}{l}\right)^{s+1}\right\}$$
$$\leqq\max\left[K_0(1+\delta)^sC_1,K_{q-1}C_2\right][p_m'(u)]^{q_0}$$
$$\times(1+\delta)^{q_0}\left(\frac{\varphi(q_0+s)}{e\rho}\right)^{\frac{(q_0+s)}{\rho}}\left(\frac{e}{q_0+s}\right)^{q_0+s}$$
$$\times\left\{1+\sum_{\substack{l+k=q_0\\ l\neq q_0}}K_\delta(1+\delta)^s\,l\left[\frac{\varphi(l)^l\,\varphi(k+s)^{k+s}}{\varphi(l+k+s)^{l+k+s}}\right]\frac{(l+k+s)^{l+k+s}}{l^l(k+s)^{k+s}}\right\}\left(\frac{\varphi(q)}{q}\right)^{s+1}.$$

We assume that the function $\zeta(r)=r^{1-\rho(r)}$ increases. Since this holds eventually, we lose no generality. For simplicity, we let $i=k+s$, $j=q_0+s$, $\alpha=\frac{\rho(1-\rho)^{-1}}{2}$. Now since $j=r_j^{\rho(r_j)}$ and $\varphi(j)=r_j$,

$$\left[\frac{\varphi(l)^l\varphi(i)^i}{\varphi(j)^j}\right]\frac{j^j}{l^li^i}=\left[\frac{\zeta(r_j)}{\zeta(r_l)}\right]^{-l}\left[\frac{\zeta(r_j)}{\zeta(r_i)}\right]^{-i}\qquad\text{if } l+i=j.$$

Suppose for the moment that $i\leqq 3j/4$. Then

$$\zeta(r_j)^\alpha-\zeta(r_i)^\alpha=\int_{r_i}^{r_j}\frac{d}{dr}\zeta(r)^\alpha dr\geqq\int_{r_{3j/4}}^{r_j}\frac{d}{dr}\zeta(r)^\alpha dr$$
$$\geqq\int_{r_{3j/4}}^{r_j}\frac{d}{dr}r^{\rho(r)/4}dr$$

for j sufficiently large, since for r large,

$$\frac{d}{dr}\zeta(r)^\alpha\geqq r^{3\rho/8-1}\quad\text{and}\quad\lim_{r\to\infty}r^{1-\frac{\rho(r)}{4}}\frac{d}{dr}r^{\rho(r)/4}=\frac{\rho}{4}.$$

Thus

$$\zeta(r_j)^\alpha-\zeta(r_i)^\alpha\geqq j^{1/4}[1-(3/4)^{1/4}]=Tj^{1/4}.$$

For $i \geqq 24\alpha$, we have

$$\left[\frac{\zeta(r_j)}{\zeta(r_i)}\right]^i \geqq \left[1+\frac{Tj^{1/4}}{\zeta(r_i)^\alpha}\right]^{i\alpha^{-1}} \geqq \left[1+\frac{iTj^{1/4}}{\zeta(r_i)^\alpha}+\ldots+KT^\gamma j^\gamma\right]^{\alpha^{-1}} \geqq T'j^3$$

where $\gamma \geqq 3\alpha$ (since $\zeta(r_i)^\alpha = O(i^{1/2+\varepsilon})$ for $\varepsilon>0$). For $i \leqq 24\alpha+1$ and $\beta = 2\max_{i \leqq 24\alpha} \zeta(r_i)$, we have $\left[\frac{\zeta(r_j)}{\zeta(r_i)}\right]^i \geqq \frac{\zeta(r_j)}{\beta} \geqq 1$ for q_0, and hence j, sufficiently large, since $\rho<1$. By symmetry, similar inequalities hold when we replace i by l. We choose q_0 so large that $\frac{K_\delta(1+\delta)^s}{T'(q_0+s)^3} \leqq (3q_0^2)^{-1}$. Thus, since for q_0 sufficiently large, either l or $(k+s)$ is greater that 24α if $l+k=q_0$, we obtain

$$\left\{1+\sum_{\substack{l+k=q_0\\ l\neq q_0}} K^\delta(1+\delta)^s l\left[\frac{\varphi(l)^l\varphi(k+s)^{k+s}}{\varphi(l+k+s)^{l+k+s}}\right]\frac{(l+k+s)^{l+k+s}}{l^l(k+s)^{k+s}}\right\} \leqq 2 \leqq q_0$$

which completes the induction. Thus

$$\limsup_{q\to\infty}\left\{\frac{q}{e}\left|\frac{R_q(u)}{A_q}\right|^{1/q}\right\} \leqq p_k'(u) \quad \text{for } k \geqq m,$$

which proves the theorem. □

Theorem 9.11. *Suppose f and g are two entire functions of finite type with respect to the proximate order $\rho(r)$ with g of minimal type. If $k=f/g$ is an entire function then $h_k^\star(z)=h_f^\star(z)$.*

Proof. Clearly $f=kg$. Since an entire function of minimal type is always of regular growth (cf. Chapter 4), it follows from Theorem 4.3 that

$$h_f^\star(z)=h_k^\star(z)+h_g^\star(z)=h_k^\star(z). \qquad \square$$

§ 3. Applications to Convolution Operators in the Spaces $E_p^{\rho(r)}$ and E^0

Our strategy has been to use the principle of duality (Proposition 9.8). We now proceed with the proof.

Theorem 9.12. *Let $\hat{\mu}$ be a convolution on $E_p^{\rho(r)}$ (resp. E^0) ($\rho \neq 1$). Then the equation $\hat{\mu}(x)=f$ for $f \in E_p^{\rho(r)}$ always admits a solution $g \in E_p^{\rho(r)}$ (resp. E^0).*

Proof. The transpose map of μ is given by $v \to \mu\star v$. Suppose $\mu\star v_1=\mu\star v_2$. Then $\tilde{\mu}(u)\tilde{v}_1(u)=\tilde{\mu}(u)\tilde{v}_2(u)$ and hence $\tilde{\mu}(u)(\tilde{v}_1(u)-\tilde{v}_2(u))\equiv 0$ as a formal power series. But the product of two power series is zero if and only if one of the factors is zero, and if $\mu\equiv 0$ then $\tilde{\mu}(u)\equiv 0$, since the polynomials are dense in

$E_p^{\rho(r)}$ (resp. E^0). Thus $\tilde{v}_1(u)=\tilde{v}_2(u)$, and since $v_1=v_2$ for the polynomials, $v_1\equiv v_2$. Thus the map is one-to-one. We now show that image is weakly closed.

First we suppose $\rho<1$. Let $\{\alpha_\lambda\}=\{\mu\star v_\lambda\}$ be a sequence in the image such that $\alpha_\lambda\to\alpha_0$ weakly. Then $\tilde{\alpha}_\lambda(u)=\tilde{\mu}(u)\tilde{v}_\lambda(u)$ as formal power series. Denote

$$\tilde{\alpha}_\lambda(u)=\sum_{q=0}^{\infty} P_q^\lambda(u),\qquad \tilde{\mu}(u)=\sum_{q=0}^{\infty} T_q(u),\qquad \tilde{v}_\lambda(u)=\sum_{q=0}^{\infty} R_q^\lambda(u).$$

Suppose s is the smallest integer for which $T_s\not\equiv 0$. Then

$$R_q^\lambda(u)=[T_s(u)]^{-1}[P_{q+s}^\lambda(u)-\sum_{\substack{l+k=q\\ l\neq q}} R_l^\lambda(u)\,T_{k+s}(u)].$$

We now show by induction on q that $R_q^\lambda(u)$ converges to a polynomial $R_q(u)$. For $q=0$, this follows from Lemma 9.9, and if this is true for all $q\leqq q_0-1$, the result then follows for q_0 by again applying Lemma 9.9. Thus $\tilde{\alpha}_0(u)=\tilde{\mu}(u)\,\tilde{\tau}(u)$, and it follows from Theorem 9.10 that $\tilde{\tau}(u)\in Q_p^{\rho(r)}$ (resp. Q_0) so that $\alpha_0=\mu\star\tau$ with $\tau\in Q_p^{\rho(r)}$. Thus the image is closed.

Now we suppose $\rho>1$ and $\alpha_\lambda=\mu\star v_\lambda$ a sequence in the image such that $\alpha_\lambda\to\check{\alpha}$ weakly. Then $\tilde{\alpha}_\lambda(u)=\tilde{\mu}(u)\tilde{v}_\lambda(u)$ as entire functions. For $z\in\mathbb{C}^n$, let

$$\sum_{q=0}^{\infty} T_q(z,u'),\qquad \sum_{q=0}^{\infty} P_q^\lambda(z,u'),\qquad \sum_{q=0}^{\infty} R_q^\lambda(z,u')$$

be the Taylor series expansions of $\tilde{\mu}(u)$, $\tilde{\alpha}_\lambda(u)$ and $\tilde{v}_\lambda(u)$ respectively at the point z. Suppose s (depending on z) is the smallest integer for which $T_s(z,u')\not\equiv 0$. Then as above, $R_q^\lambda(z,u')$ approaches a limit $R_q(z,u')$ for all λ and

$$R_q(z,u')=[T_s(z,u')]^{-1}[P_{q+s}(z,u')-\sum_{\substack{l+k=q\\ l\neq q}} R_l(z,u')\,T_{k+s}(z,u')].$$

We now show that $\sum R_q(z,u')$ converges in a neighborhood of $u'=0$.

Let $\Delta(z,r)$ be a polydisc with center z and polyradius r such that on

$$A=\{u\colon |u_j-z_j|\leqq r_j,\ j=1,\dots,n-1,\ |u_n-z_n|=r_n\},$$

$T_s(z,u')\neq 0$ and set $\dfrac{1}{K}=\min\limits_{A}|T_s(z,u')|$. There exists a constant $C\geqq 1$ such that for $u\in\Delta(z,r)$, $|P_q(z,u')|\leqq C2^{-q}$, $|T_q(z,u')|\leqq C2^{-q}$. We show by induction that on $\Delta'=\Delta(z,r/2)$, $|R_q(z,u')|\leqq(4CK)^q$. We have by the Cauchy Formula

$$\begin{aligned}|R_q(z,u')|\leqq \int_{|\xi_n-u_n|=r_n} &|[P_{q+s}(z,u_1',\dots,u_{n-1}',\xi_n)\\ &-\sum_{\substack{l+k=q\\ l\neq k}} R_l(z,u_1',\dots,u_{n-1}',\xi_n)]\\ &\times T_s(u_1',\dots,u_{n-1}',\xi_n)^{-1}(\xi_n-u_n')^{-1}\,d\xi_n|\end{aligned}$$

so for $q=0$ the result is immediate, and once it is established for $q-1$ we see that

$$|R_q(z,u')| \leqq 2K(2CK)^{q-1} \sum_{n=0}^{\infty} (\tfrac{1}{2})^n \leqq (4CK)^q.$$

Thus $\dfrac{\tilde{\alpha}(u)}{\tilde{\mu}(u)} = F(u)$ is actually an entire function, and by Theorem 9.11 and Theorem 9.5, $F(u) = \tilde{v}(u)$ for $v \in (E_p^{\rho(r)})$, so the image is closed and the Theorem is proved. □

§ 4. Supplementary Results for Proximate Orders with $\rho > 1$

We will show that for strong proximate orders that we can improve the precision of our results. This will stem from Proposition 1.22 which says that for a *strong* proximate order, $r^{\rho(r)}$ is a convex increasing function of r, so if we compose with a plurisubharmonic function, the result remains plurisubharmonic. In particular if $p(z) \geqq 0$ is a support function (i.e. $p(tz) = t\,p(z)$, $t > 0$, $p(z_1+z_2) \leqq p(z_1) + p(z_2)$), then we can write $p(z) = \sup_{u \in K} \mathbb{R}e\langle z, u\rangle$ for some convex compact set K, so $p(z)$ is plurisubharmonic and $p(z)^{\rho(p(z))}$ is plurisubharmonic also. Note that for $p(z)$ a complex norm, then $\log p(z)$ is plurisubharmonic and $(p(z))^{\rho(p(z))}$ is also plurisubharmonic *for every* ρ (by Proposition 1.22). This goes a long way in explaining why one must take a complex norm for $\rho < 1$ but only a positive support function for $\rho > 1$.

Let $p_m(z) = p(z) + \dfrac{1}{m}\|z\|$. Then $K_m = \{z : p_m(z) \leqq 1\}$ is a compact convex set and so $p'_m = \sup_{u \in K_m} \mathbb{R}e\langle u, z\rangle$ is also a positive support function and $p'_k \geqq p'_m$ for $k \geqq m$. We let $E_p^{\rho(r)} = \bigcap_{m=1}^{\infty} B^{\star}_{w_m}$ with $w_m(z) = p_m(z)^{\rho(p_m(z))}$ and $F_{p'}^{\rho(r)} = \bigcup B^{\star}_{w'_m}$ with $w'_m = p'^{\rho(p'_m)}_m$. We equip $E_p^{\rho(r)}$ with the projective limit topology, so that it becomes a Fréchet space, and we equip $F_{p'}^{\rho(r)}$ with the inductive limit topology. Then $(F_{p'}^{\rho(r)})'$, the dual space of continuous linear functionals, is just $(F_{p'}^{\rho(r)}) = \bigcap_{m=1}^{\infty} (B^{\star}_{w'_m})$, and if we equip $(B^{\star}_{w'_m})'$ with the dual topology

$$\|v\|_m = \sup_{\substack{f \in B^{\star}_{w_m} \\ \|f\|_{B_{w_m}} = 1}} |v(f)|, \tag{9,6}$$

then we can give $(F_{p'}^{\rho(r)})'$ the projective limit topology, under which it becomes a Fréchet space.

Lemma 9.13. *Every element $\alpha \in (E_{p'}^{\rho(r)})'$ can be represented by a measure μ such that $\int \exp w'_m d|\mu| < +\infty$ for every m.*

Proof. We recall that $C^{\star}_{w'_m}$ is the space of continuous functions $k(z)$ such that $\lim_{\|z\| \to \infty} |k(z) \exp -w'_m(z)| = 0$. A Cauchy measure v_γ is integration on a rectifiable curve γ contained in some complex line. We note that the closure of the linear subspace spanned by the Cauchy measures is just $(B^{\star}_{w_m})^{\perp}$, since if f is continuous and $v_\gamma(f)=0$ for every Cauchy measure, f is holomorphic in every complex line by Morera's Theorem and hence f is globally holomorphic by Hartog's Theorem (cf. [B]).

Note that $\|v\|_{m+1}\geqq\|v\|_m$ in general. Let μ_1 represent α in $(B_{w_1})'$ and let μ_2' represent α in $(B_{w_2'})'$. Then the measure $(\mu_2'-\mu_1)$ is orthogonal to $B_{w_1'}$, so we can find a finite linear combination of Cauchy measures v_2 such that $\|\mu_2'-\mu_1-v_2\|_1<1/2$. Set $\mu_2=\mu_2'-v_2$. We choose by induction $\mu_1,\ldots,\mu_{m-1}$ such that $\|\mu_{m-1}-\mu_{m-2}\|_{m-2}<1/2^{m-2}$. Then we can find μ_m' which represents α in $B_{w'_m}$, and we can find a finite linear combination of Cauchy measures v_m such that $\|\mu_m-v_m-\mu_{m-1}\|_{m-1}<1/2^{m-1}$. We set $\mu_m=\mu_m'-v_m$. Then we can extend μ_m to $\tilde{\mu}_m$ on all $C^{\star}_{w_m'}$ by the Hahn-Banach Theorem and this will not increase the norm. Then $\lim\limits_{m\to\infty}\tilde{\mu}_m$ exists in each $C^{\star}_{w_m'}$, and if $\mu=\lim\limits_{m\to\infty}\tilde{\mu}_m$, it has the desired properties. □

Corollary 9.14. *If we equip $(E^{\rho(r)}_{p'})'$ with its Fréchet space topology, then $((E^{\rho(r)}_{p'})')'=(F^{\rho(r)}_{p'})$.*

Proof. Let $\alpha\in\mathscr{C}_0^\infty(B(0,1))$ such that $0\leqq\alpha$, $\int\alpha(z)d\tau(z)=1$, and α depends only on $\|z\|$. For $\mu\in(F^{\rho(r)}_{p'})'$, a measure by Lemma 9.13, we set $\tilde{\mu}=\mu\star\alpha$ so that $\tilde{\mu}$ is a $\mathscr{C}^\infty$ function and $\tilde{\mu}(f)=\mu(f\star\alpha)=\mu(f)$ if f is holomorphic, since $f\star\alpha=f$ for holomorphic functions. Hence $(F^{\rho(r)}_{p'})'=\bigcap\limits_{m=1}^{\infty}Q_m$, where Q_m is the space of functions k in $\mathbb{C}^n$ such that $\int\exp w_m'(z)k(z)d\tau(z)<\infty$ for every m. Since the dual space of Q_m is just a space of functions, it remains to show that these functions are actually in $F^{\rho(r)}_{p'}$.

Suppose $h(z)$ is such a function. Since $h\in(Q_m)'$, we have $h(z)\exp-w_m'(z)$ essentially bounded, and so for $\hat{m}>m$, $\lim\limits_{\|z\|\to\infty}|h(z)\exp-w_{\hat{m}}'(z)|=0$ if we exclude a set of measure zero.

In fact, $h(z)$ is holomorphic, since if α_γ is any Cauchy measure, then $\alpha_\gamma(f)=0$ for every $f\in F^{\rho(r)}_{p'}$ so $\alpha_\gamma(h)=0$ also. Thus by Morera's Theorem, h is holomorphic. This completes the proof. □

Lemma 9.15. *Let $\rho(r)$ be a strong proximate order with $\rho>1$. If $\eta(r)$ is a non-negative function such that $\lim\limits_{r\to\infty}\eta(r)r^{-\rho(r)}=0$, then there exists a positive function $\xi(r)$ with non-negative first and second derivatives such that $\xi(r)\geqq\eta(r)$ and $\lim\limits_{r\to\infty}\xi(r)r^{-\rho(r)}=0$.*

Proof. Let $\{\varepsilon_m\}$ be a decreasing sequence of positive numbers with $\lim\limits_{m\to\infty}\varepsilon_m=0$ and $\{r_m\}$ an increasing sequence such that $\eta(r)\leqq\varepsilon_{m+1}r^{\rho(r)}$ for $r\geqq r_m$. We assume without loss of generality that both $\dfrac{d}{dr}r^{\rho(r)}$ and $\dfrac{d^2}{dr^2}(r^{\rho(r)})$ are everywhere non-negative.

We construct a function $\xi_1(r)$ to be piecewise linear. The construction will be done by induction. For $m=1$, we let $\xi_1(r)$ be a constant such that $\xi_1(r)=\max_{r\leqq r_2}(\eta(r), \varepsilon_1 r^{\rho(r)})$. Having constructed $\xi_1(r)$ for $r\leqq r_m$ with the property that $\xi_1(r)\geqq\varepsilon_{m-1}r^{\rho(r)}$ for $r_{m-1}\leqq r\leqq r_m$, we construct $\xi_1(r)$ for $r_m\leqq r\leqq r_{m+1}$. We continue $\xi_1(r)$ linearly unless there exists an R_m with $r_m\leqq R_m\leqq r_{m+1}$ such that $\xi_1(R_m)=\varepsilon_{m-1}R_m^{\rho(R_m)}$. It this occurs, we continue $\xi_1(r)$ past R_m by taking $\delta>0$ and taking the tangent to the curve $\varepsilon_{m-1}r^{\rho(r)}$ at R_m; at $R_m+q\delta$ for q an integer, we extend this continuation by making a linear extension with slope $\frac{d}{dr}\{\varepsilon_{m-1}r^{\rho(r)}\}|_{R_m+q\gamma}$. By choosing δ sufficiently small, we shall have $\xi_1(r)\geqq\varepsilon_m r^{\rho(r)}$ in the interval $r_m\leqq r\leqq r_{m+1}$. This establishes the induction. Furthermore, it is clear that $\xi_1(r)\geqq\eta(r)$ and that for any m and r sufficiently large, $\xi_1(r)\leqq\varepsilon_m r^{\rho(r)}$. Suppose $\alpha(r)$ is the function from Corollary 9.14 and let $\xi(r)=\int\xi_1(r')\alpha(r-r')dr'$. Since $\xi_1(r')$ is convex, $\xi(r)\geqq\xi_1(r)$ and so $\xi(r)$ satisfies the requirements of the Lemma. □

Before proceeding, we note that if $\rho(r)$ is a strong proximate order ($\rho>1$) and $\rho^\star(r)$ is its conjugate proximate order, then $\rho^\star(r)$ is also a strong proximate order. We leave this simple calculation to the reader.

Theorem 9.16. *The Fourier-Borel transform establishes an isomorphism between the spaces*

i) $(E_p^{\rho(r)})'$ *and* $F_{\tau p'}^{\rho^\star(r)}$ *and between the spaces*

ii) $(F_{p'}^{\rho(r)})'$ *and* $E_{\tau p}^{\rho^\star(r)}$ *where* $\tau=\dfrac{\rho}{(\rho-1)^{(\rho-1)/\rho}}=A_\rho^{-1}$.

Proof. Let $v\in(E_p^{\rho(r)})'$. Then by Lemma 9.13, there exists an m such that $|v(f)|\leqq C_m\sup_z|f(z)\exp-p_m(z)^{\rho(p_m(z))}|$. Thus

$$\begin{aligned}|f_v(u)|&\leqq C_m\sup_z|\exp\langle u,z\rangle-p_m(z)^{\rho(p_m(z))}|\\&\leqq C_m\exp(\sup_{t\geqq0}(\sup_{p_m(z)=t}\{\mathbb{R}e\langle u,z\rangle t-t^{\rho(t)}\})\\&\leqq C_m\exp\sup_{t\geqq0}(p'_m(u)t-t^{\rho(t)}).\end{aligned}$$

Now $\frac{d}{dt}(p'_m(u)t-t^{\rho(t)})=p'_m(u)-\left(\rho'(t)\log t+\frac{\rho(t)}{t}\right)t^{\rho(t)}$ and since $\rho(t)\to\rho$ and $t\rho'(t)\log t\to0$, it follows that for large values of $\|u\|$, this function takes on an absolute maximum. For $\delta>0$ and $\|u\|$ sufficiently large (depending on δ), the maximum occurs at $t_u^{\rho(t_u)-1}=\dfrac{p'_m(u)}{\rho+\xi(u)}$ for $|\xi(u)|<\delta$ and equals

$$p'_m(u)^{\frac{\rho(t_u)}{\rho(t_u)-1}}\left\{\left(\frac{1}{\rho+\xi(u)}\right)^{\frac{1}{\rho(t_u)-1}}-\left(\frac{1}{\rho+\xi(u)}\right)^{\frac{\rho(t_u)}{\rho(t_u)-1}}\right\}$$

which is less that or equal to $[(\tau+\varepsilon)p'_m(u)]^{\rho^\star(k(u)p'_m(u))}$ where $\varepsilon \to 0$ as $\delta \to 0$ and $0 < a \leqq k(u) \leqq b < +\infty$. Thus the maximum for large $\|u\|$ is less than $[\tau p'_{m+1}(u)]^{\rho^\star(p'_{m+1}(u))}$ by Theorem 1.18. So the mapping $v \to f_v(u)$ of $(E_p^{\rho(r)})'$ is into $F_{\tau p'}^{\rho^\star(r)}$. Similarly, one shows that the mapping $v \to f_v(u)$ of $(E_{p'}^{\rho(r)})$ is into $E_{\tau p}^{\rho^\star(r)}$.

We now show that the mapping is onto. We let x, a $2n$-tuple, represent the real coordinates of z, and we define

$$\varphi(v) = \sup_{x \in K} (x_1 \operatorname{Im} v_1 + \ldots + x_{2n} \operatorname{Im} v_{2n}),$$

where $K = \{z \colon \mathbb{R}e\langle z, u\rangle \leqq p(u),\ u \in \mathbb{C}^n\}$. Then $\varphi(v)$ is a plurisubharmonic function of the variable v, and so $\theta(v) = (\varphi(v))^{\rho(\varphi(v))}$ is also plurisubharmonic.

Let $F(u) \in E_p^{\rho(r)}$, and set $\eta(r) = \sup_{\|u\| \leqq r} (\sup(\log|F(u)| - p(u)^{\rho(p(u))}, 0))$. It follows from Hartog's Lemma (more precisely, Corollary 1.32) that on the compact set $\|u\| = 1$, $\dfrac{\log|F(ru)|}{r^{\rho(r)}} \leqq p_m(u)^\rho$ for every $r > R_m$. This implies by Theorem 1.18 that $\lim_{r \to \infty} \eta(r) r^{-\rho(r)} = 0$. So by Lemma 9.15, there exists a positive function $\xi(r)$, convex and increasing in r, such that $\lim_{r \to \infty} \xi(r) r^{-\rho(r)} = 0$ and $\xi(r) \geqq \eta(r)$. Let $\varphi^\star(v) = \sup_{\|x\| \leqq 1} (x_1 \operatorname{Im} v_1 + \ldots + x_{2n} \operatorname{Im} v_{2n})$ and set $\xi^\star(v) = \xi(\varphi^\star(v))$, which is plurisubharmonic.

Let Σ be the n-dimensional subspace $v = (iu_1, -u_1, \ldots, iu_n, -u_n)$ in $\mathbb{C}^{2n}$ and let

$$w(iu_1, -u_1, \ldots, iu_n, -u_n) = F(u_1, \ldots, u_n)$$

on Σ. Then $|w(v)| \leqq C_0 \exp(\theta(v) + \xi^\star(v))$ on Σ, so if $\varepsilon > 0$ and

$$\theta'(v) = \theta(v) + \xi^\star(v) + \log(1 + \|v\|^2)^{n+\varepsilon},$$

$$\int_\Sigma |w(v)|^2 \exp(-2\theta'(v)) d\tau(v) < +\infty.$$

Thus, by Theorem 7.1 we can find an entire function W in $\mathbb{C}^{2n}$ such that $W = w$ on Σ and

$$\int |W(v)|^2 \exp(-2\theta''(v)) (1 + \|v\|^2)^{-3n-\varepsilon} d\tau(v) < +\infty,$$

where $\theta''(v) = \sup_{\|v - v'\| \leqq 2n} \theta'(v)$. From this we conclude via Schwarz' Lemma (cf. Lemma 3.47) that there exists a constant C'_0 such that

$$|W(v)| = C'_0 (1 + \|v\|)^{3n+\varepsilon} \exp \theta'''(v),$$

where $\theta'''(v) = \sup_{\|v - v'\| \leqq 1} \theta''(v')$.

Let α be the regularizing function introduced before and set

$$\tilde{\alpha}(v) = \int \alpha(x) \exp - i(x_1 v_1 + \ldots + x_{2n} v_{2n}) d\tau(x). \tag{9,7}$$

Hence $\tilde{\alpha}(0) = \int \alpha(x) d\tau(x) = 1$, and since α depends only on $\|x\|$, $\tilde{\alpha}(v)$ is a function of $v_1^2 + \ldots + v_{2n}^2$, so $\tilde{\alpha}(v) \equiv 1$ on Σ. By repeated integration by parts applied to (9,7) we see that $|\tilde{\alpha}(v)| \leqq \frac{C_k}{(1+\|v\|)^k} \exp \varepsilon(|\operatorname{Im} v_1|+\ldots+|\operatorname{Im} v_{2n}|)$. Hence if we set $\tilde{W}=\tilde{\alpha}\cdot W$, then $\tilde{W}=W$ on Σ and

$$(9,8) \qquad |\tilde{W}(v)| \leqq \frac{\tilde{C}}{(1+\|v\|)^{2n-1}} \exp\left(\theta'''(v)+\varepsilon \sum_{i=1}^{2n} |\operatorname{Im} v_i|\right).$$

By the Paley-Wiener Theorem, if

$$\mu(x)=\frac{1}{(2\pi)^{2n}} \int_{\mathbb{R}^{2n}} \exp i\langle x, v+iv'\rangle \tilde{W}(v+iv')\,d\tau(v),$$

then $\mu(x)$ is continuous and independent of v' and the Fourier-Laplace transform of $\mu(x)$ is $W'(v)$

$$(\text{i.e. } W'(v)=\int \exp\{-i(x_1 v_1+\ldots+x_{2n}v_{2n})\}\,\mu(x)\,d\tau(x)).$$

Thus the Fourier-Borel transform of $\mu(x)$ is just $w(v)=F(u)$, and it follows from (9,8) that

$$\begin{aligned} \mu(x) &\leqq K_m \exp(\inf_u (p_m(u)^{\rho(p_m(u))} - \mathbb{R}e\langle u, z\rangle) \\ &\leqq K_m \exp\{-(\tau p'_{m-1}(u))^{\rho^\star(p_{m-1}(u))}\} \end{aligned}$$

if we repeat the calculations made above. Hence $\mu(x) \in (F_{\tau p'}^{\rho^\star(r)})'$, and the map $(E_{\tau p'}^{\rho^\star(r)})' \to E_p^{\rho(r)}$ is onto. Since

$$\tau^\star = \frac{\rho^\star}{(\rho^\star-1)^{\frac{(\rho^\star-1)}{\rho^\star}}} = \frac{1}{\tau} \qquad \left(\text{where } \frac{1}{\rho^\star}+\frac{1}{\rho}=1\right)$$

and $\rho^{\star\star}(r)=\rho(r)$, we have $(F_{p'}^{\rho(r)}) \to E_{\tau p}^{\rho^\star(r)}$ is onto. Similarily, one shows that the mapping of $(E_p^{\rho(r)})'$ into $F_{\tau p'}^{\rho^\star(r)}$ is onto.

Thus, the mapping $v \to f_v$ is a continuous mapping of the Fréchet space $(F_{p'}^{\rho(r)})'$ onto $E_{\tau p}^{\rho^\star(r)}$, which implies by Proposition 9.8 that the transpose map of $(E_{\tau p}^{\rho^\star(r)})'$ into $F_{p'}^{\rho(r)}$ is one-to-one with closed image. In fact, we know that the map is onto, which implies in turn that the map $(F_{p'}^{\rho(r)})$ is one-to-one onto $E_{\tau p}^{\rho^\star}(r)$, which establishes the desired isomorphisms. □

Corollary 9.17. *In the space* $E_p^{\rho(r)}$, *the subspaces spanned by*

i) $\exp\langle u, z\rangle$ *for* $u \in K$ *with* $\overset{\circ}{K} \neq \{\phi\}$ *or*

ii) $z^\alpha \exp\langle u_0, z\rangle$ *for all multi-indices* α

are dense (in particular, the exponential functions and the polynomials are dense).

Proof. For every $v \in (E_p^{\rho(r)})'$, if $v(\exp\langle u, z\rangle)=0$ for $u \in K$ then $f_v \equiv 0$ from which (i) follows. We have $v(z^\alpha \exp\langle z, u_0\rangle = c_\alpha$ where c_α is the coefficient of $(u-u_0)^\alpha$ in the Taylor series expansion of f_v at u_0. Thus, if $c_\alpha=0$ for all α, $f_v \equiv 0$ from which (ii) follows. □

Theorem 9.18. *Let $\mu\in(E_p^{\rho(r)})'$ for a strong proximate order $\rho(r)$ with $\rho>1$ and $p(z)$ a non-negative support function and suppose that f_μ has minimal order with respect to $\rho^\star(r)$. Then the convolution equation $\hat{\mu}(x)=f$ has a solution $g\in(E_p^{\rho(r)})'$.*

Proof. The map $v\to\mu\star v$ is one-to-one and has closed image as one sees easily by repeating the second half of Theorem 9.12. We need only apply Proposition 9.8. □

§5. The Case $\rho=1$

In what preceeds, we have considered proximate orders for $\rho<1$ and $\rho>1$. The case of proximate orders for $\rho=1$ is extremely delicate since for $\rho\neq1$, either $r^{\rho(r)-1}$ increases or decreases but for $\rho=1$ this becomes problematical, and the theory and calculations become impossible without making additional assumptions. In a certain sense, this also translates the central role that the exponential functions play in the theory. We shall not consider proximate orders but we shall assume that $\rho\equiv1$.

Let $p(z)$ be a support function and E_p the Fréchet space of functions that we get by setting $w_m=p+\frac{1}{m}\|z\|$. If $\mu\in E_p'$ we define its Fourier-Borel transform by $f_\mu(u)=\mu(\exp\langle z,u\rangle)$. If $\tau=v\star\mu$ is the convolution of the measures μ and v, then $f_\tau(u)=f_v(u)\,f_\mu(u)$.

Suppose that g is any function holomorphic in a neighborhood of K. Then g defines a continuous linear operator S_g from E_p into E_p by Theorem 8.9:

$$S_g(F(u))=\frac{-(-1)^{\frac{n(n+1)}{2}}}{(2\pi i)^n}\int_{\Sigma(w)} g(z)\exp\langle z,u\rangle\frac{\partial^{n-1}}{\partial\xi_0^{n-1}}\left(\frac{\mathfrak{L}_F(\bar{\xi})}{\xi_0}\right)\tilde{\Omega}(z,\bar{\xi}),\qquad K\subset\omega$$

via the projective Laplace transform.

Lemma 9.19. *Let $\psi_{z_0}=\mathfrak{L}_{\exp\langle z_0,u\rangle}$ for $z_0\in K$. Then the linear functional on $\mathscr{H}(K)$ determined by ψ_{z_0} is $T_{\psi_{z_0}}=\delta(z_0)$, the Dirac measure with support z_0.*

Proof. Let f be a representative of $\bar{f}\in\mathscr{H}(K)$ defined in some strictly convex neighborhood ω of K. Since ω is a Runge domain, f can be uniformly approximated by polynomials in an open neighborhood of K, and since $z_i=\lim\limits_{|\lambda|\to0}\frac{e^{z_i\lambda}-1}{\lambda}$, $\lambda\in\mathbb{C}$, f can be uniformly approximated by exponentials. Since

$$F(u)=\frac{-(-1)^{\frac{n(n+1)}{2}}}{(2\pi i)^n}\int_{\Sigma(\omega)}\exp\langle z,u\rangle\frac{\partial^{n-1}}{\partial\xi_0^{n-1}}\left(\frac{\mathfrak{L}_F(\bar{\xi})}{\xi_0}\right)\tilde{\Omega}(z,\bar{\xi}),$$

$T_{\psi_{z_0}}$ is just $f(z_0)$ for the exponentials. It now follows from the uniform convergence in a neighborhood of K that $T_{\psi_{z_0}} = f(z_0)$ for all $f \in \mathscr{H}(K)$. □

Lemma 9.20. *Let $v \in E_p'$. If f_v is its Fourier-Borel transform, then the linear operator $Q_{f_v}\colon E_p \to E_p$ is just the transpose of the convolution $v \star \mu$ (i.e. $(Q_{f_v}(F), \mu) = (F, v \star \mu)$).*

Proof. We can represent μ by a measure such that $\mu \exp(p(u) + \varepsilon \|u\|)$ has bounded mass. Thus from Fubini's Theorem,

$$\mu(F(u)) = \frac{-(-1)^{\frac{n(n+1)}{2}}}{(2\pi i)^n} \int_{\Sigma(\omega)} \mu(\exp\langle z, u\rangle) \frac{\partial^{n-1}}{\partial \xi_0^{n-1}} \left(\frac{\mathfrak{L}_F(\xi)}{\xi_0}\right) \tilde{\Omega}(z, \xi)$$

for ω a small strictly convex neighborhood of K. Thus μ is completely determined by its values on a set of exponentials $\exp\langle z, u\rangle$ defined for z in a neighborhood of K. We choose ω so small that f_v is defined and bounded on ω. Then for $z_0 \in \omega$

$$\begin{aligned}(Q_{f_v}(\exp\langle z_0, u\rangle), \mu) &= \mu\left(\frac{-(-1)^{\frac{n(n+1)}{2}}}{(2\pi^i)^n} \int_{\Sigma(\omega)} \exp\langle z, u\rangle f_v(z) \frac{\partial^{n-1}}{\partial \xi_0^{n-1}} \left(\frac{\psi_{z_0}(\xi)}{\xi_0}\right) \tilde{\Omega}(z, \xi)\right) \\ &= f_v(z_0)\mu(\exp\langle z_0, u\rangle) = f_v(z_0) f_\mu(z_0).\end{aligned}$$

□

Theorem 9.21. *Let $v \in E_p'$ and $\hat{v}$ the associated convolution operator. Then for $F \in E_p$, there exists a $G \in E_p$ which is a solution of the equation $\hat{v}(x) = F$.*

Proof. The mapping $\mu \to f_\mu$ is a one-to-one linear mapping of E_p' into $\mathscr{H}(K)$. We give $\mathscr{H}(K)$ the topology of convergence of the Taylor series coefficients at each point of K. This is at least as weak as the equivalence on $\mathscr{H}(K)$ of the weak topology on E_p', since for a multi-index α,

$$\mu(u^\alpha \exp\langle z_0, u\rangle) = \frac{\partial^\alpha f_\mu(z_0)}{\partial z^\alpha}.$$

If $f_v f_{\mu_\gamma}$ is a filter converging to $g \in \mathscr{H}(K)$, then we must have $g = f_\gamma \cdot f_\mu$ as in the proof of Theorem 9.12, so the mapping $f_\mu \to f_v \cdot f_\mu$ is one-to-one with closed image, and $\mu \to v \star \mu$ is one-to-one with closed image. Thus $\hat{v}$ is onto by Proposition 9.8. □

We recall that a function $g(z)$ will be said to be *subadditive* if $g(z_1 + z_2) \leqq g(z_1) + g(z_2)$. A support function is our first example of a subadditive function. We know that the dual space of $B_g^\star$ is just the set of measures in $\mathbb{C}^n$ for which $\mu \exp g(z)$ is a bounded measure. The space B_g' is in general much more complicated, and we do not attempt a description here, but we note that the space of Baire measures for which $\mu \exp g(z)$ is a bounded measure generates a (not necessarily closed) subspace of B_g', which we shall note $\tilde{B}_g'$, and $\bar{B}_g'$ will be its closure in B_g'.

Suppose $\mu \in \tilde{B}'_g$ with norm $\|\mu\|$ and α as defined in Lemma 9.14, and let $\tilde{\mu}=\mu\star\alpha$, which is a $\mathscr{C}^\infty$ function. If $g(z)$ is a subadditive, then

$$\begin{aligned}\int_{\mathbb{C}^n} |\tilde{\mu}(z)| \exp g(z) d\tau(z) &\leqq \int_{\mathbb{C}^n}\int_{\mathbb{C}^n} \alpha(z') d|\mu|(z-z') \exp g(z) d\tau(z)\\ &= \int \alpha(z') d\tau(z') \int \exp g(w-z') d|\mu|(w)\\ &\leqq K\|\mu\| \text{ by subadditivity,}\end{aligned}$$

where $K=\sup\limits_{\|z'\|=1} \exp g(z')$.

Thus, we can associate $\tilde{B}'_g$, and hence $\bar{B}'_g$, with a closed subspace of the Banach space $L^1(\exp g(z) d\tau(z))=\{f\colon \int |f(z)| \exp g(z) d\tau(z)<+\infty\}$.

Lemma 9.22. *If $g(z)$ is subadditive, then the dual space of $\tilde{B}'_g$ is just B_g and the dual space of $B^{\star\prime}_g$ is a space of entire functions.*

Proof. Since B'_g is a subspace of $L^1(\exp g(z) d\tau(z))$, its dual space can be associated with the space of functions H such that $H \exp g(z)$ is essentially bounded in $\mathbb{C}^n$.

If v_γ is any Cauchy measure, then $v_\gamma(f)=0$ for $f\in B_g$, and so $v_\gamma(f) = v_\gamma\star\alpha(f)=0$. Thus, $\tilde{v}_\gamma(H)=0$, so H is holomorphic in every complex line by Morera's Theorem, and hence is globaly holomorphic by Hartog's Theorem (cf. [B]). □

Lemma 9.23. *If $g(z)$ is subadditive, then for $\alpha, \beta\in B'_g$ (resp. $B^{\star\prime}_g$) the operation of convolution $\alpha\star\beta(f)=\alpha_z[\beta_w f(z+w)]$ defines a continuous linear functional on B_g (resp. $B^\star_g$) and*

$$\|\alpha\star\beta\| \leqq \|\alpha\|\,\|\beta\|.$$

Proof. It is a simple consequence of the subadditivity of $g(z)$ and the Cauchy Integral Formula that for $f\in B_g$ (resp. $B^\star_g$), all derivatives of f are in B_g (resp. $B^\star_g$). Let $h^i=(0,\ldots,h,O,\ldots,0)$, h in the ith place. By Taylor's Theorem with remainder we have for $|h|<1$

$$f(z+w+h^i)-f(z+w)=h\frac{\partial f(z+w)}{\partial z_i}+h^2\frac{1}{2\pi i}\int_{|\xi|=1}\frac{f(z+w+\xi^i)}{\xi^2(\xi-h)}d\xi.$$

The integral on the right-hand side is an entire function in B_g (resp. $B^\star_g$) for fixed w. Upon dividing by h and letting h approach zero, we see that $\tilde{f}(z) =\beta_w f(z+w)$ is again an entire function, and it follows from the subadditivity of $g(z)$ that $\tilde{f}(z)\in B_g$. Thus we can apply α to it. If $\|f\|=1$, then by the subadditivity of g,

$$\|f(z+w)\| \leqq \exp[g(z)+g(w)] \quad \text{and} \quad |\tilde{f}(z)| \leqq \|\beta\| \exp g(z),$$

so $\|\tilde{f}(z)\| \leqq \|\beta\|$ and hence $|\alpha(\tilde{f}(z))| \leqq \|\alpha\|\,\|\beta\|$. □

Thus $\bar{B}'_g$ (resp. $B^{\star\prime}_g$) becomes a Banach algebra under the operation of convolution. In fact it is a commutative Banach algebra with identity, the identity element being given by $\delta(0)$, the Dirac measure.

Lemma 9.24. *The maximal ideal space M of non-zero homomorphisms in $\bar{B}'_g$ is just the space of exponentials* $\exp\langle u, z\rangle$ *in B'_g.*

Proof. A homomorphism on a commutative Banach algebra with identity is always continuous, so by Lemma 9.23, such a homomorphism is given by some entire function in B_g. Let f be such an entire function. For two points z_0 and w_0, we consider $(f, \delta(z_0)\star\delta(w_0)) = f(z_0+w_0) = f(z_0)f(w_0)$. This formula says that if $f(z)$ is zero at any point, it is zero everywhere. Since we assume that f is not the trivial homomorphism, $f(z)\neq 0$, and we can define a branch of $\log f(z)$ in $\mathbb{C}^n$. Since $f(0)=1$, $\log|f(z)|$ determines a real linear function which has a linear conjugate function uniquely determined by the condition $f(0)=1$. Then $f(z)=\exp\langle u, z\rangle$. □

Corollary 9.25. *The maximal ideal space of $B^{\star\prime}_g$ is a subset of the exponentials* $\exp\langle u, z\rangle$ *in B'_g.*

We note that $B^\star_g$ can consist only of the zero function even if B_g does not. For example, if we take $g(z)=\mathbb{R}e\langle z, u_0\rangle$ so that B_g contains $\exp\langle z, u_0\rangle$, then $B^\star_g$ is empty since $f\in B^\star_g$ and $\lim_{\|z\|\to\infty} |f(z)\exp -g(z)| = 0$ implies that $f(z)\exp -\langle z, u_0\rangle \equiv 0$ and so $f(z)\equiv 0$. This is why we lose some precision in describing these spaces.

Suppose now that $g(z)=p(z)$ is a support function. Then for $\alpha\in\tilde{B}'_g$ (resp. $B^{\star\prime}_g$), we define the function

$$F_\alpha(u) = \alpha(\exp\langle u, z\rangle)$$

which is continuous on $K_p = [z\colon \operatorname{Re}\langle z, u\rangle \leqq p(u), \text{ for } u\in\mathbb{C}^n]$, and holomorphic in the interior, and

$$F_{\alpha\star\beta}(u) = F_\alpha(u)\cdot F_\beta(u).$$

Theorem 9.26. *For* $\alpha\in\tilde{B}'_p$ (resp. $B^{\star\prime}_p$), *if* $F_\alpha(u)\neq 0$ *on* K_p, *then for* $f\in B_p$ *(resp.* $B^\star_p$*) there exists a unique solution* $\tilde{f}\in B_p$ *(resp.* $B^\star_p$*) of the equation* $\hat{\alpha}(x)=f$.

Proof. Consider the ideal M_α in $\bar{B}'_p$ (resp. $B^\star_{p'}$) generated by α. By Lemma 9.24, there is no non-zero homomorphism which vanishes on M_α, so M_α is not contained in any proper maximal ideal; hence $M_\alpha = \bar{B}'_g$ *(resp.* $B^{\star\prime}_g$). Thus the mapping $\beta\to\alpha\star\beta$ is one-to-one and onto, so by Proposition 9.8 the map $\hat{\alpha}$ is also one-to-one and onto. □

If $p_m(z)$ is pointwise decreasing sequence of support functions, we set $F = \bigcap_{m=1}^{\infty} B_{p_m}$, which becomes a Fréchet space when we equip it with the projective limit topology. The dual space F' of F is just the union of the dual spaces of B_{pm}. Let $\tilde{F}'=\bigcup_m \tilde{B}'_{pm}$ and $K=\bigcap K_{pm}$. We also let $F^\star=\bigcap B^\star_{pm}$.

Corollary 9.27. *If $f\in F$ (resp. $F^\star$) and if for $\alpha\in\tilde{F}'$ (resp. $F^{\star\prime}$) $F_\alpha(u)\neq 0$ on K, then there exists a unique solution $\tilde{f}\in F$ (resp. $F^\star$) of the equation $\hat{\alpha}(x)=f$.*

Proof. If $F_\alpha(u)\neq 0$ on K, by continuity $F_\alpha(u)\neq 0$ on K_{pm} for $m\geqq M_0$ and hence there exists a unique solution $\tilde{f}$ in B_{pm} (resp. $B^\star_{pm}$). Thus $\tilde{f}\in\bigcap B_{pm}$ (resp. $B^\star_{pm}$). □

The following elementary example shows that some condition on $F_\alpha(u)$ is needed. Let $n=1$, $g(z)=r$, $f(z)=\exp z$ and $\hat{\alpha}=D-1$ where $D=\dfrac{d}{dz}$. Then $f\in B_g$, but the solutions of $\hat{\alpha}(x)=f$, namely $(z+1)\exp z+C\exp z$ are not in B_g.

Let $z=x+iy$, where $x=(x_1,\ldots,x_n)$ and $y=(y_1,\ldots,y_n)$, and let $dx=dx_1,\ldots,dx_n$ be the Lebesgue measure on $\mathbb{R}^n$. We will suppose that $g(z)=g(y_1,\ldots,y_n)$ is a support function. We define the set B^p_g (resp. $B^{p\star}_g$) to be the Banach space of all $f\in B_g$ (resp. $B^\star_g$) which satisfy

$$\|f\|^p_p=\int_{\mathbb{R}^n}|f(x)|^p dx<\infty \qquad 1\leqq p<+\infty,$$

where we give B^p_g (resp. $B^{p\star}_g$) the norm $\|f\|=\|f\|_p+\|f\|_g$. (We can think of B^∞_g as B_g.)

We now characterize the dual space $(B^p_g)'$. Consider the space $B_g\times L^p$ of doubles (f,h) with $f\in B_g$ and $h\in L^p(\mathbb{R}^n)$. This is a Banach space when we give it the norm $\|(f,h)\|=\|f\|_g+\|h\|_p$. The dual is just the set of doubles (α,β) with $\alpha\in B'_g$ and $\beta\in(L^p)'$, where $(\alpha,\beta)(f,h)=\alpha(f)+\beta(h)$. The subspace of $B_g\times L^p$ composed of those (f,h) for which $h(x)=f(x)$ is a closed subspace of $B_g\times L^p$ which is isomorphic in the obvious way to B^p_g, and so by the Hahn-Banach Theorem they have the same dual spaces. The characterization of $(B^{p\star}_g)'$ is the same. We now develop the essential facts that we shall need about these spaces.

Proposition 9.28. *For $n=1$, suppose $f(z)$ is an entire function of order 1 with $h^\star_f(z)\leqq g(y)$ and suppose further that sor some $p>0$, $\int_{-\infty}^{+\infty}|f(x)|^p dx<+\infty$. Then*

$$\int_{-\infty}^{+\infty}|f(x+iy)|^p dx\leqq\exp pg(y)\int_{-\infty}^{+\infty}|f(x)|^p dx.$$

Proof. First we note that the hypothesis $\int_{-\infty}^{+\infty}|f(x)|^p dx<+\infty$ implies that $\int_{-\infty}^{+\infty}\dfrac{\log^+|f(x)|}{1+x^2}dx<+\infty$. Indeed if $\varphi(t)=\exp(pt)$, then φ is a positive non-decreasing convex function of t and so

$$\varphi\left\{\frac{1}{\pi}\int_{-\infty}^{+\infty}\frac{\log^+|f(x)|dx}{1+x^2}\right\}\leqq\frac{1}{\pi}\int_{-\infty}^{+\infty}\frac{\varphi(\log^+|f(x)|)dx}{1+x^2}$$
$$\leqq\frac{1}{\pi}\int_{-\infty}^{+\infty}\frac{\sup(1,|f|^p)dx}{1+x^2}<+\infty.$$

Let $\tau=h_f^{\star}(i)$ and $\tilde{f}(z)=f(z)\exp-\tau iz$ so that $\tilde{f}(z)\in L^p(\mathbb{R})$ and $h_{\tilde{f}}^{\star}(z)<0$ for $\operatorname{Im}z>0$. Suppose $M>-\infty$ and set $V_M(z)=\sup(\log|\tilde{f}|)+M,0)$, which is subharmonic and positive in $\mathbb{C}$. Let $A_r=\{z\colon \operatorname{Im}z>0,\ \|z\|<r\}$. For every $\varepsilon>0$, $\log|\tilde{f}(z)|\leqq\varepsilon r$ for $r>R_\varepsilon$, $z\in A_r$.

Suppose D is a domain with $\mathscr{C}^2$ boundary, $D\subset A_1$ and $bdD\cap\{y\colon y=0\}$ contains the interval $[-1/2,\ +1/2]$ on the x-axis, and let $D_r=\{rz\colon z\in D\}$. For z with $\operatorname{Im}z>0$, we let $H_M(z)=\dfrac{y}{\pi}\displaystyle\int_{-\infty}^{+\infty}\frac{V_M(t)dt}{(t-x)^2+y^2}>0$, and if $P_r(w,z)$ is the Poisson kernel in D_r then we set, $H_r^\varepsilon(z)=\displaystyle\int_{bdD_r\cap\{y>0\}}\varepsilon r P_r(w,z)dS_r(w)$, where dS is the surface measure on bdD_r. Then $H_M(z)+H_r^\varepsilon(z)$ is a harmonic function in D_r, and $H_M(z)+H_r^\varepsilon(z)\geqq V_M(z)$ in D_r for $r>R_\varepsilon$ by the Maximum Principle. Furthermore, $H_r^\varepsilon(z)=H_1^\varepsilon\left(\dfrac{z}{r}\right)\cdot r$ and by standard estimates for the Poisson kernel in a domain with $\mathscr{C}^2$ boundary, $P_1(w,z)\leqq\dfrac{c\,d(z)}{|w-z|}$, where $d(z)$ is the distance to bdD. Hence $H_1^\varepsilon\left(\dfrac{z}{r}\right)\leqq\dfrac{\varepsilon K}{r}$, and $H_M(z)+\varepsilon K\geqq V_M(z)$. Since $\varepsilon>0$ was arbitrary, $V_M(z)\leqq H_M(z)$. But

$$\frac{y}{\pi}\int_{-\infty}^{+\infty}\frac{M\,dt}{(t-x)^2+y^2}=M$$

so

$$\frac{y}{\pi}\int_{-\infty}^{+\infty}\frac{\sup(\log|\tilde{f}(t)|,M)dt}{(t-x)^2+y^2}\geqq\sup(\log|\tilde{f}(z)|,M),$$

which holds for all M, and

$$\int_{-\infty}^{+\infty}\frac{\log|\tilde{f}(t)|}{1+t^2}dt<+\infty.$$

Thus

$$\frac{y}{\pi}\int_{-\infty}^{+\infty}\frac{\log|\tilde{f}(t)|}{(t-x^2)+y^2}dt\geqq\log|\tilde{f}(z)|$$

holds for all z with $\operatorname{Im}z>0$. Again by the convexity of $\varphi(t)=\exp pt$, we have

$$|\tilde{f}(z)|^p\leqq\frac{1}{\pi}\int_{-\infty}^{+\infty}|\tilde{f}(t)|^p\frac{y}{(x-t)^2+y^2}dt$$

so

$$\int_{-\infty}^{+\infty} |\tilde{f}(x+iy)|^p dx \leqq \frac{1}{\pi} \int_{-\infty}^{+\infty} |\tilde{f}(t)|^p dt \int_{-\infty}^{+\infty} \frac{y}{(x-t)^2+y^2} dx$$

$$\leqq \int_{-\infty}^{+\infty} |f(t)|^p dt$$

and hence

$$\int_{-\infty}^{+\infty} |f(x+iy)|^p dy \leqq \exp p\tau y \int_{-\infty}^{+\infty} |f(t)|^p dt.$$

A similar result holds in the lower half-plane. □

Lemma 9.29. *Suppose* $f \in B_g^p$ *(resp.* $B_g^{p\star}$*). Then*

$$f(x+iy) \in L^p(\mathbb{R}^n) \quad \text{and} \quad \|f(x+iy)\|_p \leqq \exp g(y)\, \|f(x)\|_p.$$

Proof. Let $\frac{iy}{\|y\|} = (it_1^1, \ldots, it_n^1)$, $t_i \in \mathbb{R}$. Set $\alpha^1 = (t_1^1, \ldots, t_n^1)$, which we complete to an orthonormal system of real vectors α^j, $j=2, \ldots, n$, and set

$$A = \begin{bmatrix} t_1^1 \ldots t_n^1 \\ \vdots \quad\quad \vdots \\ t_1^n \ldots t_n^n \end{bmatrix}$$

with inverse A^{-1}, so that the Jacobian of A is 1. Let $z' = Az$, and set $\tilde{f}(z') = f(A^{-1}z')$ so that $\int_{\mathbb{R}^n} |\tilde{f}(z')|^p dx = \int_{\mathbb{R}^n} |f(x)|^p dx$ and for $y_1' = \operatorname{Im} z_1'$, $z' = Az$, $g(y_1') = g(y)$. Then for fixed $x_2', \ldots, x_n'$, by Proposition 9.28,

$$\int_{-\infty}^{+\infty} |\tilde{f}(x_1'+iy_1', x_2', \ldots, x_n')|^p dx_1' \leqq \exp pg(y) \int_{-\infty}^{+\infty} |\tilde{f}(x_1', \ldots, x_n')|^p dx_1',$$

and so

$$\int_{\mathbb{R}^n} |\tilde{f}(x'+iy_1')|^p dx' \leqq \exp pg(y) \int_{\mathbb{R}^n} |\tilde{f}(x')|^p dx'$$

$$= \exp pg(y) \int_{\mathbb{R}^n} |f(x)|^p dx.$$ □

Lemma 9.30. *If* $f \in B_g^p$ *(resp.* $B_g^{p\star}$*) then* $\frac{\partial f}{\partial z_j} \in B_g^p$ *(resp.* $B_g^{p\star}$*).*

Proof. We have already shown that $\frac{\partial f}{\partial z_j} \in B_g$ (resp. $B_g^\star$). If $\xi^j = (0, \ldots, 0, \xi, 0, \ldots, 0)$, ξ in the jth place, we have from the Cauchy Integral Formula $\frac{\partial f(x)}{\partial z_j} = \frac{1}{2\pi i} \int_{|\xi|=1} \frac{f(x+\xi^j)d\xi}{\xi^2}$ so that for $p=1$, the result follows directly from Lemma 9.29 and Fubini's Theorem.

If $p>1$, let $q=1-1/p$ and let $\mu(x)\in L^q(\mathbb{R}^n)$. Then

$$\left|\int_{\mathbb{R}^n}\frac{\partial f(x)}{\partial z_j}\mu(x)dx\right|\leqq\frac{1}{2\pi}\left|\int_{\mathbb{R}^n}\int_{|\xi|=1}\frac{f(x+\xi^j)}{\xi^2}d\xi\mu(x)dx\right|$$
$$\leqq\sup_{|\xi|=1}\int_{\mathbb{R}^n}|f(x+\xi^j)\mu(x)dx|\leqq K\|f\|_p\|\mu\|_q$$

by Lemma 9.29 and Hölder's Inequality. Thus $\frac{\partial f}{\partial z_j}(x)$ is a continuous linear functional on $L^q(\mathbb{R}^n)$ and so is in $L^p(\mathbb{R}^n)$. □

Lemma 9.31. *If $f\in B_g^p$ (resp. $B_g^{p\star}$) and $\mu(x)\in L^q(\mathbb{R}^n)$, $1/q=1-1/p$ ($q=\infty$ if $p=1$), then $k(z)=\int_{\mathbb{R}^n}f(z+x)\mu(x)dx\in B_g$.*

Proof. By Lemma 9.29 and Hölder's Inequality,

$$|\int_{\mathbb{R}^n}f(z+x)\mu(x)dx|\leqq\exp g(z)\|f\|_p\|\mu\|_q,$$

so it remains to verify that $k(z)$ is entire.

We again set $h^j=(0,\ldots,0,h,0,\ldots,0)$ (h in the jth place). By Taylor's Theorem with remainder, we have

$$f(z+w+h^j)=f(z+w)+h\frac{\partial f(z+w)}{z_j}+h^2\frac{1}{2\pi i}\int_{|\xi|=1}\frac{f(z+w+\xi^j)d\xi}{\xi^2(\xi-h)}.$$

Setting $w=x$, applying Lemmas 9.29 and 9.30 to the right-hand side, dividing by h and letting h approach zero, we see that

$$\frac{\partial k(z)}{\partial z_j}=\int\frac{\partial f(z+x)}{\partial z_j}\mu(x)dx. \qquad □$$

Of course, in general $k(z)\notin B_g^p$, so the dual space will not be a Banach algebra, but it will be a left module over those elements of B_g' (resp. $(B_g^\star)'$). For $\alpha\in B_g'$ and $\gamma\in B_g^{p'}$ (resp. $B_g^{p\star\prime}$), we define

$$\alpha\star\gamma(f)=\alpha_z[\gamma_w(f(z+w))].$$

By Lemmas 9.23 and 9.31, we see that this defines a continuous linear functional on B_g^p. The operation of convolution is associative for elements of B_g' (resp. $B_g^{\star\prime}$) (i.e. $\alpha'\star(\alpha\star\gamma)=(\alpha'\star\alpha)\star\gamma$). Thus we have:

Theorem 9.32. *Let $\alpha\in\tilde{B}_g'$ (resp. $B_g^{\star\prime}$) be such that $F_\alpha(u)\neq0$ on K_g for g a support function such that $g(z)=g(y)$. Then for $f\in B_g^p$ (resp. $B_g^{p\star}$), there exists a unique solution $\tilde{f}\in B_g^p$ (resp. $B_g^{p\star}$) of the equation $\hat{\alpha}(x)=f$.*

Proof. If $F_\alpha(u)\neq0$, then α is invertible in B_g' (resp. $B_g^{\star\prime}$), so the map $\gamma\to\alpha\star\gamma$ is one-to-one map of $B_g^{p'}$ (resp. $B_g^{p\star\prime}$) onto itself. The result now follows from Proposition 9.8. □

In the same way as we had Corollary 9.27, we have:

Corollary 9.33. *Suppose that $g_m(z)$ is a decreasing sequence of functions of the form $g_m(z)=g_m(y)$, and let $F^p=\bigcap B^p_{g_m}$ (resp. $F^{p\star}=\bigcap B^{p\star}_{g_m}$) and $K=\bigcap K_{g_m}$. If $F_\alpha(u)\neq 0$ on K, $\alpha\in\bigcup B'_{g_m}$, then for $f\in F^p$, there exists a unique $\tilde{f}\in F^p$ solution of the equation $\hat{\alpha}(x)=f$.*

The following example shows that some condition on $F_\alpha(u)$ is required. Let $n=1$, $\tau=1$ and $f(z)=\dfrac{\sin z}{z}\in L^2(\mathbb{R})$. Then if $\alpha=\dfrac{d}{dz}$, there is no solution in $B^2_{|y|}$ of the equation $\dfrac{d\tilde{f}}{dz}=f$. To see this, we write the general solution as

$$k(z)=C+\int_0^z \frac{\sin\xi}{\xi}\,d\xi,$$

so that for $x>0$, $k(x)=C+\int_0^x \frac{\sin t}{t}\,dt$ and $k(x)\to C+C_0$ as $x\to+\infty$. But for $x>0$,

$$k(-x)=C+\int_0^{-x}\frac{\sin t}{t}\,dt=C-\int_0^x\frac{\sin t}{t}\,dt$$

and $k(-x)\to C-C_0$, so $g(x)\notin L^2$ for any choice of C since $C_0\neq 0$.

§6. More on Functions of Order less than One

If $\rho(r)$ is a proximate order with $\rho<1$, then for $r>R_0$, we have $r^{\rho(r)}$ increasing and

$$\frac{d}{dr}(r^{\rho(r)})=(r\rho'(r)\log r+\rho(r))r^{\rho(r)-1}<r^{\rho(r)-1}.$$

We can thus assume that this holds everywhere. If $g(z)$ is a positive subadditive function, then $g(z)^{\rho(g(z))}$ is also subadditive since if $b\leqq a$, $\dfrac{f(b+a)-f(a)}{b}=f'(\xi)$, $\xi\geqq a$, so letting $f(r)=r^{\rho(r)}$,

$$f(b+a)-f(a)=bf'(\xi)\leqq b\cdot\xi^{\rho(\xi)-1}\leqq b^{\rho(b)}=f(b),$$

and

$$f(g(z_1+z_2))\leqq f(g(z_1)+g(z_2))\leqq f(g(z_1))+f(g(z_2)).$$

Suppose now that $p(z)$ is a norm and $\alpha\in(B_{p^{\rho(p)}})'$ is such that $\alpha(1)\neq 0$. Since the only exponential in $B_{p^{\rho(p)}}$ is the function 1, it follows as above that α has an inverse α' in $B_{p^{\rho(p)}}$. We shall now exploit this observation.

Theorem 9.34. *Let $\rho(r)$ be a proximate order with $\rho<1$. Suppose $\alpha\in\bigcap_{A>0} B_{A\|z\|^{\rho(\|z\|)}}$ such that $\alpha(1)\neq 0$. Then for every f of normal type with* *respect to $\rho(r)$, there exists a unique $\tilde{f}$ solution of the equation $\hat{\alpha}(x)=f$ such that*

i) $h^\star_{\tilde{f}}(z)=h^\star_f(z)$

ii) *if f is of regular growth for the ray (tz_0), $t>0$, then $\tilde{f}$ is also of regular growth for the ray (tz_0).*

Proof. i) We assume that f is of type $B/2$. Let $w\in S^{2n-1}$ be fixed, and let $a=h^\star(w)$. Given $\varepsilon>0$, there exists δ such that $\dfrac{\log|f(rz)|}{r^{\rho(r)}}\leqq h^\star(w)+\varepsilon/3$ for $r>R_\varepsilon$ and $\|z-w\|<\delta$. Let $\eta>0$ be given, and choose $A=\eta^{-1}(3(1+B+a))^{2/\rho}$. Suppose that $\mu=\alpha^{-1}$ in the space $B'_{(A\|z\|)^{\rho(A\|z\|)}}$ (the choice of μ might well depend on A) so that $\int \exp A\|z\|^{\rho(A\|z\|)}d|\mu|<+\infty$. Then $f(rw)=\int\tilde{f}(rw+z)\,d\mu(z)$ and

$$|f(rw)|\leqq \int_{\|z\|\leqq\eta r}|\tilde{f}(rw+z)|\,d|\mu|(z)+\int_{\|z\|>\eta r}|\tilde{f}(z+w)|\,d|\mu|(z)=I_1+I_2.$$

Now $|\tilde{f}(z)|\leqq C\exp B\|z\|^{\rho(B\|z\|)}$ since $\tilde{f}\in E_{\frac{B}{2}\|z\|^{\rho(\|z\|)}}$, so

$$\begin{aligned}
I_2&=\int_{\|z\|>\eta r}|\tilde{f}(z+rw)|\,d|\mu|(z)\leqq C\int_{\|z\|>\eta r}\exp B\|z+rw\|^{\rho(B\|2+rw\|)}d|\mu|(z)\\
&\leqq C\exp Br^{\rho(Br)}\int_{\|z\|>\eta r}\exp B\|z\|^{\rho(B\|z\|)}d|\mu|(z)\\
&\leqq C\exp Br^{\rho(r)}\int_{\|z\|>\eta r}\exp\big[B\|z\|^{\rho(B\|z\|)}\\
&\quad-(A\|z\|)^{\rho(A\|z\|)}\big]\exp(A\|z\|)^{\rho(A\|z\|)}d|\mu|(z)\\
&\leqq C\exp Br^{\rho(r)}\big[\exp B(\eta r)^{\rho(B(\eta r))}-3(1+B+a)r^{\rho(Br)}\big]\,C_\eta\\
&\leqq C\tilde{C}_\eta\exp-2a^{\rho(r)}
\end{aligned}$$

for $r>R_\eta$ by Definition 1.16 and Theorem 1.18. Suppose that $h^\star_{\tilde{f}}(w)<a$. Then there exists $\xi>0$, $\eta_\xi>0$, and R_η such that $r^{-\rho(r)}\log|\tilde{f}(rz)|\leqq a-\xi$ for $\|z-w\|<\eta_\xi$ and $r>R_\eta$ by Theorem 1.31. Let $w'\in S^{2n-1}$ be a point such that $\|w'-w\|<2^{-1}\eta_\xi$ and $h_f(w')\geqq h^\star_f(w)-\varepsilon/3$, and let r_m be a sequence which increases to infinity such that

$$\lim_{m\to\infty}\frac{\log|f(r_m w')|}{r_m^{\rho(r_m)}}=h_f(w').$$

Then since $\int d|\mu|(z)<+\infty$ and $I_2\leqq C\tilde{C}_{\eta/2}\exp-2ar^{\rho(r)}$, there exists a point w'' such that $\|w''-w\|<\eta$ and

$$\frac{\log|\tilde{f}(r_m w'')|}{r_m^{\rho(r_m)}}\geqq a-\xi/2.$$

But this is a contradiction, so $h^\star_{\tilde{f}}(z)\geqq h^\star_f(z)$. By noting that

$$\tilde{f}(z)=\int f(z+w)\,d\alpha(w),$$

we can reverse the roles of f and $\tilde{f}$ in the above calculations. Thus $h_f^\star(z)=h_{\tilde{f}}^\star(z)$.

ii) Suppose now that f is of regular growth along the ray tw, $w\in S^{2n-1}$. Let η be so small that for $r>R_\eta$

$$(9,9)\qquad |I_{\tilde{f}}^r(z',\delta)-I_{\tilde{f}}^r(w,\delta)|\leqq \xi/8 \quad \text{for } \|z'-w\|<2\eta \text{ (Lemma 4.2).}$$

By Definition 4.1, there exists R_ξ such that for $r>R_\xi$, we can find w_r' with $\|w_r'-w\|<\eta$ and $r^{-\rho(r)}\log|f(rw_r')|\geqq h_f^\star(w)-\xi/4$. By the argument of (i) above, we can find w_r'' with

$$\|w_r''-w_r'\|<\eta \quad \text{and} \quad r^{-\rho(r)}\log|f(rw_r'')|\geqq r^{-\rho(r)}\log|f(rw_r')|-\xi/4.$$

Since

$$I_{\tilde{f}}^r(w_r'',\delta)\geqq r^{-\rho(r)}\log|\tilde{f}(rw_r'')|\geqq h_f^\star(w)-\xi/2,$$

by the Mean Value Property for subharmonic functions, we see by (9,9) that

$$I_{\tilde{f}}^r(w,\delta)\geqq h_f^\star(w)-\xi=h_{\tilde{f}}^\star(w)-\xi \qquad \text{for } r>\sup(R_\xi,\tilde{R}_\xi).$$

By Theorem 1.31, there exists $\hat{R}_\xi$ such that

$$I_{\tilde{f}}^r(w,\delta)\leqq h_{\tilde{f}}^\star(w)+\xi \qquad \text{for } r>\hat{R}_\xi \quad \text{and} \quad \delta<\delta_\xi,$$

so $\tilde{f}$ is of regular growth for the ray (tw). □

§7. Convolution Operators in $\mathbb{C}^n$

We finish this chapter with an application to convolution operators in $\mathbb{C}^n$. Let Ω be an open convex subset of $\mathbb{C}^n$ and K a compact subset of $\mathbb{C}^n$. Suppose that $\mu\in\mathscr{H}(\mathbb{C}^n)'$ is carried by K. Let

$$\Omega+K=\{z=z'+z'':\, z'\in\Omega,\, z''\in K\},$$

which is an open convex subset of $\mathbb{C}^n$. For $f\in\mathscr{H}(\Omega+K)$, $\mu\in\mathscr{H}(K)'$, we define the operator $\hat{\mu}$: $\mathscr{H}(\Omega+K)\to\mathscr{H}(\Omega)$ by $\hat{\mu}(f)(z)=\mu_w(f(z+w))$, $z\in\Omega$, $w\in K$. For $\varepsilon>0$, we can find a measure μ_ε with support in $K^\varepsilon=\{z':\mathscr{F}z\in K\ni\|z'-z\|<\varepsilon\}$ such that $\hat{\mu}(f)(z)=\int f(z+w)\,d\mu_\varepsilon(w)$, so $\hat{\mu}(f)(z)$ is holomorphic in $\Omega_\varepsilon=\{z:\, d_\Omega(z)>\varepsilon\}$. Since this is true for all $\varepsilon>0$, by the uniqueness of analytic continuation, $\hat{\mu}(f)$ is holomorphic in Ω.

Theorem 9.35. *Let $\mu\in\mathscr{H}(\mathbb{C}^n)'$ be carried by the compact convex set K and let $\mathscr{F}_\mu$ be the Fourier-Borel transform of μ. If $h_{\mathscr{F}_\mu}^\star(\xi)=h_K(\xi)$ and $\mathscr{F}_\mu(\xi)$ is of regular growth in $\mathbb{C}^n$ (with respect to $\rho\equiv 1$), then for $g\in\mathscr{H}(\Omega)$, there exists a solution $f\in\mathscr{H}(\Omega+K)$ of the equation $\hat{\mu}(\alpha)=f$. If $\Omega\subset\mathbb{C}^n$ is a bounded strictly convex domain with $\mathscr{C}^2$ boundary, then $\hat{\mu}$: $\mathscr{H}(\Omega+K)\to\mathscr{H}(K)$ only if $h_{\mathscr{F}_\mu}^\star(\xi)=h_K(\xi)$ and $\mathscr{F}_\mu(\xi)$ is of regular growth in $\mathbb{C}^n$.*

Proof. For $\alpha\in\mathscr{H}(\Omega)'$, let $\mathscr{F}_\alpha(\xi)$ be its Fourier-Borel transform. Then if $\hat{\mu}^t(\alpha)$ is the transpose operator of μ, $\mathscr{F}_{\hat{\mu}^t(\alpha)}(\xi)=\mathscr{F}_\mu(\xi)\mathscr{F}_\alpha(\xi)$.

Suppose that $\hat{\mu}^t(\alpha_\nu)$ converges weakly to an element $\beta\in\mathscr{H}(\Omega+K)'$. Then $\mathscr{F}_\beta(\xi)=\mathscr{F}_\mu(\xi)\cdot G(\xi)$ for some entire function $G(\xi)$, since the Taylor series of $\mathscr{F}_\beta(\xi)$ at each point $\xi\in\mathbb{C}^n$ is divisible by the Taylor series of $\mathscr{F}_\mu(\xi)$ (cf. Theorem 9.12). Since $h_{K_1+K_2}(\xi)=h_{K_1}(\xi)+h_{K_2}(\xi)$, there exists a compact convex subset $\tilde{K}$ of Ω such that β is carried by $\tilde{K}+K$ and $h^\star_{\mathscr{F}_\beta}(\xi)\leqq h_{K+\tilde{K}}(\xi)\leqq h_K(\xi)+h_{\tilde{K}}(\xi)$. Since $\mathscr{F}_\mu(\xi)$ is of regular growth, by Theorem 4.3

$$h^\star_G(\xi)=h^\star_{\mathscr{F}_\beta}(\xi)-h^\star_{\mathscr{F}_\mu}(\xi)\leqq h_K(\xi)+h_{\tilde{K}}(\xi)-h_{\tilde{K}}(\xi)=h_K(\xi).$$

Thus by Theorem 8.9, we can find $\gamma\in\mathscr{H}(\Omega)'$ with $\mathscr{F}_\gamma(\xi)=G(\xi)$, and so the image is weakly closed. By Proposition 9.8, μ is surjective, which proves the first part of our Theorem.

Suppose now that Ω is bounded, strictly convex with $\mathscr{C}^2$ boundary and that there exists $\xi_0\in S^{2n-1}$ with $\mathscr{F}_\mu(r\,\xi_0)$ not of regular growth or $h^\star_{\mathscr{F}}(\xi_0)<h_K(\xi_0)$. Since Ω is strictly convex, there exists a unique $z_0\in bd\Omega$ such that $h_K(\xi_0)=\mathbb{R}e\langle\xi_0,z_0\rangle$.

The construction depends heavily on Theorem 4.9, and we use the notation of that theorem. Let $B(\tilde{z},s)$ be a ball contained in Ω and such that $\bar{B}\cap bd\Omega=\{z_0\}$. We assume for simplicity that $\tilde{z}=0$. Let r_m be the sequence of Theorem 4.9,

$$\tilde{A}(\xi)=\|\xi\|+\sum_{m=1}^{\infty}\zeta r_m\psi\left(\frac{\xi-r_m\xi_0}{\eta_2 r_m}\right).$$

If a is the type of $\tilde{A}(\xi)$, we set $A(\xi)=sa^{-1}\tilde{A}(\xi)$ and define $V_1(\xi)$ and $\varphi(\xi)$ as before and

$$\alpha_\nu(\xi)=\sum_{m=1}^{\nu}\psi(z-r_m\xi_0)\exp V_1(r_m\xi_0).$$

Then we can find a sequence δ_ν decreasing to zero such that $\beta_\nu=\bar{\partial}\alpha_\nu$ satisfies $\int_{\mathbb{C}^n}|\beta_\nu|^2\exp-V_2(\xi)d\tau(\xi)<C$ independent of ν for $V_2(\xi)=(1-\delta_\nu)V_1(\xi)+2\eta\varphi+\|\xi\|^{\rho'}$.

By Appendix III, we can find γ_ν such that $\bar{\partial}\gamma_\nu=\beta_\nu$ and

$$\int_{\mathbb{C}^n}|\gamma_\nu|^2\exp-[V_2(\xi)+\log(1+\|z\|^2)]d\tau(\xi)<C$$

independent of ν.

We let $g_\nu(\xi)=\alpha_\nu(\xi)-\gamma_\nu(\xi)$. Then $g_\nu(\xi)$ is an entire function of order strictly less than s and $|g_\nu(\xi)|\leqq C''\exp V_3(\xi)$ independent of ν by Lemma 3.47. Thus, we can choose a subsequence which we shall also denote by $g_\nu(\xi)$ which converges uniformly on compact subsets to a function g. It is easy to verify by construction that $|\mathscr{F}_\mu(\xi)g_\nu(\xi)|\leqq C'''\exp(h_K(\xi)+h_{\tilde{K}}(\xi))$ for $\tilde{K}$ a compact convex subset of Ω, where C''' is independent of ν. Thus, if ν is the functional carried by the ball of radius $(1-\delta_\nu)s$ whose Fourier-Borel transform is $g_\nu(\xi)$, then $\mu\star\nu$ converges weakly in $\mathscr{H}(K+\Omega)'$. But, since $\mathscr{F}_\mu(\xi)\cdot g(\xi)$ is not the Fourier-Borel transform of an element in the image space, the image is not weakly closed. □

Corollary 9.36. *If μ is carried by the origin in $\mathbb{C}^n$, then for Ω an open convex subset of $\mathbb{C}^n$, given $f \in \mathscr{H}(\Omega)$, we can find $\tilde{f} \in \mathscr{H}(\Omega)$ solution of the equation $\hat{\mu}(x) = f$.*

Proof. If μ is carried by the origin, then $\mathscr{F}_\mu(\xi)$ is of minimal type with respect to the proximate order $\rho \equiv 1$. Since a function of minimal type is always of regular growth, the corollary follows from Theorem 9.36. □

Historical Notes

The work on convolution equations in spaces of entire functions is largely inspired by the work of Malgrange [1] and Ehrenpreis [1]. Martineau [8] and Taylor [1] were the first to study convolution operators in spaces of entire functions, but they studied spaces of functions of order at least one in order to have the Fourier-Borel transform defined as a holomorphic function. The treatment of the case $\rho < 1$ by the use of formal power series with estimates on the growth of the coefficients (as presented in Sections 1–3) is due to Gruman [3], although this idea is contained implicitly in the work of Taylor. The refinement (in Section 4) to real norms and strong proximate orders for $\rho < 1$ can be found in Gruman [4]; this exploits an idea of Hörmander [*B*]. The results in Section 5 using Banach algebra technics are due to Gruman [5]. Theorem 9.32 and 9.34 are refinements of previous results and have never been published elsewhere. With respect to Section 7, we refer the reader to the interesting results of Morzakov [1].

Appendix I. Subharmonic and Plurisubharmonic Functions

Subharmonic functions and potential theory are often used in the theory of one complex variable. For holomorphic functions of several complex variables defined in a domain $\Omega\subset\mathbb{C}^n$, this same important role is played by another class of real valued functions, the class $\mathrm{PSH}(\Omega)$ of functions plurisubharmonic in Ω. For $f\in\mathscr{H}(\Omega)$, both $|f|$ and $\log|f|$ belong to $\mathrm{PSH}(\Omega)$; for $\varphi_i\in\mathrm{PSH}(\Omega)$, $i=1,\ldots,N$, $\sup\limits_{1\leqq i\leqq N}\varphi_i$ is in $\mathrm{PSH}(\Omega)$. Thus the class $\mathrm{PSH}(\Omega)$ is a natural extension of the set $\{\log|f|, f\in\mathscr{H}(\Omega)\}$ and gives general methods for the study of this set.

Definition I.1. Let $\Omega\subset\mathbb{R}^m$ be a domain. A real valued function $\varphi(x)$ with values in $[-\infty,+\infty)$ is said to be *subharmonic* in Ω if

i) $\varphi(x)$ is upper semi-continuous and $\varphi(x)\not\equiv-\infty$;

ii) for every $x\in\Omega$ and every $r<d_\Omega(x)=\inf\{\|x-x'\|: x'\in\complement\Omega\}$

$$\varphi(x)\leqq\omega_m^{-1}\int\limits_{\|\alpha\|=1}\varphi(x+r\alpha)d\omega_m(\alpha)=\lambda(x,r,\varphi)$$

where $d\omega_m$ is the Lebesgue measure on the unit sphere S^{m-1} and ω_m is the total mass of S^{m-1}. We denote by $S(\Omega)$ the family of subharmonic functions in Ω. If φ and $-\varphi$ are subharmonic in Ω, we say that φ is *harmonic* in Ω.

Remark. If φ is subharmonic in Ω and $r<d_\Omega(x)$, then

$$\varphi(x)\leqq\tau_m^{-1}r^{-2m}\int\limits_{\|x\|\leqq r}\varphi(x+x')d\tau_m(x')=A(x,r,\varphi)$$

where $d\tau_m$ is the Lebesgue measure in $\mathbb{R}^m$ and τ_m is the total mass of the unit ball $B(0,1)$.

Definition I.2. Let $\Omega\subset\mathbb{C}^n$ be a domain. A real valued function $\varphi(x)$ with values in $[-\infty,+\infty)$ is said to be *plurisubharmonic* if

i) $\varphi(z)$ is upper semi-continuous and $\varphi(z)\not\equiv-\infty$;

ii) for every r such that $\{z+uw: |u|\leqq r, u\in\mathbb{C}\}\subset\Omega$

$$\varphi(z)\leqq(2\pi)^{-1}\int\limits_0^{2\pi}\varphi(z+re^{i\theta}w)d\theta.$$

If φ and $-\varphi$ are plurisubharmonic, we say that φ is *pluriharmonic* in Ω. We denote by $\mathrm{PSH}(\Omega)$ the family of functions plurisubharmonic in Ω.

Remark 1. If $\Omega \subset \mathbb{C}^n$, then $\mathrm{PSH}(\Omega) \subset S(\Omega)$ and if $n=1$, $\mathrm{PSH}(\Omega)=S(\Omega)$.

Remark 2. $\varphi \subset \mathrm{PSH}(\Omega)$ if and only if it is upper semi-continuous, not identically $-\infty$, and if its restriction to every complex line L^1 meeting Ω is either subharmonic or the constant $-\infty$ on each connected L^1-open set of $L^1 \cap \Omega$.

Examples. 1) Any continuous convex function in Ω (in terms of the underlying real variables) is in $\mathrm{PSH}(\Omega)$, for then we have

$$\varphi(x) \leqq \tfrac{1}{2}[\varphi(x+y)+\varphi(x-y)],$$

and if we replace y by $ye^{i\theta}$ and integrate with respect to $\dfrac{d\theta}{2\pi}$ we obtain ii) of Definition I.2;

2) if $f \in \mathscr{H}(\Omega)$, then $\log|f| \in \mathrm{PSH}(\Omega)$; it is enough to show that $\varphi(u)= f(z+uw)$ satisfies $\log|\varphi(0)| \leqq \dfrac{1}{2\pi}\int_0^{2\pi} \varphi(re^{i\theta})d\theta$ when the disc $\{z+uw\colon |u| \leqq r\}$ is in Ω.

If $\varphi(u) \equiv 0$, then the inequality is trivial. If not, we let $\varphi(u)=\prod_{j=1}^{k}(u-a_j)\,g(u)$ for $|u| \leqq r$, where $g(u)$ is holomorphic and has no zeros for $|u| \leqq r$. Then $\log|g(u)|$ is a harmonic function, and since

$$(2\pi)^{-1}\int_0^{2\pi} \log|re^{i\theta}-a_j|\,d\theta = \sup(\log|a_j|, \log r) \geqq \log|a_j|$$

we have

$$(2\pi)^{-1}\int_0^{2\pi} \log|\varphi(re^{i\theta})|\,d\theta \geqq \sum_{j=1}^{k} \log|a_j| + \log|g(0)| = \log|\varphi(0)|.$$

Proposition I.3. i) *If* $\varphi \in \mathrm{PSH}(\Omega)$ *and* $c>0$, *then* $c\varphi \in \mathrm{PSH}(\Omega)$.

ii) *If* φ_1 *and* φ_2 *are in* $\mathrm{PSH}(\Omega)$, *then* $\sup(\varphi_1, \varphi_2) \in \mathrm{PSH}(\Omega)$.

iii) *If* φ_ν *is a decreasing sequence of plurisubharmonic functions in* Ω, *then either* $\lim_{\nu\to\infty} \varphi_\nu(z) \equiv -\infty$ *or* $\varphi(z)=\lim_{\nu\to\infty}\varphi_\nu(z) \in \mathrm{PSH}(\Omega)$.

Proof. These are immediate consequences of Definition I.2. □

Definition I.4. A function $\varphi \in S(\Omega)$ will be said to be *continuous* if it is continuous for the completion of the Euclidean topology on $\mathbb{R}$ to the point $-\infty$.

Remark. $\varphi \in S(\Omega)$ is continuous if and only if $\exp\varphi(x)$ is continuous for the Euclidean topology on $\mathbb{R}$.

Proposition I.5. *If* $\Omega \subset \mathbb{R}^m$ *and* $\varphi \in \mathscr{C}^2(\Omega)$, *then* $\varphi \in S(\Omega)$ *if and only if* $\Delta\varphi(x) \geqq 0$, *where* Δ *is the Laplacian* $\Delta = \sum_{i=1}^{m} \dfrac{\partial^2}{\partial x_i^2}$. *If* $\Omega \subset \mathbb{C}^n$ *and* $\varphi \in \mathscr{C}^2(\Omega)$ *then*

$\varphi\in \mathrm{PSH}(\Omega)$ *if and only if*

$$L(\varphi, w)=\sum_{j,k=1}^{n}\frac{\partial^2\varphi(z)}{\partial z_j\partial\bar z_k}w_j\bar w_k\geqq 0 \qquad \text{for all } w\in\mathbb{C}^n.$$

Proof. We write the Taylor series expansion of $\varphi(x)$

$$\varphi(x')=\varphi(x)+\sum_{j=1}^{m}\frac{\partial\varphi(x)}{\partial x_j}(x'_j-x_j)$$
$$+1/2\sum_{j,k=1}^{m}\frac{\partial^2\varphi}{\partial x_j\partial x_k}(x)(x'_j-x_j)(x'_k-x_k)+o(\|x'-x\|^2).$$

Then

$$\lambda(x,r,\varphi)=\varphi(x)+\frac{1}{2m}\Delta\varphi(x)r^2+o(r^2),$$

since $\lambda(x,r,x'_j-x_j)=0$ for all j by symmetry (this is an odd function) and $\lambda(x,r,(x'_j-x_j)(x'_k-x_k))=0$ for $j\neq k$, again by symmetry.

Thus

$$\lim_{r\to 0}[\lambda(x,r,\varphi)-\varphi(x)]r^{-2}=\frac{1}{2m}\Delta\varphi(x)\geqq 0.$$

On the other hand, if ω_m is the measure of the unit sphere S^{m-1} in $\mathbb{R}^m$, $\omega_m=2\pi^{m/2}[\Gamma(m/2)]^{-1}$, then we obtain from Gauss' Theorem (or Green's Theorem)

$$\lambda(x,r,\varphi)=\varphi(x)+\int_0^r t^{-m+1}dt\int_{B(x,t)}\Delta\varphi(x')d\tau_m(x'). \tag{I,1}$$

Thus $\Delta\varphi(x)\geqq 0$ implies that $\lambda(x,r,\varphi)\geqq\varphi(x)$.

It follows from Remark 2 that $\varphi\in\mathrm{PSH}(\Omega)\cap\mathscr{C}^2(\Omega)$ is plurisubharmonic if and only if

$$\Delta_u\varphi(z+uw)|_{u=0}=4\sum_{j,k=1}^{n}\frac{\partial^2\varphi(z)}{\partial z_j\partial\bar z_k}w_j\bar w_k\geqq 0 \tag{I,2}$$

for every $w\in\mathbb{C}^n$. □

Proposition I.6. *If $\Omega\subset\mathbb{R}^m$ and $\varphi\in S(\Omega)\cap\mathscr{C}^2(\Omega)$, then $\lambda(x,r,\varphi)$ and $A(x,r,\varphi)$ are increasing in r and convex functions of $u_m(r)=-r^{2-m}$ for $m>2$, of $u_2(r)=\log r$ if $m=2$. If $\Omega\subset\mathbb{C}^n$ and $\varphi\in\mathrm{PSH}(\Omega)\cap\mathscr{C}^2(\Omega)$, then $\lambda(z,r,\varphi)$ and $A(z,r,\varphi)$ are increasing with r and convex in $\log r$.*

Proof. It follows from (I,1) that

$$(m-2)r^{m-1}\frac{\partial\lambda}{\partial r}(x,r,\varphi)=\frac{\partial\lambda(x,r,\varphi)}{\partial u_m(r)}$$

is increasing, which shows the first part for $m>2$. For $m=2$, we have

$$r\frac{\partial\lambda}{\partial r}(x,r,\varphi)=\frac{\partial\lambda(x,r,\varphi)}{\partial u_2(r)}.$$

If $\Omega \subset \mathbb{C}^n$ and $\varphi \in \mathrm{PSH}(\Omega) \cap \mathscr{C}^2(\Omega)$, then it follows from (I, 1) for $m=2$ and (I, 2) that

$$v(z, z', r) = \frac{\partial}{\partial \log r} \int_0^{2\pi} \varphi(z + re^{i\theta} z') d\theta$$

is increasing in r. Since this is true for all z',

$$v(z, r) = \omega_{2n}^{-1} \int_{\|z'\|=r} v(z, z', r) d\omega_{2n}(z') = \frac{\partial}{\partial \log r} \lambda(z, r, \varphi)$$

is an increasing function of $\log r$, hence $\lambda(z, r, \varphi)$ is convex in $\log r$. □

Remark. The result that $\lambda(z, r, \varphi)$ is increasing and convex in $\log r$ is an important property of plurisubharmonic functions and is not in general true for $\mathbb{R}^{2n}$-subharmonic functions.

In the definition of subharmonic and plurisubharmonic functions, we require only upper semi-continuity, whereas in Propositions I.5 and I.6, we made assumptions on the regularity of the functions. We now extend these properties in a way to drop the regularity assumptions.

Lemma I.7. *Let $\Omega \subset \mathbb{R}^m$ be a domain and $0 < c \leqq 1$. Suppose that $\Omega' \subset \Omega$ and for $x \in \Omega'$, $B(x, c d_\Omega(x)) \subset \Omega'$. Then either $\Omega' = \Omega$ or $\Omega' = \emptyset$.*

Proof. By the hypothesis, Ω' is open. Suppose that $x_0 \in \bar{\Omega}' \cap \Omega$ and let $d = d_\Omega(x_0)$. Then there exists a point $x' \in \Omega'$ such that $\|x' - x_0\| \leqq cd/4$. This implies that $d_\Omega(x') \geqq 3d/4$ and $x_0 \in B(x', c d_\Omega(x')) \subset \Omega'$. Thus Ω' is also closed, and since Ω is connected, $\Omega' = \Omega$ or $\Omega' = \emptyset$. □

In $\mathbb{C}^n$, we denote by $D_{z,w}$ the disc

$$D_{z,w} = \{z' \in \mathbb{C}^n \colon z' = z + uw, u \in \mathbb{C}, |u| \leqq 1\}.$$

Lemma I.8. *Let $\Omega \subset \mathbb{C}^n$ be a domain. For $z \in \Omega$, we set $S(z, \Omega) = \bigcup_{D_{z,w} \subset \Omega} D_{z,w}$. Then $S(z, \Omega)$ is open and if $\Omega' \subset \Omega$ has the property that $z \in \Omega'$ implies that $S(z, \Omega) \subset \Omega'$, then $\Omega' = \Omega$ or $\Omega' = \emptyset$.*

Proof. Obviously, $S(z, \Omega)$ is a disked neighborhood of z. Let $z_0 \in S(z, \Omega)$ be such that $z_0 \neq z$. Then $z_0 = z + z'$ with $z' \neq 0$ and the disc. $D_{z,z'}$ is compact in Ω. There exists a disked open neighborhood U of the origin such that $D_{z,z'} + U \subset \Omega$. But $D_{z,z'} + U$ is a union of discs centered at z, so $D_{z,z'} + U \subset S(z, \Omega)$. Thus $S(z, \Omega)$ contains an open neighborhood of z_0 and hence is open.

To prove the second part of the lemma, we note that $S(z, \Omega)$ contains the ball $B(z, d_\Omega(z))$ and so the conclusion follows from Lemma I.7. □

Proposition I.9. *If $\Omega \subset \mathbb{C}^n = \mathbb{R}^{2n}$ is a domain, then* $\mathrm{PSH}(\Omega) \subset S(\Omega) \subset L^1_{\mathrm{loc}}(\Omega)$.

Proof. Let N be the set of points in Ω such that $\int_U \varphi \, d\tau_{2n} = -\infty$ for every neighborhood U of z, $U \Subset \Omega$. For $z \in N$ and $\|z' - z\| < \frac{1}{6} d_\Omega(z)$, the ball $B' = B(z', \frac{1}{3} d_\Omega(z))$ is compact in Ω and is a neighborhood of z. Therefore $\varphi(z') \leqq A(z', \frac{1}{3} d_\Omega(z), \varphi) = -\infty$. Thus for $z \in N$, $\varphi(z') \equiv -\infty$ for $\|z' - z\| < c d_\Omega(z)$ with $c > 0$. This implies that $B(z, c d_\Omega(z)) \subset N$. From Lemma I.7 and Definition I.1, we deduce $N = \emptyset$. □

Corollary I.10. *For Ω a domain in $\mathbb{R}^m$ (respectively $\mathbb{C}^n$), $S(\Omega)$ [respectively $\mathrm{PSH}(\Omega)$] is a convex cone over $\mathbb{R}^+$; it is closed under the operation $(\varphi_1, \varphi_2) \to \varphi_3 = \sup(\varphi_1, \varphi_2)$.*

Proof. For $\varphi_1, \varphi_2 \in \mathrm{PSH}(\Omega)$ (or $S(\Omega)$), the set $\{z\colon \varphi_1(z) = -\infty \text{ or } \varphi_2(z) = -\infty\}$ is of measure zero by Proposition I.9. Thus, $t\varphi_1 + (1-t)\varphi_2$ is not identically $-\infty$, $0 \leqq t \leqq 1$, and hence is in $\mathrm{PSH}(\Omega)$ or $S(\Omega)$. □

Definition I.11. A subset $E \subset \Omega$, a domain in $\mathbb{R}^m$ (resp. $\mathbb{C}^n$) is said to be *polar* (resp. *pluripolar*) if there exists $\varphi \in S(\Omega)$ (resp. $\mathrm{PSH}(\Omega)$) such that $E \subset \{x\colon \varphi(x) = -\infty\}$.

Corollary I.12. *A (pluri)polar set in a domain $\Omega \subset \mathbb{C}^n$ is of Lebesgue measure zero.*

Proposition I.13 (Maximum Principle). *Let $\Omega \subset \mathbb{R}^m$ be a domain and $\varphi \subset S(\Omega)$. Let $m = \sup_\Omega \varphi$. If there exists $x_0 \in \Omega$ such that $\varphi(x_0) = m$, then $\varphi \equiv m$.*

Proof. If $B(x_0, r) \subset \Omega$, then $m = \varphi(x_0) \leqq A(x, r, \varphi) \leqq m$. Thus, $\varphi(x) \equiv m$ in $B(x_0, r)$, for otherwise, by the upper semi-continuity of φ there would exist $\varepsilon > 0$ and an open subset U of $B(x_0, r)$ for which $\varphi(x) < m - \varepsilon$ on U and so $A(x, r, \varphi) < m$. Thus, the set $M = \{x \in \Omega\colon \varphi(x) \geqq m\}$ is open, and it is closed since φ is upper semi-continuous. Since $M \neq \emptyset$, $M = \Omega$. □

Proposition I.14. *Let $\varphi(z, t)$ be a real valued function of $z \in \Omega \subset \mathbb{C}^n$ and $t \in T$, T a locally compact space. Let μ be a positive measure on T. Suppose that*

i) *$z \to \varphi(z, t)$ is plurisubharmonic in Ω and $(\theta, t) \to \varphi(z + w e^{i\theta}, t)$ is $d\theta x d\mu$ measurable.*

ii) *For every compact subset $K \subset \Omega$, there exists a constant $M(K)$ such that $\varphi(z, t) \leqq M(K)$ for every $t \in T$.*

Then $\psi(z) = \int \varphi(z, t) \, d\mu(t) \in \mathrm{PSH}(\Omega)$ or is identically $-\infty$.

Proof. We shall verify properties i) and ii) of Definition I.2. Let $\psi^\star(z) = \limsup_{z' \to z} \psi(z)$ be the upper regularization of $\psi(z)$.

Then there exists a sequence w_q such that $w_q \to 0$ and

$$\psi^\star(z) = \lim_{q \to \infty} \psi(z + w_q) = \limsup_{q \to \infty} \int \varphi(z + w_q, t) \, d\mu(t).$$

From Fatou's Lemma and the uniform bound for $\varphi(z+w_q, t)$, we obtain

$$\psi^\star(z) \leqq \int \limsup_{q\to\infty} \varphi(z+w_q, t)d\mu(t) \leqq \int \varphi(z,t)d\mu(t) = \psi(z),$$

where the second inequality stems froms the semi-continuity of $\varphi(z,t)$ for fixed t. Thus, $\psi^\star(z)=\psi(z)$, and so $\psi(z)$ is upper semi-continuous. To show ii) we observe that

$$\psi(z) = \int \varphi(z,t)d\mu(t) \leqq \int d\mu(t) \int_0^{2\pi} \varphi(z+we^{i\theta}, t)\frac{d\theta}{2\pi}$$

for every disc $\{z+uw\colon |u|\leqq r\}$ contained in Ω. We then conclude from the measurability in $d\mu x d\theta$ that

$$\psi(z) \leqq \int\int d\mu(t)\frac{d\theta}{2\pi}\varphi(t+we^{i\theta}, t) = \frac{1}{2\pi}\int_0^{2\pi} \psi(z+we^{i\theta})d\theta. \qquad \square$$

Remark. Proposition I.14 remains valid for $\varphi(x,t)$ subharmonic in $x\in\Omega$ for $t\in T$, where we replace condition ii) by condition ii'): $(\alpha,t)\to\varphi(x+r\alpha,t)$ is $d\omega_m x d\mu$ measurable.

Let $\alpha(x)\in\mathscr{C}_0^\infty(B(0,1))$ such that $\alpha(x)\geqq 0$, α depends only on $\|x\|$ and $\int\alpha(x)d\tau_m=1$. We consider the positive functions $\alpha_\varepsilon(x)=\varepsilon^{-m}\alpha(x/\varepsilon)$, which form, as ε tends to zero, an approximation to the Dirac measure with point mass at the origin.

Proposition I.15. *Let* $\varphi\in S(\Omega)$ (*resp.* $\mathrm{PSH}(\Omega)$) and set

$$\varphi_\varepsilon(x) = \varphi\star\alpha_\varepsilon(x) = \int \varphi(x+x')\alpha_\varepsilon(x')d\tau(x').$$

Then

i) $\varphi_\varepsilon(x)\in S(\Omega_\varepsilon)\cap\mathscr{C}^\infty(\Omega_\varepsilon)$ [*resp.* $\varphi_\varepsilon(z)\in\mathrm{PSH}(\Omega_\varepsilon)\cap\mathscr{C}^\infty(\Omega_\varepsilon)$] *where* $\Omega_\varepsilon=\{x\colon \mathrm{d}_\Omega(x)>\varepsilon\}$

ii) $\varphi_\varepsilon(x)$ *is an increasing function of* ε *for* $\varepsilon<d_\Omega(x)$ *and* $\lim_{\varepsilon\to 0}\varphi_\varepsilon(x)=\varphi(x)$.

Proof. By ii) of Definitions I.1 and I.2 we obtain $\varphi_\varepsilon(x)\geqq\varphi(x)$ for $\varepsilon<d_\Omega(x)$. By Proposition I.14 and the subsequent remark $\varphi_\varepsilon(x)$ is in $S(\Omega_\varepsilon)$ (resp. $\mathrm{PSH}(\Omega_\varepsilon)$).

Let $\eta>0$ be given. By the upper semi-continuity of φ, there exists $t_\eta>0$ such that $\varphi(x+y)\leqq\varphi(x)+\eta$ for $\|y\|\leqq t_\eta$. Thus for $\varepsilon<t_\eta$

$$\varphi_\varepsilon(x) = \int\varphi(x+y)\alpha_\varepsilon(y)d\tau(v) = \int\varphi(x+\varepsilon y')\alpha(y')d\tau(y') \leqq \varphi(x)+\eta$$

and $\lim_{\varepsilon\to 0}\varphi_\varepsilon(x)=\varphi(x)$.

It follows from Proposition I.6 that the mean value $\lambda(x,r,\varphi_\varepsilon)$ of φ_ε on $S(x,r)$ is an increasing function of r for fixed ε and $d_\Omega(x)>r+\varepsilon$. Hence $\lambda(x,r,\varphi)=\lim_{\varepsilon\to 0}\lambda(x,r,\varphi_\varepsilon)$ is an increasing function of r for fixed x, $d_\Omega(x)>r$.

From this we obtain for $\varepsilon' < \varepsilon$

$$\begin{aligned}\varphi_\varepsilon(x) &= \int \varphi(x+\varepsilon y')\alpha(y')d\tau(y') = \omega_m^{-1} \int \lambda(x, \varepsilon t, \varphi)\alpha(t)dt \\ &\geqq \omega_m^{-1} \int \lambda(x, \varepsilon' t, \varphi)\alpha(t)dt = \varphi_{\varepsilon'}(x).\end{aligned}$$ □

Subharmonic and plusubharmonic functions are locally integrable. Thus, using differentiation for distributions, we extend to $S(\Omega)$ and so $\mathrm{PSH}(\Omega)$ the properties given first for differentiable subharmonic and plurisubharmonic functions. For $\varphi \in S(\Omega)$, we consider the Laplacian (defined as a distribution):

$$\Delta \varphi = \sum_{j=1}^{m} \frac{\partial^2 \varphi}{\partial x_j^2}$$

and for $\varphi \in \mathrm{PSH}(\Omega)$, the Levi form

$$L(\varphi, \omega) = \sum_{i,j=1}^{n} \frac{\partial^2 \varphi}{\partial z_i \partial \bar{z}_j} w_i \bar{w}_j \tag{I,3}$$

(I,3) is a distribution in Ω depending on the vector w.

Proposition I.16. *Let $\varphi \in S(\Omega)$. Then the distribution $\Delta\varphi$ is a positive measure. If $\varphi \in \mathrm{PSH}(\Omega)$, then $L(\varphi, w)$ is a positive measure for every $w \in \mathbb{C}^n$.*

Proof. Let $\psi \in \mathscr{C}_0^\infty(\Omega)$, $\psi \geqq 0$. By Proposition I.15, there exists a sequence φ_q of functions subharmonic and $\mathscr{C}^\infty$ in a neighborhood of support ψ such that φ_q decreases to φ. From Proposition I.5, $\int \Delta \varphi_q \psi \, d\tau = \int \varphi_q \Delta \psi \, d\tau \geqq 0$. Since φ is in $L^1_{\mathrm{loc}}(\Omega)$, we obtain from the Lebesgue Dominated Convergence Theorem that $\Delta\varphi(\psi) = \int \varphi \Delta \psi \, d\tau = \lim_{q\to\infty} \int \varphi_q \Delta\psi \, d\tau \geqq 0$. Thus, $\Delta\varphi$ is a positive measure. Similarly for $\varphi \in \mathrm{PSH}(\Omega)$, we choose φ_q to be plurisubharmonic and $\mathscr{C}^\infty$ on a neighborhood of support ψ. Then

$$\begin{aligned}L(\varphi, w)(\psi) &= \int \varphi L(\psi, w) d\tau = \lim_{q\to\infty} \int \varphi_q L(\psi, w) d\tau \\ &= \lim_{q\to\infty} \int L(\varphi_q, w)\psi \, d\tau \geqq 0.\end{aligned}$$ □

Proposition I.17. *Let $\varphi \in \mathrm{PSH}(\Omega)$. Then $\lambda(z, r, \varphi)$ and*

$$M_\varphi(z', r) = \sup_{\|z''\| < r} \varphi(z', z''), \; z' \in \mathbb{C}^m, \; z'' \in \mathbb{C}^{n-m},$$

are increasing convex functions of $\log r$.

Proof. For $\eta > 0$, let φ_q be a sequence of $\mathscr{C}^\infty$ plurisubharmonic functions such that φ_q decreases to φ in $\Omega_\eta = \{z : d_\Omega(z) > \eta\}$. Then by Proposition I.6, $A(z, r, \varphi_q)$ and $\lambda(z, r, \varphi_q)$ are increasing convex functions of $\log r$ for $r < d_\Omega(z) - \eta$. Thus, the same is true for $A(z, r, \varphi)$ and $\lambda(z, r, \varphi)$. Since η was arbitrary, this is true for all $r < d_\Omega(z)$. Since $M_\varphi(z', r) = \sup_{\|\alpha\| \leqq 1} \varphi(z' + \alpha z'')$,

$M_\varphi(z', r)$ is a plurisubharmonic function of the variable z'' for fixed z', and $M_\varphi(z', r) = \lambda_{z''}(0, r, M_\varphi(z', r))$ is an increasing convex function of $\log r$. □

Definition I.18. A function $\varphi(x)$ is *locally subharmonic* in a domain $\Omega \subset \mathbb{R}^m$ (resp. locally plurisubharmonic in a domain $\Omega \subset \mathbb{C}^n$) if φ is upper semicontinuous, $\varphi \not\equiv -\infty$, and if for every $x \in \Omega$, there exists $\rho(x) > 0$ such that $\varphi(x) \leqq \lambda(x, r, \varphi)$ for $r < \rho(x)$ (resp. for every $z \in \Omega$, there exists $\rho(z)$ such that $\varphi(z) \leqq (2\pi)^{-1} \int_0^{2\pi} \varphi(z + we^{i\theta}) d\theta$ for $\|w\| < \rho(z)$).

Proposition I.19. *If φ is a locally subharmonic in $\Omega \subset \mathbb{R}^m$ (resp. locally plurisubharmonic in $\Omega \subset \mathbb{C}^n$), then $\varphi \in S(\Omega)$ (resp. $\varphi \in \mathrm{PSH}(\Omega)$).*

Proof. Let φ be locally subharmonic. Suppose that it is not subharmonic in Ω. Then there exists a ball $B(\xi, r) \subset \Omega$ with $\varphi(\xi) = M > -\infty$ such that $\lambda(\xi, r, \varphi) \leqq M - \varepsilon$ for some $\varepsilon > 0$. By the semi-continuity of φ we can find a function χ continuous on $bdB(\xi, r)$ such that $\chi \geqq \varphi$ and $\lambda(\xi, r, \chi) < M - \varepsilon/2$. Let $\psi(x)$ be the function harmonic in $B(\xi, r)$ and equal to χ on $bdB(\xi, r)$.

Then $\varphi_1 = \varphi - \psi$ is locally subharmonic in $B(\xi, r)$ and $\varphi_1(\xi) > \varepsilon/2 > 0$. Furthermore, $\varphi_1(x) < 0$ on $bdB(\xi, r)$. Thus, the set $\{x : \varphi_1(x) \geqq \varepsilon/2\}$ is compact in $B(\xi, r)$. Since a locally subharmonic function satisfies the maximum principle (cf. Proposition I.13 and its proof), we obtain $\varphi_1 = \text{const.} > 0$, which contradicts the fact that $\varphi_1 < 0$ on $bdB(\xi, r)$. If φ is locally plurisubharmonic, then there exists x_0 such $\varphi(x_0) \neq -\infty$, and on every component ω of $L^1 \cap \Omega$ for every complex line L^1, φ is $-\infty$ or is locally subharmonic; then it is $-\infty$ or subharmonic on ω. Thus, it is plurisubharmonic in Ω (cf. Remark 2 following Definition I.2). □

Corollary I.20. *Let Ω be a domain in $\mathbb{C}^n$ and $\{U_i\}_{i=1}^\infty$ a covering of Ω by domains U_i $\left(\text{i.e. } \Omega = \bigcup_{i=1}^\infty U_i\right)$. If φ is defined on Ω and $\varphi \in \mathrm{PSH}(U_i)$ for every i, then $\varphi \in \mathrm{PSH}(\Omega)$.*

Corollary I.21. *Let $M \subset \Omega$ be closed and $M \subset \{x : \varphi(x) = -\infty, \varphi \in S(\Omega)\}$. Then $\complement M$ is connected. The same conclusion holds if $\varphi \in \mathrm{PSH}(\Omega)$. If M is an analytic variety in Ω, $\complement M$ is connected.*

Proof. Suppose not. Then we have a decomposition $\Omega = \Omega_1 \cup \Omega_2 \cup M$ with $\Omega_1 \cap \Omega_2 = \emptyset$. It follows from Proposition I.9 that $\mathring{M} = \emptyset$. Since $M \cap \Omega_1 = M \cap \Omega_2 = \emptyset$, there exists $\xi \in M$ such that $\xi \in bd\Omega_1 \cap bd\Omega_2$. Let r be chosen such that $r < d_\Omega(\xi)$. Then $B(\xi, r/2) \cap \Omega_1 \neq \emptyset$ and $B(\xi, r/2) \cap \Omega_2 \neq \emptyset$.

Let φ be defined by $\varphi = -1$ on Ω_1 and $\varphi = 0$ on Ω_2. Let $\psi \in S(\Omega)$ be such that $\psi = -\infty$ on M and $\psi < 0$ on $B(\xi, r)$. Let $\varphi_q = \varphi + q^{-1}\psi$, $q \in \mathbb{Z}$ and $g(x) = \lim_{q \to \infty} \varphi_q(x)$. Since φ_q is locally subharmonic in Ω, $\varphi_q \in S(\Omega)$ and so

$g^\star(x) = \limsup_{x' \to x} g(x')$ is subharmonic. But $g^\star(\xi) = 0 > A(\xi, r/2, g^\star)$, which is a contradiction.

If M is an analytic variety, take $r < d_\Omega(\xi)$ such that $M \cap B(\xi, r) = \bigcap_j F_j^{-1}(0)$, with F_j holomorphic in $B(\xi, r)$ and define $\psi = \log \sum_j |F_j|^2$. □

Proposition I.22. *Let $\Omega \subset \mathbb{R}^m$ be a domain (resp. $\Omega \subset \mathbb{C}^n$) and let M be a closed polar set in Ω (resp. pluripolar set). Suppose that $\varphi \in S(\Omega \cap \complement M)$ [resp. $\varphi \in \mathrm{PSH}(\Omega \cap \complement M)$] such that φ is bounded above in a neighborhood of every point $x \in M$. Then there exists a unique extension $\tilde{\varphi} \in S(\Omega)$ (resp. $\mathrm{PSH}(\Omega)$) such that $\varphi = \tilde{\varphi}$ on $\Omega \cap \complement M$.*

Proof. For $\xi \in M$ and $r < d_\Omega(\xi)$, we choose $\psi \in S(\Omega)$ such that $\psi(x) = -\infty$ on M and $\psi < 0$ on $B(\xi, r)$. We can suppose $\varphi \leqq 0$ on $B(\xi, r)$ since $\bar{B}(\xi, r) \cap M$ is compact. We set $\varphi_q = \varphi + q^{-1}\psi$ and $g(x) = \lim_{q \to \infty} \varphi_q$. Since φ_q is locally subharmonic in $B(\xi, r)$, it is subharmonic in $B(\xi, r)$ thus $g^\star(x) \in S(\Omega)$ and $g^\star(x) = \varphi(x)$ for $x \in \Omega \cap \complement M$.

If $\tilde{\varphi}$ is an extension of φ to $B(\xi, r)$, then, since M is of measure zero, $\tilde{\varphi}(x) = \lim_{t \to 0} A(x, t, \tilde{\varphi}) = \lim_{t \to 0} A(x, t, \varphi)$, so the extension is unique. The result now follows from Proposition I.19. The proof for the plurisubharmonic case is identical. □

Corollary I.23. *Suppose that $\Omega \subset \mathbb{C}^n$ and that M is a closed $\mathbb{R}^{2n}$-polar set in Ω. Let f be a holomorphic single valued function in $\Omega \cap \complement M$ such that $|f|$ is bounded in a neighborhood of every point $\xi \in M$. Then there exists a unique $f \in \mathscr{H}(\Omega)$ such that $f = \tilde{f}$ on $\Omega \cap \complement M$.*

Proof. Let $f(z) = u(z) + iv(z)$. We apply first Proposition I.22 to $u(z)$ and $-u(z)$, which are subharmonic in $\Omega \cap \complement M$. Thus, there are subharmonic functions in Ω, $\tilde{u}(z)$ and $-\tilde{u}(z)$ which extend $u(z)$ and $-u(z)$ as subharmonic functions in Ω. For $\xi \in M$, $\tilde{u}(\xi) = \lim_{t \to 0} A(\xi, t, u) = -\lim_{t \to 0} A(\xi, t, -u) = -[-u(\xi)]$. Hence $\tilde{u}(z)$ is harmonic. Similarly, $v(z)$ is harmonic and so $f(z) = \tilde{u}(z) + i\tilde{v}(z) \in \mathscr{C}^\infty(\Omega)$. Hence $\bar{\partial} f = 0$ in Ω by continuity. □

Remark. If M is an analytic subvariety, Corollary I.23 is the classical first Continuation Theorem of Riemann.

Proposition I.24. *Let $\psi(t)$ be an increasing convex function defined on $[-\infty, +\infty)$ and let $\varphi \in \mathrm{PSH}(\Omega)$. Then $\psi \circ \varphi \in \mathrm{PSH}(\Omega)$.*

Proof. Let $\chi \in \mathscr{C}_0^\infty(\Omega)$, $\chi \geqq 0$, let φ_q be a sequence of $\mathscr{C}^\infty$ plurisubharmonic functions which decrease to φ in a neighborhood of $\operatorname{supp} \chi$, and let $\psi_\nu(t)$ be a sequence of $\mathscr{C}^\infty$ increasing convex functions which decrease to $\psi(t)$. It then follows from the continuity of $\psi(t)$ that $\Phi_q(z)=\psi_q(\varphi_q(t))$ decreases to $\Phi(z)=\psi(\varphi(z)$ for every $z\in\operatorname{supp}\chi$. A simple calculation shows that

$$L(\Phi_q, w)(\chi)=\int L(\Phi_q, w)(\psi_q'\circ\varphi_q)\chi\,d\tau+\int|\langle\operatorname{grad}\varphi_q, \bar{w}\rangle|^2(\psi_q''\circ\varphi_q)\chi\,d\tau\geqq 0.$$

Thus $\Phi_q(z)$ is plurisubharmonic by Proposition I.3 and I.5. The result now follows from Corollary I.20. □

Proposition I.25. *Suppose Ω is a domain in $\mathbb{C}^n$ with the following property: for $z=(z_k)=(x_k+iy_k)\in\Omega$ and $0\leqq t\leqq 1$, we have $z'=(x_k+ity_k)\in\Omega$. Then if $\varphi\in\mathrm{PSH}(\Omega)$ depends only on the x_k, it is a continuous convex function of $x=(x_1, \ldots, x_n)$.*

Proof. Let $\pi\colon z\to x\in\mathbb{R}^n$ be the natural projection onto the real coordinates. Then φ extends in a natural way to a plurisubharmonic function on $\Omega'=\omega x\mathbb{R}^n$, where $\omega=\pi(\Omega)$. Let $\varepsilon>0$ and let $\Omega_\varepsilon'=\{z\in\Omega'\colon d_{\Omega'}(z)>\varepsilon\}$. Then $\varphi_\varepsilon(z)\in\mathrm{PSH}(\Omega_\varepsilon')\cap\mathscr{C}^\infty(\Omega_\varepsilon')$ and $\varphi_\varepsilon(z)$ depends only on x. Furthermore, $L(\varphi_\varepsilon, w)=\sum_{j,k=1}^{n}\frac{\partial^2\varphi_\varepsilon}{\partial x_k\partial x_j}w_k\bar{w}_j$, and if $w\in\mathbb{R}^n$, φ_ε is seen to be convex. Since a decreasing sequence of convex functions is convex, φ is convex, and since a convex function locally bounded from above is continuous, φ is continuous. □

Corollary I.26. *Let $\Omega\subset\mathbb{C}^n$ be the domain $\Omega=\{z\colon 0\leqq r_j'<|z_j|<r_j''\}$. A function $\varphi(r)$, $r=(r_1, \ldots, r_n)$, $r_j=|z_j|$ defined in Ω is in $\mathrm{PSH}(\Omega)$ if and only if it is a convex function of the variable $v=(v_1, \ldots, v_n)$, $v_j=\log r_j$.*

Proof. Let $z=(z_1, \ldots, z_n)\in\Omega$. Then we can find a neighborhood ω_z of z such that we can define a branch $\log z_k=v_k+iv_k'$ of $\log z_k$ in ω_z for every k. For $\varphi\in\mathrm{PSH}(\Omega)$, $\psi(v_k)=\tilde{\psi}(v_k+iv_k')=\varphi(e^{v_1}, \ldots, e^{v_n})$ is a plurisubharmonic function of the variable $w=(v_1+iv_1', \ldots, v_n+iv_n')$. By Proposition I.25 it is a convex function of $v=(v_1, \ldots, v_n)$. Conversely, if $\psi(v)$ is defined in the open set $\omega=\{v\colon \log r_j'<v_j<\log r_j''\}$ and is a convex function of the variable v, we extend ψ as a convex function on $\omega+i\mathbb{R}^n$ by $\tilde{\psi}(v_k+iv_k')=\psi(v_k)$.

By the remark following Definition I.2, $\tilde{\psi}\in\mathrm{PSH}(\omega+i\mathbb{R}^n)$. □

Theorem I.27. *Let $\Omega\subset\mathbb{R}^m$ (resp. $\mathbb{C}^n$) and let $\varphi_\nu\in S(\Omega)$ [resp. $\varphi_\nu\in\mathrm{PSH}(\Omega)$] be an increasing sequence uniformly bounded above on every compact subset of Ω. Set $\varphi=\lim_{\nu\to\infty}\varphi_\nu$, $\varphi^\star(x)=\limsup_{x'\to x}\varphi(x')$ and $A(x, r, \varphi)$ the mean value of φ on the ball $B(x, r)$. Then:*

i) *$\varphi^\star(x)=\lim_{r\to 0}A(x, r, \varphi)$ and $\varphi^\star\in S(\Omega)$ [resp. $\varphi^\star\in\mathrm{PSH}(\Omega)$]*

ii) *$\{x\in\Omega\colon \varphi(x)<\varphi^\star(x)\}$ is of Lebesgue measure zero in Ω.*

Proof. Clearly $\varphi\in L^1_{\mathrm{loc}}(\Omega)$, and by the Lebesgue Dominated Convergence Theorem, $A(x, r, \varphi)=\lim_{\nu\to\infty}A(x, r, \varphi_\nu)$. Thus $A(x, r, \varphi)$ is a continuous function of x for $r>0$, $A(x,r,\varphi)\in S(\Omega)$ [resp. $A(x,r,\varphi)\in \mathrm{PSH}(\Omega)$] and is a convex and increasing function of r, since this is true for $A(x,r,\varphi_\nu)$. Hence $\psi(x)=\lim\limits_{r\to 0} A(x,r,\varphi)$ is an upper semi-continuous function of x and $\psi\in S(\Omega)$ [resp. $\psi\in\mathrm{PSH}(\Omega)$]. Moreover, for $r>0$ and all ν,

$$\varphi_\nu(x)\leqq\varphi(x)\leqq A(x,r,\varphi)$$

so $\varphi^\star(x)\leqq A(x,r,\varphi)$ by the continuity of $A(x,r,\varphi)$ and $\varphi^\star(x)\leqq\psi(x)$. Since $\lim\limits_{r\to 0} A(x,r,\varphi^\star)=\varphi^\star$ everywhere by upper semi-continuity,

$$\psi(x)=\lim_{r\to 0} A(x,r,\varphi)\leqq\lim_{r\to 0} A(x,r,\varphi^\star)=\varphi^\star \quad\text{and}\quad \psi(x)=\varphi^\star(x).$$

It is a classical property of a function in $L^1_{\mathrm{loc}}(\Omega)$ that $\varphi(x)=\lim\limits_{r\to 0} A(x,r,\varphi)$ for almost all x, which proves (ii). □

Remark. From Theorem I.27, we deduce.

(1) Given a sequence $\varphi_\nu(x)\in S(\Omega)$ [resp. $\varphi_\nu\in\mathrm{PSH}(\Omega)$] locally bounded above, and $\psi(x)=\limsup\limits_{\nu}\varphi_\nu(x)\not\equiv-\infty$, then $\psi^\star(x)\in S(\Omega)$ [resp. $\mathrm{PSH}(\Omega)$] and the set $\psi(x)<\psi^\star(x)$ is of Lebesgue measure zero in Ω.

(2) The cones $S(\Omega)$ and $\mathrm{PSH}(\Omega)$ are closed sets in $L^1_{\mathrm{loc}}(\Omega)$ and given a Cauchy sequence $\varphi_\nu\in S(\Omega)$ [resp. in $\mathrm{PSH}(\Omega)$] which converges to $\varphi\in L^1_{\mathrm{loc}}(\Omega)$, $\psi^\star(x)=[\limsup\limits_{\nu\to\infty}\varphi_\nu(x)]^\star$ is a limit of φ_ν in $L^1_{\mathrm{loc}}(\Omega)$ and $\varphi=\psi=\psi^\star$ almost everywhere.

To see 1), set $\varphi_{n,p}(x)=\sup\varphi_\nu(x)$ for $n\leqq\nu\leqq n+p$; $\varphi_{n,p}\in S(\Omega)$ [resp. $\mathrm{PSH}(\Omega)$]. By Theorem I.27, if $\varphi_n=\lim\limits_{p\to\infty}\varphi_{n,p}\leqq\varphi_n^\star$, the set $e_n=[x\colon \varphi_n(x)<\varphi_n^\star(x)]$ is of Levesgue measure zero in Ω. Then $\lim\limits_{n\to\infty}\varphi_n(x)=\psi(x)$ and if $g=\lim\limits_{n}\varphi_n^\star$, $\psi(x)\leqq g(x)$ and the set $[x\colon\psi(x)<g(x)]\subset\bigcup\limits_n e_n$ is of measure zero; therefore, $g(x)=\psi^\star(x)$ (for the proof that $\psi^\star\in S(\Omega)$ [resp. $\psi^\star\in\mathrm{PSH}(\Omega)$] see Chapter 1, Theorem 1.27).

To see 2), we note that by the hypothesis $\varphi_\nu(x)\leqq A(x,r,\varphi_\nu)$, the sequence is locally bounded above, and $\lim\limits_{\nu\to\infty} A(x,r,\varphi_\nu)=A(x,r,\varphi)$. Then, by Fatou's Lemma and 1) we write

$$\psi(x)\leqq\limsup_{\nu\to\infty} A(x,r,\varphi_\nu)=A(x,r,\varphi)\leqq A(x,r,\psi)=A(x,r,\psi^\star)$$

and

$$\psi^\star(x)\leqq A(x,r,\varphi)\leqq A(x,r,\psi^\star).$$

Theorem I.28 (Inverse Function Theorem for Plurisubharmonic Functions). *Let $\Omega\in\mathbb{C}^n$ be a domain and set $\Delta=\Omega\times\mathbb{C}$. For $\varphi\in\mathrm{PSH}(\Delta)$, we let $M_\varphi(z,r)=\sup\limits_{|\lambda|\leqq r}\varphi(z,\lambda)$, $z\in\mathbb{C}^n$, $\lambda\in\mathbb{C}$. Then if there exists $z_0\in\Omega$ such that $\varphi(z_0,\lambda)\not\equiv\varphi(z_0,0)$:*

i) *Given $z\in\Omega$, either $M_\varphi(z,r)$ is the constant $\varphi(z,0)$ or $M_\varphi(z,r)$ is an increasing convex function of $\log r$ and $\lim\limits_{r\to\infty}(\log r)^{-1}M_\varphi(z,r)>0$;*

ii) *for* $z\in\Omega$, *we set* $\delta(z,m)=\{\sup r: r>0,\ M_\varphi(z,r)<m\}$, *which is defined for* $m>\varphi(z,0)$. *Then* $\delta(z,m)>1$ *in* $\Omega_m=\{z\in\Omega: M_\varphi(z,1)<m\}$, *the function* $\psi(z,m)=-\log\delta(z,m)$ *is a negative plurisubharmonic function on every connected component of* Ω_q *for* $m>q$, [*or* $\psi(z,m)\equiv-\infty$ *in* Ω_q *if* $\varphi(z,\lambda)$ *does not depend on* λ *in* $\Omega_q\times\mathbb{C}$]. *Moreover* $\psi(z,m)$ *is decreasing in* m *and* $\lim_{m\to\infty}\psi(z,m)\equiv-\infty$.

Proof. Part i) follows from Proposition I.17. Thus, the points $z\in\Omega$ fall into two classes, those for which $M_\varphi(z,r)$ is constant, and then $\varphi(z,\lambda)\equiv\varphi(z,0)$, or those for which $\varphi(z,\lambda)$ is non-constant.

The upper semi-continuity of $M_\varphi(z,r)$ as a function of $(z,r)\in\mathbb{C}^n\times\mathbb{R}^+$ implies that $\delta(z,m)$ is lower semi-continuous, and hence $\psi(z,m)$ is upper semicontinuous. If $M_\varphi(z,r)$ is increasing and $\lim_{r\to\infty}M_\varphi(z,r)=+\infty$, $\psi(z,m)$ is decreasing and $\lim_{m\to\infty}\psi(z,m)=-\infty$; if $\varphi(z,\lambda)\equiv\varphi(z,0)$, then $\psi(z,m)$ is defined with value $\psi(z,m)=-\infty$ for every $m>\varphi(z,0)$.

We first consider the case where $M_\varphi\in\mathscr{C}^\infty(\Omega\times\mathbb{R})$. Set $u=u_1+iu_2$ with $u_1=\log|\lambda|=\log r$. Then a simple calculation shows that $M_\varphi(z,u)$ is a plurisubharmonic function of the variable (z,u), since

$$\frac{\partial M_\varphi}{\partial u}=\frac{\partial M_\varphi}{\partial u_1}=\frac{r}{\left(\dfrac{\partial r}{\partial\lambda}\right)}\frac{\partial M_\varphi}{\partial\lambda}=\frac{r}{\left(\dfrac{\partial r}{\partial\lambda}\right)}\frac{\partial M_\varphi}{\partial\bar\lambda}.$$

We have $\dfrac{\partial M_\varphi}{\partial u_2}=0$, $\dfrac{\partial M_\varphi}{\partial u_1}>0$. Since $M_\varphi(z,u_1)=m$, we obtain $u_1=\log\delta(z,m)$. From the Implicit Function Theorem, we obtain

$$\begin{aligned}\frac{\partial^2 u_1}{\partial z_j\partial\bar z_k}=-\left(\frac{\partial M_\varphi}{\partial u_1}\right)^{-3}&\left\{\frac{\partial^2 M_\varphi}{\partial z_j\partial\bar z_k}\left(\frac{\partial M_\varphi}{\partial u_1}\right)^2-\frac{\partial^2 M_\varphi}{\partial u_i\partial z_j}\frac{\partial M_\varphi}{\partial\bar z_k}\frac{\partial M_\varphi}{\partial u_1}\right.\\&\left.+\frac{\partial^2 M_\varphi}{\partial u_1^2}\frac{\partial M_\varphi}{\partial\bar z_k}\frac{\partial M_\varphi}{\partial z_j}-\frac{\partial^2 M_\varphi}{\partial u_1\partial\bar z_k}\frac{\partial M_\varphi}{\partial z_j}\frac{\partial M_\varphi}{\partial u_1}\right\}\end{aligned}$$

and hence

$$\sum_{j,k=1}^n\frac{\partial^2 u_1}{\partial z_j\partial\bar z_k}w_j\bar w_k=-\left(\frac{\partial M_\varphi}{\partial u_1}\right)^{-3}L(M_\varphi,W),$$

where

$$W=\left(w_1\frac{\partial M_\varphi}{\partial u_1},\ w_2\frac{\partial M_\varphi}{\partial u_1},\ \dots,\ w_n\frac{\partial M_\varphi}{\partial u_1},\ -\sum_{j=1}^n\frac{\partial M_\varphi}{\partial z_j}w_j\right)\in\mathbb{C}^{n+1},$$

so in this case, $-\log\delta(z,m)$ is plurisubharmonic. For the general case, we let $\varphi_\nu(z)$ be a sequence of $\mathscr{C}^\infty$ plurisubharmonic functions which decrease to φ and let $\delta_\nu(z,m)$ be the associated functions on a domain $\Omega'\Subset\Omega$, for $z_0\in\Omega'$. Then $-\log\delta_\nu(z,m)$ decreases to $-\log\delta(z,m)$, which is plurisubharmonic in Ω' by Proposition I.3. □

Appendix II. The Existence of Proximate Orders

Theorem II.1. *Let $M(r)$ be a continuous positive function for $r>0$ such that $\limsup_{r\to\infty}\frac{\log M(r)}{\log r}=\rho<+\infty$. Then there exists a strong proximate order $\rho(r)$ such that $M(r)\leqq r^{\rho(r)}$ for all $r>0$ and $M(r_m)=r_m^{\rho(r_m)}$ for an increasing sequence of values r_m tending to $+\infty$.*

Proof. Let $\varphi(r)=M(r)\cdot r^{-\rho}$ so that $\limsup_{r\to\infty}\frac{\log\varphi(r)}{\log r}=0$. We change variables by letting $x=\log r$ and $y=\log\varphi(r)$, so that

$$y=\varphi_1(x)=\log\varphi(\exp x) \quad\text{and}\quad \limsup_{x\to\infty}\frac{\varphi_1(x)}{x}=0.$$

The idea of the proof is to construct piecewise a concave majorant to the curve $y=\varphi_1(x)$ which coincides for a sequences of points x_m tending to infinity. This majorant will then be successively modified so as to have the differentiability properties required by the definition of a strong proximate order. The proof is divided into several steps.

1. First we construct a function $\psi_1(x)$ with the following properties:

i) $\psi_1(x)$ is concave

ii) $\lim_{x\to\infty}\frac{\psi_1(x)}{x}=0$, $\lim_{x\to\infty}\psi_1(x)=+\infty$

iii) $\lim_{x\to\infty}\psi_1'(x)=0$

iv) $\limsup_{x\to\infty}[\psi_1(x)+\varphi_1(x)]=+\infty$.

Let ε_m be a sequence which decreases to zero. We choose by induction an increasing sequence of points x_m tending to $+\infty$ and linear functions $\alpha_m(x)$ of slope ε_m such that $\alpha_m(x_m)=\alpha_{m+1}(x_m)\leqq -m$ and $\varphi_1(x)>-\alpha_m(x)+m$ for $x\geqq x_m$: let $\alpha_1(x)=-\varepsilon_1 x$; we choose a point x_1 such that $\varepsilon_1 x_1\geqq 1$ and $\varphi_1(x_1)>-\varepsilon_1 x_1+1$; having chosen x_m and α_m, we let $\alpha_{m+1}(x)=\alpha_m(x)-\varepsilon_m(x-x_m)$; we choose x_{m+1} so large that $\alpha_{m+1}(x_{m+1})\leqq-(m+1)$ and $\varphi_1(x_{m+1})>-\alpha_m(x_{m+1})+m+1$ and set $\tilde\psi_1(x)=-\alpha_m(x)$ for $x_{m-1}\leqq x\leqq x_m$.

Then $\tilde\psi_1(x)$ satisfies (i), (ii), (iv), and (iii) except for the points x_m where $\tilde\psi_1'(x)$ does not exist. Thus, we modify $\tilde\psi_1$ in the following way: let l_m be the bisector of the obtuse angle formed by the lines $y=\alpha_m(x)$ and $y=\alpha_{m+1}(x)$

and let $\tilde{\delta}_m$ be the circle of radius δ_m centered on l_m and tangent to α_m and α_{m+1}; we use an arc on the circle

$$\tilde{\delta}_m = \{(x, y): |x - x_m|^2 + |y - y_m|^2 = \delta_m^2\} \quad \text{and} \quad \psi_1(x) = y_m + \sqrt{\delta_m^2 - (x - x_m)^2}$$

for x between the x-coordinates of the two points of tangency, then for δ_m sufficiently small, (i), (ii), and (iv) still hold; $\psi_1(x) \in \mathscr{C}^1(x)$ and (iii) also holds.

2. Suppose that $\psi(x)$ is a function which satisfies (i), (ii) and (iii) of 1. Then there exists a function $\theta(x)$ such that

(iv′) $\lim_{x\to\infty} \theta(x) = +\infty$

(v) $\lim_{x\to\infty} \dfrac{\theta(x)}{x} = 0, \ \lim_{x\to\infty} \theta'(x) = 0$

(vi) $\lim_{x\to\infty} \dfrac{\theta''(x)}{\theta'(x)} = 0$

(vii) $\theta(x) \geqq \psi(x)$

(viii) there exists an increasing sequence x_m tending to $+\infty$ such that x_m is an extremal point for the curve $y = \psi(x)$, and furthermore $\theta(x_m) = \psi(x_m)$.

Let ε_m be a sequence monotonically decreasing to zero. By induction, we shall find a sequence of points x_m increasing to infinity and functions $\tilde{\theta}_m(x)$ defined on $x_m \leqq x \leqq x_{m+1}$ such that $\tilde{\theta}_m(x_m) = \tilde{\theta}_{m-1}(x_m)$ and $\tilde{\theta}'_m(x_m) = \tilde{\theta}'_{m-1}(x_m)$, $\left|\dfrac{\tilde{\theta}''_m(x)}{\tilde{\theta}'_m(x)}\right| < \varepsilon_m$ for $x_m \leqq x \leqq x_{m+1}$, $\tilde{\theta}_m(x) \geqq \psi(x)$ for $x_m \leqq x \leqq x_{m+1}$ and such that there exists a point x'_m, $x_m \leqq x'_m \leqq x_{m+1}$ for which $[x'_m, \tilde{\theta}_m(x'_m)]$ is an extremal point for the curve $y = \psi(x)$.

Let $\theta_1(x)$ be a linear function with slope ε_1 whose graph is tangent to the curve $y = \psi(x)$. The line $y = a + \varepsilon_1 x$ lies above the curve for a large by (iii). If we decrease a in a continuous manner, we find an a_0 for which $y = a_0 + \varepsilon_1 x$ is tangent to the curve at $[x'_1, \psi(x'_1)]$ which is an extremal point by (iii). Let $x_0 = 0$ and

$$\theta_m(x, c_1^{(m)}) = c_0^{(m)} + c_1^{(m)}(x - x_m) - c_2^{(m)} \exp - \varepsilon_m(x - x_m).$$

Then $\theta_m(x, c_1^{(m)})$ approaches the function $c_0^{(m)} + c_1^{(m)}(x - x_m)$ asymptotically. Let $\xi_m = \theta'_{m-1}(x_m)$, which approaches $c_1^{(m-1)}$ for large x_m. We choose $c_0^{(m)}$ and $c_2^{(m)}$ (depending on the parameter $c_1^{(m)}$) so that $\theta_m(x_m) = \theta_{m-1}(x_m)$ and $\theta'_m(x_m) = \theta'_{m-1}(x_m)$, that is if $y_{m-1} = \theta_{m-1}(x_{m-1})$

$$c_2^{(m)} = \frac{1}{\varepsilon_m}(\xi_m - c_1^{(m)}) \quad \text{and} \quad c_2^{(m)} = y_{m-1} + \frac{1}{\varepsilon_m}(\xi_m - c_1^{(m)}),$$

and we choose $0 < c_1^{(m)} < \xi_m$ so that $c_2^{(m)} > 0$. Then

$$\left|\frac{\theta''_m(x)}{\theta'_m(x)}\right| = \frac{c_2^{(m)} \varepsilon_m^2}{\varepsilon_m c_2^{(m)} + c_1^{(m)} \exp \varepsilon_m(x - x_m)} < \varepsilon_m \quad \text{for } x > x_m.$$

We choose x_m so large that $\xi_m \leqq 2c_1^{(m-1)}$ and $\theta_m[x, \frac{1}{2}c_1^{(m-1)}] > \psi(x)$ for $x \geqq x_m$, which is possible by (iii). Then there exists a $c_1^{(m)} < \frac{1}{2}c_1^{(m-1)}$ such that the curve $y=\theta_m(x, c_1^{(m)})$ meets $y=\psi(x)$ tangentially at $[\psi_m(x'_m), x'_m]$. Since $y=\theta_m(x, c_1^{(m)})$ contains no line segments, $[\psi_m(x'_m), x'_m]$ is an extremal point for $y=\psi(x)$. Furthermore, $c_1^{(m)}<2^{-m}c_1^{(1)}$, so $c_1^{(m)}$ goes to zero monotonically. Let $\tilde{\theta}_m(x)=\theta_m(x, c_1^{(m)})$ for $x_m \leqq x \leqq x_{m+1}$. Then $\tilde{\theta}(x)$ satisfies conditions (iv')–(viii) except for the points x_m where θ'' is not continuous. Since the points x_m do not lie on the curve $y=\psi(x)$, by changing $\tilde{\theta}(x)$ in a small neighborhood of x_m, we can construct a new function $\theta(x)$ whose second derivative lies between the upper and lower limits of $\tilde{\theta}''(x)$ at x_m and which still possesses the properties (iv'), (v) and (viii). Then $\left|\dfrac{\theta''(x)}{\theta'(x)}\right|<\varepsilon_{m-1}$ for $x_m \leqq x \leqq x_{m+1}$, so $\theta(x)$ satisfies (vii) also.

Let $\theta_1(x)$ be the function so constructed in (2) for $\psi_1(x)$. Let $\tilde{\psi}_2(x)$ be the smallest concave majorant of $\varphi_2(x)=\varphi_1(x)+\theta_1(x)$, and let $\theta_2(x)$ be the function constructed in (2) for $\tilde{\psi}_2(x)$. Then $\theta_2(x) \geqq \tilde{\psi}_2(x) \geqq \varphi_1(x)+\theta_1(x)$, and for a sequence x'_m of extremal points tending to infinity, $\theta_2(x'_m)=\tilde{\psi}_2(x'_m)$. But for every extremal point, $\tilde{\psi}_2(x)=\varphi_1(x)+\theta_1(x)$ since $\tilde{\psi}_2$ is the smallest concave majorant, so $\theta_2(x'_m)=\varphi_1(x'_m)+\theta_1(x'_m)$.

Let

$$\rho(r)=\rho+\frac{\theta_2(\log r)-\theta_1(\log r)}{\log r}.$$

Since

$$\varphi_1(x) \leqq \theta_2(x)-\theta_1(x), \qquad \log\varphi(r) \leqq \theta_2(\log r)-\theta_1(\log r),$$

and we have

$$M(r) \leqq r^{\rho}\exp(\theta_2(\log r)-\theta_1(\log r))=r^{\rho(r)}$$

with equality for a sequence of points tending to infinity. By (v), we obtain that $\lim_{r\to\infty}\rho(r)=\rho$ and

$$\lim_{r\to\infty}\rho'(r)r\log r$$

$$=\lim_{r\to\infty} r\log r\left[\frac{\theta'_2(\log r)-\theta'_1(\log r)-(\theta_2(\log r)-\theta_1(\log r))}{r\log r}\right]=0$$

Furthermore,

$$\rho''(r)=\frac{\theta''_2(\log r)-\theta''_1(\log r)}{r^2\log r}$$

$$+\left\{\frac{[\theta'_1(\log r)-\theta'_2(\log r)]}{r^2\log r}+\frac{[\theta_2(\log r)-\theta_1(\log r)]}{r^2(\log r)^2}\right\}$$

$$\times\left\{1+\frac{2}{\log r}\right\}.$$

It follows from (vii) and (vi) that $|\theta''_i(\log r)|<|\theta'_i(\log r)|=o(1)$ and $|\theta_i(\log r)| =o(\log r)$, $i=1,2$, so $\lim_{r\to\infty} r^2\log r\,\rho''(r)=0$. □

Appendix III. Solution of the $\bar{\partial}$-Equation with Growth Conditions

The basic technique in the theory of functions of several complex variables is the solution of the $\bar{\partial}$-equation, since a continuous function f defined in a domain $\Omega \subset \mathbb{C}^n$ is holomorphic if and only if $\bar{\partial} f = 0$ for the current $\bar{\partial} f$. We recall here the solution given by Hörmander [B] using L^2 estimates and Hilbert space techniques for the equation $\bar{\partial} u = g$ with $\bar{\partial} g = 0$. We consider only (0, 1) forms for g in this Appendix, which is sufficient for the problems treated in this book, and we refer the reader to [B] for the general solution.

1. Basic Lemmas on Non-bounded Operators Between Hilbert Spaces

Let H_1 and H_2 be two complex Hilbert spaces with inner-products $\langle\ ,\ \rangle_1$ and $\langle\ ,\ \rangle_2$ respectively. We will consider an operator A from H_1 to H_2 defined on a linear subspace D_A of H_1, called the domain of A, and a linear mapping A of D_A into H_2. We would like to find the transpose of the operator $A^\star$ with domain $D_{A^\star}$ in H_2 such that for $x \in D_A$, $y \in D_{A^\star}$

$$\langle Ax, y\rangle_2 = \langle x, A^\star y\rangle_1. \tag{III,1}$$

Since $A^\star y$ will be uniquely determined only if D_A is dense in H_1, we shall always assume that this is the case. From (III,1) we then obtain

$$|\langle Ax, y\rangle_2| \leqq \|A^\star y\|_1 \, \|x\|_1 = C_y \|x\|_1.$$

Thus, to the operator A with domain D_A, we associate the subspace $D_{A^\star}$ defined to be the set of the elements $y \in H_2$ for which there exists a constant C_y such that for every $x \in D_A$

$$|\langle Ax, y\rangle_2| \leqq C_y \|x\|_1. \tag{III,2}$$

For $y \in D_{A^\star}$, we consider the mapping $x \to \langle Ax, y\rangle_2$. From (III,2) it is a continuous linear functional on D_A equipped with the norm $\|\ \|_1$. It extends uniquely to a continuous linear functional defined on H_1, and thus there exists a unique element $A^\star y$ in H_1 such that $\langle Ax, y\rangle_2 = \langle x, A^\star y\rangle_1$. The application $y \to A^\star y$ from $D_{A^\star}$ into H_1 is linear. The operator $A^\star$ with domain $D_{A^\star}$ is the *transpose operator* of A.

Proposition III.1. *The operator $A^\star$ is closed, that is if $y_n \to y_0$, $y_n \in D_{A^\star}$, and $z_n = A^\star y_n \to z_0 \in H_1$, then $y_0 \in D_{A^\star}$ and $z_0 = A^\star y_0$.*

Proof. Since z_n is a Cauchy sequence in H_1, there exists an M such that $\|z_n\| \leqq M$, and hence $|\langle x, A^\star y_n \rangle_1| = |\langle x, z_n \rangle_1| \leqq M \|x\|_1$ for $x \in D_A$. Thus, $|\langle Ax, y_n \rangle_2| \leqq M \|x\|_1$ when $y_n \to y_0$, and since $\langle Ax, y_n \rangle_2 \to \langle Ax, y_0 \rangle_2$, we see that $|\langle Ax, y_0 \rangle_2| \leqq M \|x\|_1$. Hence $y_0 \in D_{A^\star}$ by (III,2). Furthermore, $\langle x, z_n \rangle_1 \to \langle x, z_0 \rangle_1$, hence $\langle Ax, y_0 \rangle_2 = \langle x, z_0 \rangle_1$ and $z_0 = A^\star y_0$ by uniqueness. □

We would like to define $(A^\star)^\star$ and verify that $(A^\star)^\star = A$. To do so, we introduce the additional hypothesis that A *is closed* and show that $D_{A^\star}$ is dense in H_2. Then $(A^\star)^\star$ can be defined, and $(A^\star)^\star = A$ (which implies that $D_A = D_{(A^\star)^\star}$ when A is closed).

We introduce the product space $H = H_1 \times H_2$ and $\tilde{H} = H_2 \times H_1$ and the mappings B_1 and B_2 of H into $\tilde{H}$ and $\tilde{H}$ into H respectively given by $B_1(x, y) = (y, -x)$ and $B_2(y, x) = (x, -y)$.

The graph G_A of A is the set (x, Ax) in H with $x \in D_A$ and the graph $G_{A^\star}$ of $A^\star$ is the set $(y, A^\star y)$ in $\tilde{H}$ with $y \in D_{A^\star}$.

We equip H with the inner product $\langle (x, y), (u, v) \rangle = \langle x, u \rangle_1 + \langle y, v \rangle_2$ and $\tilde{H}$ with the inner product $\langle (y, x), (v, u) \rangle = \langle x, u \rangle_1 + \langle y, v \rangle_2$.

Thus, an operator A is closed if and only if its graph is closed.

Proposition III.2. *We have*

$$G_{A^\star} = [B_1(G_A)]^\perp \quad \text{in } \tilde{H}. \tag{III,3}$$

Proof. The relation $\langle Ax, y \rangle_2 - \langle x, z \rangle_1 = 0$ for all $x \in D_A$ implies $y \in D_{A^\star}$ and $z = A^\star y$ by the definition of $A^\star$. But this is equivalent to $\langle (Ax, -x), (y, z) \rangle = 0$ in $\tilde{H}$. □

Proposition III.3. *If G_A is closed, then $A^\star$ is dense in H_2, $D_A = D_{(A^\star)^\star}$, and $A = (A^\star)^\star$.*

Proof. From the closure of G_A, we see that $B_1(G_A) = (B_1(G_A))^{\perp\perp}$, and from Proposition III.2, we see that $(B_1(G_A))^\perp = (G_{A^\star})^\perp$. Hence

$$G_A = B_2(B_1(G_A)) = B_2(G_{A^\star}^\perp) = (B_2(G_{A^\star})^\perp) \quad \text{and}$$

$$G_A = (B_2(G_{A^\star}))^\perp \quad \text{in } H. \tag{III,4}$$

Let $u \in H_2 \cap (D_{A^\star})^\perp$. Then $\langle (0, u), (A^\star y, -y) \rangle = 0$ in H for all $y \in D_{A^\star}$, and so $(0, u) \in B_2(G_{A^\star})^\perp$. Thus, by (III,4), $(0, u) \in G_A$ and $u = A(0) = 0$, from which it follows that $A^\star$ is dense in H_2. By applying (III,3) to $A^\star$, we obtain $G_{(A^\star)^\star} = (B_2(G_{A^\star}))^\perp$. Thus by (III,4), $G_A = G_{(A^\star)^\star}$, from which it follows that $D_A = D_{(A^\star)^\star}$ and $A = (A^\star)^\star$. □

Lemma III.4. *Let A be a closed operator from a dense subspace D_A of a Hilbert space H_1 into a closed subspace F of a Hilbert space H_2. Then $F = A(D_A)$ if and only if there exists a constant $C \geqq 0$ such that for every $y \in F \cap D_A^\star$, $\|y\|_2 \leqq C\|A^\star y\|_1$.*

Proof. Let $z \in F$. The solution of the equation $Ax = z$ is equivalent to the existence of an x such that $\langle x, A^\star y\rangle_1 = \langle z, y\rangle_2$ for every $y \in D_{A^\star}$, since $(A^\star)^\star = A$ by Proposition III.3.

Suppose that there exists a constant C for which the conclusion of Lemma III.4 is valid. If $y \in F^\perp$, then $\langle z, y\rangle_2 = 0$.

If $y \in F \cap D_{A^\star}$, then by hypothesis

$$|\langle z, y\rangle_2| \leqq \|z\|_2 \|y\|_2 \leqq C\|z\|_2 \|A^\star y\|_1, \tag{III,5}$$

so the linear functional $l(y) = \langle z, y\rangle_2$ is continuous on $A^\star(D_{A^\star})$. Hence there exists an $x \in H_1$, such that

$$\langle x, A^\star y\rangle_1 = \langle z, y\rangle_2. \quad \text{Thus } Ax = z \text{ and } F = A(D_A).$$

Suppose that $F = A(D_A)$. Let $B = \{y \colon y \in F \cap D_{A^\star}, \|A^\star y\|_1 \leqq 1\}$.

We shall show that B is a bounded set in H_2. The space $A(D_A)$ is closed and therefore is a Hilbert space; then for $y \in B$ and $z \in F = A(D_A)$ we obtain:

$$|\langle y, z\rangle_2| = |\langle Ay, x\rangle_1| \leqq \|A^\star y\|_1 \|x\|_1 \leqq \|x\|_1.$$

Thus the family B is pointwise bounded on F. By the Banach-Steinhaus Theorem it is equicontinuous and uniformly bounded on the unit ball in F by a constant C. This implies that $\left|\left\langle y, \frac{y}{\|y\|}\right\rangle\right| \leqq C$ or $\|y\|_2 \leqq C$. □

Lemma III.5. *Let A be a closed operator defined on a domain D_A dense in H_1 and let F be a closed subspace of the Hilbert space H_2 such that $A(D_A) \subset F$. Suppose that there exists $C > 0$ such that $\|y\|_2 \leqq C\|A^\star y\|_1$ for every $y \in F \cap D_{A^\star}$. Then for $v \in H_1 \cap [A^{-1}(0)]^\perp$, there exists $w \in D_{A^\star}$ such that $A^\star w = v$ and $\|w\|_2 \leqq C\|v\|_1$.*

Proof. Since A is closed, if $x_n \in D_A$, $x_n \to x_0$, and $Ax_n = 0$, then $x_0 \in D_A$ and $Ax_0 = 0$, so $A^{-1}(0)$ is a closed subspace of H_1. Now $Ax = 0$ is equivalent to $\langle Ax, y\rangle_2 = 0$ for all $y \in D_{A^\star}$, since $D_{A^\star}$ is dense in H_2, and is hence equivalent to $\langle x, A^\star y\rangle_1 = 0$ for all $y \in D_{A^\star}$. On the other hand, $\langle u, A^\star y\rangle_1 = 0$ for all $y \in D_{A^\star}$, implies $u \in D_{(A^\star)^\star} = D_A$ and $\langle Au, y\rangle_2 = \langle u, A^\star y\rangle_1 = 0$ for all $y \in D_{A^\star}$, so $Au = 0$. Thus $A^{-1}(0) = A^\star(D_{A^\star})^\perp$ and similarly $A^{\star -1}(0) = A(D_A)^\perp$ by symmetry.

By hypothesis $F^\perp$ is included in $A(D_A)^\perp = A^{\star -1}(0)$. Thus, the restriction of $A^\star$ to $F \cap D_{A^\star}$ is equal to $A^\star(D_{A^\star})$ in H_1. The inequality $\|y\|_2 \leqq C\|A^\star y\|_1$ implies that this image is closed in H_1. Thus, $v \in (A^\star(D_{A^\star}))^{\perp\perp} = A^\star(D_{A^\star})$. Hence, we can find $w \in F \cap D_{A^\star}$ such that $A^\star w = v$, and so $\|w\|_2 \leqq C\|v\|_1$. □

2. Inequalities for the $\bar{\partial}$-equation

Let φ be a continuous function on $\Omega \subset \mathbb{C}^n$ and let $L^2(\varphi)$ be the completion of $\mathscr{C}_0^\infty(\Omega)$ for the norm induced by the inner product

$$\langle f, g \rangle_\varphi = \int_\Omega f \bar{g} e^{-\varphi(z)} d\tau(z).$$

The space $L^2(\varphi)$, considered as a subspace of distributions, is composed precisely of those distributions T for which

$$|T(f)|^2 \leqq C_T \int_\Omega |f|^2 e^{-\varphi(z)} d\tau(z) \qquad \text{for } f \in \mathscr{C}_0^\infty(\Omega).$$

By $L^2_{(0,1)}(\varphi)$ (resp. $L^2_{(0,2)}(\varphi)$) we will mean the space of $(0,1)$ forms $f = \sum_{i=1}^n f_i d\bar{z}_i$ (resp. the space of $(0,2)$-forms $f = \sum_{i<j} f_{ij} d\bar{z}_i \wedge d\bar{z}_j$) such that $f_i \in L^2(\varphi)$ (resp. $f_{ij} \in L^2(\varphi)$), and we equip $L^2_{(0,1)}$ with the norm $\|f\|_\varphi^2 = \sum_{i=1}^n \|f_i\|_\varphi^2$ (resp. we equip $L^2_{(0,2)}(\varphi)$ with the norm $\|f\|^2 = \sum_{i<j} \|f_{ij}\|_\varphi^2$); $L^2_{(0,1)}(\varphi)$ (resp. $L^2_{(0,2)}(\varphi)$) is the completion with this norm of the $(0,1)$ forms with coefficients in $\mathscr{C}_0^\infty(\Omega)$ (resp. $L^2_{(0,2)}(\varphi)$ is the completion of the $(0,2)$ forms with coefficients in $\mathscr{C}_0^\infty(\Omega)$).

If φ_1 and φ_2 are two continuous functions in Ω, we define the operator $A = \bar{\partial}$ on a subset of $L^2(\varphi)$ into $L^2_{(0,1)}(\varphi_2)$ by $\bar{\partial} f = \sum_{i=1}^n \frac{\partial f}{\partial \bar{z}_i} d\bar{z}_i$ taken as a current. The domain of A is the set of $f \in L^2(\varphi_1)$ such that $\bar{\partial} f \in L^2_{(0,1)}(\varphi_2)$.

Proposition III.6. *The operator $A = \bar{\partial}$ is a closed densely defined operator.*

Proof. Since $\mathscr{C}_0^\infty(\Omega)$ is dense in $L^2(\varphi_1)$, A is densely defined. If $f_n \to f_0$ in $L^2(\varphi_1)$, then $f_n \to f_0$ in $L^1_{\text{loc}}(\Omega)$. Since derivation is a continuous operation in the space of distributions, $Af_n \to Af_0$ as a current. Thus, if $Af_n \to g_0$ in $L^2_{(0,1)}(\varphi_2)$, $Af_0 = g_0$ for the currents, and A is closed. □

We shall also consider the operator $B = \bar{\partial}$ which maps the space $L^2_{(0,1)}(\varphi_2)$ into $L^2_{(0,2)}(\varphi_3)$ given by

$$\bar{\partial}\left(\sum_{i=1}^n f_i d\bar{z}_i\right) = \sum_{i<j} \left(\frac{\partial f_j}{\partial \bar{z}_i} - \frac{\partial f_i}{\partial \bar{z}_j}\right) d\bar{z}_i \wedge d\bar{z}_j.$$

For what follows, we choose a function $\alpha(z) \in \mathscr{C}_0^\infty(B_0)$ defined in the unit ball B_0 of $\mathbb{C}^n$ and such that $\int \alpha(z) d\tau(z) = 1$, and we set $\alpha_\varepsilon(z) = \varepsilon^{-2n} \alpha\left(\frac{z}{\varepsilon}\right)$.

Lemma III.7. *Given $g \in L^2(\Omega)$ with compact support, then $g_\varepsilon = g \star \alpha_\varepsilon$ is in $\mathscr{C}_0^\infty$ and* $\lim_{\varepsilon \to 0} \|g_\varepsilon - g\|_{L^2} = 0$.

Proof. Since $g_\varepsilon(z)=\int g(z+z')\alpha_\varepsilon(z')d\tau(z')$ and g has compact support, we may differentiate under the integral, which shows that g_ε is in $\mathscr{C}_0^\infty$.

If u is continuous, it follows from the formula

$$u_\varepsilon(z)-u(z)=\int (u(z-\varepsilon z')-u(z))\alpha_\varepsilon(z')d\tau(z')$$

and the uniform continuity of u that u_ε converges uniformly to u.

Since $\int \alpha_\varepsilon(z)d\tau(z)=1$, it follows from Minkowski's Inequality that in L^2 norms $\|u_\varepsilon\|\leqq\|u\|$. For any $\eta>0$, we can find a continuous function v with compact support such that $\|u-v\|<\eta$. Thus, $\|u_\varepsilon-v_\varepsilon\|<\eta$ and

$$\limsup_{\varepsilon\to 0}\|u_\varepsilon-u\|\leqq\limsup_{\varepsilon\to 0}\|u_\varepsilon-v_\varepsilon\|+\|u-v\|+\limsup_{\varepsilon\to 0}\|v_\varepsilon-v\|\leqq 2\eta,$$

since $v_\varepsilon\to v$ uniformly. □

Now we shall calculate explicitly $A^\star$. Let $g\in\mathscr{C}_0^\infty(\Omega)$ and $f=\sum_{j=1}^n f_i d\bar{z}_i$ in $L^2_{(0,1)}(\Omega)$. Then if $f\in D_{A^\star}$,

$$\int (A^\star f)\bar{g}e^{-\varphi_1}d\tau=\langle A^\star f, g\rangle_1=\langle f, Ag\rangle_2=\int\left(\sum_{j=1}^n f_j\overline{\left(\frac{\partial g}{\partial\bar{z}_j}\right)}\right)e^{-\varphi_2}d\tau$$

so

$$\text{(III,6)}\qquad A^\star f=-e^{\varphi_1}\sum_{j=1}^n\frac{\partial}{\partial z_j}e^{-\varphi_2}f_j.$$

Proposition III.8. *Let K_0 be a fixed compact set and β_m a sequence of functions in $\mathscr{C}_0^\infty(\Omega)$ such that $0\leqq\beta_m\leqq 1$, $\beta_m\equiv 1$ for $z\in K_0$, and such that for every compact subset K, there exists m_K such that $m\geqq m_K$ implies $\beta_m\equiv 1$ on K. Suppose that $\varphi_2\in\mathscr{C}^1(\Omega)$ and*

$$\text{(III,7)}\qquad e^{-\varphi_{j+1}}\sum_{k=1}^n|\partial\beta_m/\partial z_k|^2\leqq e^{-\varphi_j},\qquad j=1,2.$$

Then the (0, 1) *forms with coefficients in $\mathscr{C}_0^\infty(\Omega)$ are dense in $D_{A^\star}\cap D_B$ for the norm $|||f|||=\|A^\star f\|_1+\|f\|_2+\|Bf\|_3$, where $\|\ \|_1$ is the norm in $L^2(\varphi_1)$, $\|\ \|_2$ the norm in $L^2_{(0,1)}(\varphi_2)$, and $\|\ \|_3$ the norm in $L^2_{(0,2)}(\varphi_3)$.*

Proof. Since β_m has compact support for all m and is in $\mathscr{C}^\infty$, $\bar{\partial}\beta_m\wedge f$ and $\beta_m\bar{\partial}f$ have coefficients in $L^2_{(0,2)}(\varphi_3)$ if $f\in D_B$.

Furthermore, $B(\beta_m f)-\beta_m(Bf)=\bar{\partial}\beta_m\wedge f$, and from (III,7), we have

$$|B(\beta_m f)-\beta_m(Bf)|^2e^{-\varphi_3}\leqq|f|^2e^{-\varphi_2}.$$

Since $B(\beta_m f)-\beta_m(Bf)$ converges to zero almost everywhere, it follows from the Lebesgue Dominated Convergence Theorem that

$$\lim_{m\to\infty}\|B(\beta_m f)-\beta_m(Bf)\|_3=0\qquad\text{for } f\in D_B.$$

Suppose that $f\in D_{A^\star}$ and $g\in D_A$. Then

$$\langle\beta_m f, Ag\rangle_2=\langle f,\bar{\beta}_m(Ag)\rangle_2=\langle f, A(\beta_m g)\rangle_2+\langle f, gA\beta_m\rangle_2,$$

and

$$|\langle f, A(\beta_m g)\rangle_2| = |\langle A^\star f, \beta_m g\rangle_1| \leqq C(m,f)\|g\|_1$$

$|\langle f, gA\beta_m\rangle_2| \leqq C'(m,f)\|g\|_1$ since β_m has compact support.

Thus $|\langle \beta_m f, Ag\rangle_2| \leqq (C(m,f)+C'(m,f))\|g\|_1$ and $\beta_m f \in D_{A^\star}$. From (III,6) we see that

$$|A^\star(\beta_m f) - \beta_m(A^\star f)| = e^{(\varphi_1-\varphi_2)} \sum_{k=1}^{n} |f_j| \left|\frac{\partial \beta_m}{\partial z_j}\right|.$$

It then follows from (III,7) and Schwarz' Inequality that

$$|A^\star(\beta_m f) - \beta_m(A^\star f)|^2 \leqq \sum_j |f_j|^2 e^{(\varphi_1-\varphi_2)}.$$

Thus, since $\lim_{m\to\infty} |A^\star(\beta_m f) - \beta_m(A^\star f)| = 0$ pointwise, again by the Lebesgue Dominated Convergence Theorem, we see that

$$\lim_{m\to\infty} \|A^\star(\beta_m f) - \beta_m(A^\star f)\|_1 = 0.$$

Thus the elements of $D_B \cap D_{A^*}$ with compact support are dense for the norm $|||\ |||$.

Let $f \in D_{A^\star} \cap D_B$ have compact support. Then $f \star \alpha_\varepsilon$ for $\varepsilon < 1$ has its support contained in a fixed compact set, and since φ_2 is continuous, the $L^2_{(0,1)}(0)$ norm of f_ε is equivalent to the $L^2_{(0,1)}(\varphi_2)$ norm of f_ε. Thus, by Lemma III.7, $\lim_{\varepsilon\to 0} \|f_\varepsilon - f\|_2 = 0$. Furthermore, since $B(f\star\alpha_\varepsilon) = Bf\star\alpha_\varepsilon$ and the support of Bf is contained in that of f, the $L^2_{(0,2)}(0)$ norm is equivalent to the $L^2_{(0,2)}(\varphi_3)$ norm on the support of f and $\lim_{\varepsilon\to 0}\|Bf_\varepsilon - Bf\|_3 = 0$ by Lemma III.7.

In the same way, by (III,6), since $A^\star$ is a differential operator, $\operatorname{supp} A^\star f \subset \operatorname{supp} f$; however, since $A^\star$ is not a constant coefficient operator, it does not commute with the regularization. We have in fact $e^{(\varphi_2-\varphi_1)}A^\star = \theta + a$, where θ is a constant coefficient differential operator and a is multiplication by continuous functions. Thus

$$(\theta+a)(f\star\alpha_\varepsilon) = [(\theta+a)f]\star\alpha_\varepsilon + a(f\star\alpha_\varepsilon) - (af)\star\alpha_\varepsilon.$$

As above, the right hand side converges to $(\theta+a)f + af - af$ in $L^2(1)$ and hence $\lim_{\varepsilon\to 0}\|A^\star(f\star\alpha_\varepsilon) - A^\star f\|_1 = 0$. □

Theorem III.9. *Let K_0 be a fixed compact subset of Ω and $\beta_m \in \mathscr{C}_0^\infty(\Omega)$ such that $0 \leqq \beta_m \leqq 1$, $\beta_m \equiv 1$ for $z \in K_0$ and such that for every compact subset $K \subset \Omega$, there exists m_K such that for $m \geqq m_K$, $\beta_m \equiv 1$ on K. Let $\psi \in \mathscr{C}^2$ such that $\sum_{k=1}^{m} \left|\frac{\partial \beta_m}{\partial z_k}\right|^2 \leqq e^\psi$. Let $\varphi \in \mathrm{PSH}(\Omega) \cap \mathscr{C}^2$ such that $\sum_{j,k=1}^{n} \frac{\partial^2 \varphi(z)}{\partial z_j \partial \bar{z}_k} w_j \bar{w}_k \geqq C(z)\|w\|^2$ for a function C and for all $w \in \mathbb{C}^n$.*

Set $\varphi_1 = \varphi - 2\psi$, $\varphi_2 = \varphi - \psi$, $\phi_3 = \varphi$. Then

$$\int (C - 2|\partial\psi|\,|f|^2) e^{-\varphi} d\tau \leqq 2\|A^\star f\|_1^2 + \|Bf\|_2^2.$$

Proof. Set $\delta_j g = e^{\varphi}\dfrac{\partial}{\partial z_j}(e^{-\varphi} g) = \dfrac{\partial g}{\partial z_j} - g\dfrac{\partial \varphi}{\partial z_j}$. We then obtain the relation

$$\text{(III,8)} \qquad \delta_j \frac{\partial}{\partial \bar{z}_k} - \frac{\partial}{\partial \bar{z}_k}\delta_j = \frac{\partial^2 \varphi}{\partial \bar{z}_k \partial z_j}.$$

From (III,6), we obtain

$$e^{\psi} A^{\star} f = \sum_{j=1}^{n} \delta_j f_j + \sum_{j=1}^{n} f_j \frac{\partial \psi}{\partial z_j},$$

since $\varphi_1 - \varphi_2 = -\psi$.

Using the inequality for vectors $\|a-b\|^2 \leqq 2\|a\|^2 + 2\|b\|^2$, we see that

$$\int \sum_{j,k=1}^{n} \delta_j f_j \overline{\delta_k f_k} e^{-\varphi} d\tau \leqq 2\|A^{\star} f\|_1^2 + 2\int \sum_{j=1}^{n} |f_j|^2 \cdot |\bar{\partial}\psi|^2 e^{-\varphi} d\tau.$$

An easy calculation shows that

$$\begin{aligned} |\bar{\partial} f|^2 &= \sum_{i<j} \left[\left(\frac{\partial f_i}{\partial \bar{z}_j} - \frac{\partial f_j}{\partial \bar{z}_i}\right)\right] \overline{\left[\left(\frac{\partial f_i}{\partial \bar{z}_j} - \frac{\partial f_j}{\partial \bar{z}_i}\right)\right]} \\ &= \sum_{i<j} \left|\frac{\partial f_i}{\partial \bar{z}_j}\right|^2 + \sum_{i<j} \left|\frac{\partial f_j}{\partial \bar{z}_i}\right|^2 - \sum_{i<j} \frac{\partial f_i}{\partial \bar{z}_j} \cdot \overline{\frac{\partial f_j}{\partial \bar{z}_i}} - \sum_{i<j} \frac{\partial f_j}{\partial \bar{z}_i} \cdot \overline{\frac{\partial f_i}{\partial \bar{z}_j}} \\ &= \sum_{i,j=1}^{n} \left|\frac{\partial f_i}{\partial \bar{z}_j}\right|^2 - \sum_{i,j=1}^{n} \frac{\partial f_i}{\partial \bar{z}_j} \overline{\frac{\partial f_j}{\partial \bar{z}_i}}. \end{aligned}$$

Adding together these two results, we obtain

$$\begin{aligned} &\int \sum_{j,k=1}^{n} \left(\delta_j f_j \cdot \overline{\delta_k f_k} - \frac{\partial f_j}{\partial \bar{z}_k} \cdot \overline{\frac{\partial f_k}{\partial \bar{z}_j}}\right) e^{-\varphi} d\tau + \int \sum_{k,j=1}^{n} \left|\frac{\partial f_k}{\partial \bar{z}_j}\right|^2 e^{-\varphi} d\tau \\ &\quad \leqq 2\|A^{\star} f\|_1^2 + \|Bf\|_3^2 + 2\int |f|^2 |\bar{\partial}\psi|^2 e^{-\varphi} d\tau. \end{aligned}$$

Suppose now that the coefficients of f are in $\mathscr{C}_0^{\infty}$. An integration by parts then gives

$$\int \delta_j f_j \cdot \overline{\delta_k f_k} e^{-\varphi} d\tau = -\int f_j \overline{\frac{\partial}{\partial \bar{z}_j}(\delta_k f_k)} e^{-\varphi} d\tau$$

and

$$\begin{aligned} -\int f_j \overline{\frac{\partial}{\partial \bar{z}_j} \delta_k f_k} e^{-\varphi} d\tau &= -\int f_j \overline{\delta_k \left(\frac{\partial f_k}{\partial \bar{z}_j}\right)} e^{-\varphi} d\tau \\ &\quad + \int f_j \bar{f}_k \frac{\partial^2 \varphi}{\partial \bar{z}_k \partial z_j} d\tau \qquad \text{by (III,8).} \end{aligned}$$

Another integration by parts yields

$$-\int f_j \overline{\delta_k \left(\frac{\partial f_k}{\partial \bar{z}_j}\right)} e^{-\varphi} d\tau = \int \frac{\partial f_j}{\partial z_k} \cdot \overline{\frac{\partial f_k}{\partial z_j}} e^{-\varphi} d\tau,$$

and so

$$\int \sum_{j,k} f_j \bar{f}_k \frac{\partial^2 \varphi}{\partial z_j \partial \bar{z}_k} e^{-\varphi} d\tau + \sum_{j,k=1}^{n} \left| \frac{\partial f_j}{\partial \bar{z}_k} \right|^2 e^{-\varphi} d\tau$$
$$\leqq 2 \|A^\star f\|_1^2 + \|Bf\|_3^2 + 2 \int |f|^2 |\partial \psi|^2 e^{-\varphi} d\tau.$$

The conclusion now follows from the hypothesis on φ and Proposition III.8. □

Definition III.10. A domain $\Omega \in \mathbb{C}^n$ will be said to be *pseudoconvex* if there exists $\gamma(z) \in \mathrm{PSH}(\Omega) \cap \mathscr{C}^\infty(\Omega)$ with $\left[\frac{\partial^2 \gamma}{\partial z_j \partial \bar{z}_k}(z)\right]$ positive definite for every $z \in \Omega$ and such that for every $r \in \mathbb{R}$, $\Omega_r = \{z \in \Omega: \gamma(z) < r\} \Subset \Omega$.

Lemma III.11. *Let Ω be a pseudoconvex domain in $\mathbb{C}^n$ and let $\varphi \in \mathrm{PSH}(\Omega) \cap \mathscr{C}^2(\Omega)$. Let $C(z) > 0$ be a real valued and continuous function such that*

$$C(z) \|w\|^2 \leqq \sum_{j,k=1}^{n} \frac{\partial^2 \varphi}{\partial z_j \partial \bar{z}_k} w_j \bar{w}_k$$

for every $z \in \Omega$, $w \in \mathbb{C}^n$. Then for given $g \in L^2_{(0,1)}(\varphi)$ such that $\bar{\partial} g = 0$, we can find $u \in L^2(\varphi)$ such that $\bar{\partial} u = g$ and

$$\text{(III,9)} \qquad \int_\Omega |u|^2 e^{-\varphi} d\tau \leqq 2 \int_\Omega \sum_1^n |g_j|^2 \frac{e^{-\varphi}}{C} d\tau = 2 T_g$$

provided T_g is finite.

Proof. Let $\gamma(z)$ be the function of Definition III.10 associated with Ω. For any fixed r, we can assume that the functions β_m of Proposition III.8 are such that $\beta_m \equiv 1$ on Ω_{r+1} for all m. We can then find $\psi \geqq 0$, $\sum_{m=1}^{\infty} \sum_{k=1}^{n} \left| \frac{\partial \beta_m}{\partial \bar{z}_k} \right|^2 \leqq e^\psi$ and such that $\psi = 0$ in Ω_{r+1}.

Let $\chi(r)$ be an increasing convex function such that $\chi(\gamma(z)) \geqq 2\psi(z)$ and

$$\chi'(\gamma(z)) \sum_{j,k=1}^{n} \frac{\partial^2 \gamma}{\partial z_j \partial \bar{z}_k} w_j \bar{w}_k \geqq 2 |\partial \psi|^2 \|w\|^2.$$

Let $\varphi_1^r = \varphi + \chi(\gamma) - 2\psi$, $\varphi_2^r = \varphi + \chi(\gamma) - \psi$, and $\varphi_3^r = \varphi + \chi(\gamma)$. By Theorem III.9, for $f \in D_{A^\star} \cap D_B$ with coefficients in $\mathscr{C}_0^\infty$,

$$\int_\Omega C(z) |f(z)|^2 e^{-\varphi(z)} d\tau(z) \leqq 2 \|A^\star f\|_1^2 + \|Bf\|_3^2$$

where the norm $\| \ \|_1$ is taken in $L^2(\varphi_1^r)$ and $\| \ \|_3$ in $L^2(\varphi_3^r)$. Suppose that $T_g = 1/2$. Then by the Cauchy-Schwarz Inequality, we have

$$|\langle g, f \rangle_2|^2 \leqq 1/2 \int_\Omega C(z) |f(z)|^2 e^{-\varphi_3^r(z)} d\tau(z) \leqq \|A^\star f\|_1^2 + 1/2 \|Bf\|_3^2.$$

Let $f=f_1+f_2$, where $Bf_1=0$ and f_2 is orthogonal in $L^2_{(0,1)}(\varphi_2^r)$ to the kernel of B. Since the range of A is contained in the kernel of B, f_2 is orthogonal to the range of A, so $A^\star f_2=0$. Since $\bar{\partial}g=0$, $\langle g, f_2\rangle_2=0$ and $|\langle g, f\rangle_2|^2 =|\langle g, f_1\rangle_2|^2 \leqq \|A^\star f_1\|^2=\|A^\star f\|_1^2$.

Applying the Hahn-Banach Theorem to the anti-linear form

$$A^\star f \to \langle g, f\rangle_2, \qquad f\in D_{A^\star},$$

we find an element $u_r\in L^2(\varphi_1)$ such that $\int_\Omega |u_r|^2 e^{-\varphi_1^r}d\tau \leqq 1$ and $\langle g, f\rangle_2 =\langle u_r, A^\star f\rangle_1$. Thus $Au_r=g$. We choose a sequence r_j which increases to infinity such that $u_{r_j}\to u$ converges weakly in $L^2(\Omega_r)$ for every r. Since $\bar{\partial}u_r=g$, we have $\bar{\partial}u=g$ because differentiation is weakly continuous. Since $\varphi_1^r=\varphi$ on Ω_r, we have $\int_{\Omega_r} |u|^2 e^{-\varphi}d\tau\leqq 1$ for every r, and so $\int_\Omega |u|^2 e^{-\varphi}d\tau \leqq 1$. □

Theorem III.12. *Let Ω be a pseudoconvex domain in $\mathbb{C}^n$ and let $\varphi\in \mathrm{PSH}(\Omega)$. For every $g\in L^2_{(0,1)}(\varphi)$ such that $\bar{\partial}g=0$, there exists a function u such that $\bar{\partial}u=g$ and*

$$\text{(III,10)} \qquad \int_\Omega |u|^2(1+\|z\|^2)^{-2}e^{-\varphi}d\tau \leqq \int_\Omega |g|^2 e^{-\varphi}d\tau.$$

Proof. Let $\gamma(z)$ be the function of Definition III.10. Let $\Omega_j=\{z\in\Omega: \gamma(z)<j\}\Subset\Omega$. The open sets Ω_j are also pseudoconvex, since $(j-\gamma(z))^{-1}$ is plurisubharmonic in Ω_j.

By Proposition I.15, we can find $\varphi_j\in \mathrm{PSH}(\Omega_j)\cap\mathscr{C}^\infty(\Omega_j)$ such that φ_j decreases to φ as j tends to infinity.

Let $\psi_j=\varphi_j+2\log(1+\|z\|^2)$. Then, since

$$\sum_{j,k=1}^n \frac{\partial^2\log(1+\|z\|^2)}{\partial z_j\partial\bar{z}_k}w_j\bar{w}_k=(1+\|z\|^2)^{-2}(\|w\|^2(1+\|z\|^2)-(\langle w, z\rangle)^2)$$
$$\geqq (1+\|z\|^2)^{-2}\|w\|^2$$

we can take $C(z)=2(1+\|z\|^2)^{-2}$ in Lemma III.11.

Thus, for every j, we can find u_j defined in Ω_j such that

$$\int_{\Omega_j} |u_j|^2(1+\|z\|^2)^{-2}e^{-\varphi_j}d\tau \leq \int_{\Omega_j} |g|^2 e^{-\varphi_j}d\tau \leqq \int_\Omega |g|^2 e^{-\varphi}d\tau.$$

Since the sequence φ_j is uniformly bounded above on every compact subset of Ω, we can find a subsequence u_{j_k} such that u_{j_k} converges weakly to a function u in $L^2(\Omega_l)$ for every l. Then $\bar{\partial}u=g$, since differentiation of a distribution is continuous in L^1_{loc} and

$$\int_{\Omega_l} |u|e^{-\varphi_j}d\tau \leqq \int_\Omega |g|^2 e^{-\varphi}d\tau$$

for all j and l, so

$$\int_\Omega |u|^2 e^{-\varphi}(1+\|z\|^2)^{-2}d\tau \leqq \int_\Omega \|g\|^2 e^{-\varphi}d\tau.$$ □

Bibliography

A. Gunning, R., Rossi, H., *Analytic Functions of Several Complex Variables*, Prentice-Hall, Englewood Cliffs, N.J. (1965)

B. Hörmander, L., *An Introduction to Complex Analysis in Several Variables*, North-Holland, Amsterdam (1973)

C. Lelong, P., *Fonctions Plurisousharmoniques et Formes Différentielles Positives*, Dunod, Paris (1968) and Gordon and Breach, New York (1969)

D. Levin, B.Ja., *Distribution of Zeros of Entire Functions*, Trans. Math. Monographs, Vol. 5, American Math. Soc., Providence, R.I. (1964)

E. de Rahm, G., *Variétés différentiables*, Hermann, Paris (1960)

F. Schwartz, L., *Théorie des Distributions*, Hermann, Paris (1966)

G. Treves, F., *Linear Partial Differential Equations with Constant Coefficients*, Gordon and Breach, New York (1966)

H. Wells, R.O., *Differential Analysis on Complex Manifolds*, Springer-Verlag, New York (1980)

Agarwal, A.K.

1. On the properties of an entire function of two complex variables, Canad. J. Math. 20 (1968), 51–57
2. On the geometric means of entire functions of several complex variables, Trans. Amer. Math. Soc. 151 (1970), 651–657

Agranovič, P.Z.

1. The existence of a function holomorphic in a cone with a prescribed indicator and having proximate order (Russian), Teor. Funkcii Funkcional Anal. i. Prilozen Vyp. 24 (1975), 3–15
2. Functions of several variables with completely regular growth (Russian), Teor, Funkcii Funkcional Anal. i. Prilozen Vyp. 30 (1978), 3–13

Agranovič, P.Z., Ronkin, L.I.

1. Functions of completely regular growth of several variables, Ann. Polon. Math. 39 (1981), 239–254

Aĭzenberg, L.A.

1. The general form of a linear continuous functional in spaces of functions holomorphic in convex domains in $\mathbb{C}^n$, Soviet Math. 7 (1966), 198–202

Alexander, H.

1. On a problem of Julia, Duke Math. J. 42 (1975), 327–332
2. Projective capacity. Recent developments in several complex variables, 1–27, Princeton Univ. Press (1981), editor J. Fornaes

Andersson, M., Berndtsson, B.

1. Henkin-Ramirez formulas with weight factors, Ann. Inst. Fourier 32 (1982)

Andreotti, A., Stoll, W.

1. *Analytic and algebraic dependence of meromorphic functions*, Lecture Notes in Math. 234, Springer-Verlag, Berlin 1971

Avanissian, V.

1. Fonctions plurisousharmonique et fonctions doublement sousharmoniques, Ann. Sci. Ecole Norm. Sup. 78 (1961), 101–161

2. Fonctions plurisousharmoniques différences de deux fonctions plurisousharmoniques de type exponentiel, C.R. Acad. Sci. Paris 252 (1961), 499-500
3. Fonctions entières de p variables et fonctions plurisousharmoniques à croissance très lente. J. Analyse Math. Jerusalem 9 (1971-72), 347-361
4. Ouverts d'exclusion dans $\mathbb{C}^p$ ($p \geqq 2$) pour les fonctions entières à croissance lente, C.R. Acad. Sci. Paris, Sér. A-B, 274 (1972), 1915-1918
5. Quelques applications de la méthode des "boules d'exclusion" dans $\mathbb{C}^p$ (Armenian and Russian Summaries), Izv. Akad. Nauk. Armjan SSR, Ser. Math. 8 (1973), N° 4, 346, 306-320

Avanissian, V., Gay, R.
1. Sur les fonctions entières de plusieurs variables, C.R. Acad. Sci. Paris, Sér. A-B, 266 (1968), 1187-1190
2. Sur une transformation des fonctionnelles analytiques et ses applications aux fonctions entière de plusieurs variables, Bull. Soc. Math. France, 103 (1975), N° 3, 341-384

Bavrin, I.I.
1. The nature of a pair of analytic functions, one of which is entire, which are univalent in the space of two complex variables (Russian), Moskov. Oblast. Pedagog. Inst. Uc. Zap. 57 (1957), 33-37

Berenstein, C.A.
1. The number of zeros of an analytic function in a cone, Bull. Amer. Math. Soc. 81 (1975), 213-214

Berenstein, C.A., Dostal, M.A.
1. A lower estimate for exponential sums, Bull. Amer. Math. Soc. 80 (1974) 687-691
2. The Ritt Theorem in several variables, Ark. Mat. 12 (1974), 267-380

Berenstein, C.A., Taylor, B.A.
1. Interpolation problems in $\mathbb{C}^n$ with application to harmonic analysis, J. Analyse Math. Jerusalem 38 (1980), 188-254

Berndtsson, B.
1. Zeros of analytic functions of several variables and related topics, Thesis, Univ. of Goteborg, 1977
2. Zeros of analytic functions of several variables, Arkiv för Mat. 16 (1978)
3. A note on Pavlov-Korevaar interpolation, Nederl. Akad. Wetensch. Proc. ser. A. 81 (1978)

Bernšteĭn, S.N.
1. On entire functions of finite degree of several complex variables (Russian), Doklady Akad. Nauk. SSR (N.S) 60 (1948), 949-952

Bieberbach, L.
1. Beispiel zweier ganzer Funktionen zweier komplexer Variablen, welche eine schlicht volumentreue Abbildung des R_4 auf einen Teil seiner selbst vermitteln, Preuss. Akad. Wiss. Sitzungsber. (1933), 476-479

Bitlyan, I.F., Gol'dberg, A.A.
1. The Wiman-Valiron Theorems for integral functions of several complex variables (Russian, English summary). Vestnik Leningrad Univ. 14 (1959), Vol. 13, 27-41

Boas, R.P.
1. *Entire Functions*, Academic Press (1954), New York

Bochner, S.
1. Entire functions in several variables with constant absolute values on a circular uniqueness set. Proc. Amer. Soc. 13 (1942), 117-120

Bombieri, E.
1. Algebraic values of meromorphic maps, Invent. Math. 10 (1970), 248-263

Bombieri, E., Lang, S.
1. Analytic subgroups of group varieties, Invent. Math. 11 (1970), 1-14

Borel, E.
1. Leçons sur les séries à termes positifs, Gauthier-Villars, Paris 1902

Bose, S.K., Kumar, K.
1. On a class of Dirichlet series over $\mathbb{C}^2$, Ann. Soc. Sci. Bruxelles, Sér. I, 89 (1975), N° 4, 509-521

Bose, S.K., Sharma, D.
1. Integral functions of two complex variables, Compositio Math. 15 (1963), 210–226
Carlson, J.
1. Some degeneracy theorems for entire functions with values in an algebraic variety. Trans. Amer. Math. Soc. 168 (1972), 273–301
2. A remark on the transcendental Bezout problem, Value Distribution Theory (Part A), Proc. Tulane Univ. Program on Value Distribution Theory in Complex Analysis, Marcel-Dekker, New York (1974), 133–143
3. A moving lemma for the transcendental Bezout problem, Ann. of Math. 103 (1976), 305–330
4. A result on value distribution of holomorphic maps of $\mathbb{C}^n \to \mathbb{C}^n$, Several complex variables (Proc. Sympos. Pure Math., Vol. XXX, Part 2, Williams Coll. Williamstown, Mass. 1975), Amer. Math. Soc., Providence, R.I. (1977), 225–227
Carlson, J., Griffiths, P.A.
1. The order functions for entire holomorphic mappings, Value Distribution Theory, (Part A) Proc. Tulane Univ. Program on Value Distribution Theory in Complex Analysis, Marcel-Dekker, New York (1974), 225–248
Chern, S.S.
1. The integrated form of the first main theorem for complex analytic mappings in several complex variables, Ann. of Math. 2, 71 (1960), 536–551
Chou, C.C.
1. Sur le module minimal des fonctions entières de plusieurs variables complexes d'ordre inférieur à 1, C.R. Acad. Sci. Paris, Sér. A–B, 267 (1968), 779–780
Cornalba, M., Shiffman, B.
1. A counterexample to the "Transcendental Bezout Problem", Ann. of Math., Vol. 96 (1972), 402–406
Dalal, S.S.
1. On the order and type of integral functions of several complex variables, J. Indian Math. Soc. (N.S.) 33 (1969), 215–220
Demailly, J.P.
1. Formules de Jensen en plusieurs variables et applications arithmétiques, Bull. Soc. Math. France 110 (1982), 85–102
2. Sur les nombres de Lelong associés à l'image directe d'un courant positif fermé, Ann. Inst. Fourier Grenoble 32, N° 2 (1982), 37–66
Deny, J., Lelong, P.
1. Sur une généralisation de l'indicatrice de Phragmen-Lindelöf, C.R. Acad. Sci. Paris 224 (1947), 1046–1048
2. Etude des fonctions sousharmoniques dans un cylindre ou dans un cône, Bull. Soc. Math. France (1947), 89–112
Dikshit, G.P., Agarwal, A.K.
1. On the means of entire functions of several complex variables. Ganita 21 (1970), N° 1, 75–85
Džafarov, A.S.
1. Some inequalities for entire function of finite degree (Russian) Izv. Vyss. Ucebn. Zaved Matematika (1960), N° 1, 14, 103–115
2. Some generalizations of Berenstein's inequality for entire functions of finite degree (Azerbaijani), Akad. Nauk. Azerbaidzan. S.S.R. Trudy Inst. Math. Meh. 1 (9) (1961), 87–98
3. Inequalities between various weighted norms for entire functions of exponential type (Russian), Izv. Akad. Nauk. Azerbaidzan S.S.R. Sér. Fiz. Mat. Techn. Nauk. (1963), N° 2, 17–25, MR 28–244
4. A generalization of inequalities of Ehrenpreis, Malgrange, Hörmander, and Rosenbloom on entire functions of exponential type (Russian), Akad. Nauk Azerbaidzan SSR, Dokl. 19 (1963), N° 5, 3–6
5. Generalization of an inequality of R. Boas for entire functions of exponential type (Russian), Azerbaidzan Gos. Univ. Ucen Zap. Ser. Fiz. Mat. i Him Nauk (1964), N° 4, 3–9
6. Inequalities with a weight for entire functions of finite order (Russian), Akad. Nauk Azerbaidzan, SSR Dokl. 20 (1964), N° 12, 3–6

7. Inequalities for entire functions belonging to a certain class (Russian), Studia Sci. Math. Hungar I (1964), 17–25

Džafarov, A.S., Ibragimov, I.I.

1. Some inequalities with weight for entire functions of finite degree (Russian), Uspehi, Mat. Nauk. 19 (1964) N° 6, 120, 147–154

Džrbaǰyan, M.M.

1. On the theory of some classes of entire functions of several variables (Russian), Akad. Nauk Armyan. SSR, Izv. Fiz. Mat., Estest. Tehn. Nauk 8 (1955), N° 4, 1–23
2. On integral representation and expansion in generalized Taylor series of entire functions of several complex variables (Russian), Mat. Sb. N.S. 41, 83 (1957), 257–276

Ehrenpreis, L.

1. A fundamental principle for systems of linear differential equations with constant coefficients and some applications, Proc. Internat. Sympos. Linear Spaces (Jerusalem 1960), Jerusalem Ac. Press, Pergamon Press, Oxford (1962), 161–174

Eremine, S.A.

1. Sur des fonctions entières de deux variables (Russian, French), Ukrain Mat. Z. 9 (1957), 30–43

Evgrafov, M.A.

1. Integral representation of functions of exponential growth (Russian), Dokl. Akad. Nauk SSSR 168 (1966), 512–515

Fatou, P.

1. Sur certaines fonctions complexes de deux variables. C.R. Acad. Sci. Paris 175 (1922), 1030–1033
2. Sur les fonctions méromorphes de deux variables. C.R. Acad. Sci. Paris 175 (1922), 862–865

Favarov, S.Ja.

1. The addition of the indicator of entire and subharmonic functions of several variables (Russian), Mat. Sb. (N.S.) 105, 147 (1978), N° 1, 128–140

Filmonova, L.A.

1. A certain condition for the representability of an entire function of two complex variables by a double Newton series (Russian), Ural. Gos. Univ. Mat. Zap. 8, tetrad'4, 100–108, 136 (1974)

Fuks, B.A.

1. *Introduction to the theory of analytic functions of several complex variables*, Translation Math. Monographs 8, Amer. Math. Soc., Providence, R.I., 1963
2. *Special chapters in the Theory of analytic functions of several complex variables*, Translation Math. Monographs 14, Amer. Math. Soc., Providence R.I., 1965

Gavrilova, R.M.

1. The representation of entire functions of two complex variables by Dirichlet series (Russian), Teor. Funkcii Funkcional. Anal. i Prilozen. Vyp. 10 (1970). 71–78

Geče, F.I.

1. Systems of entire functions of several variables and their applications to the theory of differential equations (Russian), Izv. Akad. Nauk. Armjan SSR, Ser. Fiz. Mat. Nauk 17 (1964), N°2, 17–46
2. Growth characteristics of entire functions of several complex variables (Russian), Dokl. Akad. Nauk SSSR, 164 (1965), 487–490 (English translation: Soviet Math. Dokl. 6 (1965), 1242–1246
3. On a certain class of entire functions of several variables (Russian), Ukrain Mat. Z. 18 (1966), N° 3, 13–27
4. Refined growth characteristics of entire functions of several complex variables (Russian, Lithuanian, and German summaries), Litovsk Mat. Sb. 8 (1968), 461–488
5. An investigation of the growth of entire and holomorphic functions of several complex variables by means of directed characteristics (Ukranian; English and Russian summaries), Dopovidi Akad. Nauk Ukrain SR, Ser. A (1975), 105–110

Geče, F.I., Kurei, A.I.

1. The entire solutions of linear partial differential equations of infinite order (Russian; Armenian and English summaries), Izv. Akad. Nauk. Armjan. SSR, Ser. Mat. 8 (1973) N° 2, 123–143

Gol'dberg, A.A.
1. Elementary remarks on the formulas defining order and type of functions of several variables (Russian), Akad. Nauk. Armjan SSR, Dokl. 29 (1959), 145–151
Gopola, K.J., Nagaraja Rao, I.H.
1. On orders and types of an entire function, J. Austral. Math. Soc. 15 (1973), 393–408
Griffiths, P.A.
1. On the Bezout problem for entire analytic sets, Ann. of Math. 100 (1974), 533–552
Griffiths, P.A., King, J.
1. Nevanlinna theory and holomorphic mappings between algebraic varieties, Acta Math. 130 (1973), 145–220
Gromov, V.P.
1. The representation of functions by double Dirichlet sequences (Russian), Mat. Zametki 7 (1970), 53–61; English translation. Math. Notes 7, 1970, 33–37
Gross, F.
1. Entire functions all of whose derivatives are integral at the origin. Duke Math. J. 31 (1964), 617–622
2. Generalized Taylor series and orders and types of entire functions of several complex variables. Trans. Amer. Math. Ser. 120 (1945), 124–144
3. Entire functions of several variables with algebraic derivatives at certain algebraic points. Pacific J. Math. 31 (1969), 693–701
Gruman, L.
1. Entire functions of several variables and their asymptotic growth. Ark. Mat. 9 (1971), 141–163
2. The regularity of growth of entire functions whose zeros are hyperplanes. Ark. Mat. 10 (1972), 23 – 31
3. The growth of entire solutions of differential equations of finite and infinite order. Ann. Inst. Fourier (Grenoble), 22 (1972), N° 1, 211–238
4. Some precisions on the Fourier-Borel transform and infinite order differential equations. Glasgow Math. J. 14 (1973), 161–167
5. Infinite order differential equations in Banach spaces of entire functions. J. London Math. Soc., 2 (1974), 492–500
6. Interpolation in families of entire functions in $\mathbb{C}^n$, Canad. Math. Bull. Vol. 19 N° 1 (1976), 109–112
7. Les zéros des fonctions entières d'ordre fini de croissance régulière dans $\mathbb{C}^n$, C.R. Acad. Sci. Paris, t 282, 363–365 (1976)
8. The area of analytic varieties in $\mathbb{C}^n$, Math. Scand 41 (1978), 365–397
9. Value distribution for holomorphic maps in $\mathbb{C}^n$, Math. Ann. 245 (1979), 199–218
10. Propriétés arithmétiques des fonctions entières, Bull. Soc. Math. France 108 (1980), 421–440
11. La géométrie globale des ensembles analytiques dans $\mathbb{C}^n$, Séminaire P. Lelong-H. Skoda 1978–1979. Lecture Notes in Math. N° 822, Springer-Verlag (Berlin), 90–99
12. Ensembles exceptionnels pour les applications holomorphes dans $\mathbb{C}^n$, Séminaire P. Lelong-P. Dolbeault-H. Skoda 1981–1983, Lecture Notes in Math. N° 1028, Springer-Verlag (Berlin), 125–162
13. The zeros of functions of finite order in $\mathbb{C}^n$, Ann. Polon. Math. XL (1983), 161–177
Grušin, V.V.
1. On a certain theorem of Phragmen-Lindelöf type (Russian), Vestmik Muskov Univ. Ser. I. Mat. Meh. Z 1 (1966), N° 2, 15–17
Gupta, M.
1. On the class of entire functions of several complex variables having finite order point, J. Korean Math. Soc. 13 (1976), 19–25
Gurevič, D.I.
1. Closed ideals with exp. polynomial generators in rings of entire functions of two variables (Russian; Armenian and Englished summaries), Izv. Akad. Nauk Armajan. SSR Ser. Mat. 9 (1974), N° 6, 459–472, 510

Hahn, K.T.
1. A remark on integral functions of several complex variables. Pacific J. Math. 26 (1968), 509–513
Hantler, S.L.
1. Estimates for the $\bar{\partial}$-Neumann operator in weighted Hilbert spaces. Trans. Amer. Math. Soc. 217 (1976), 395–406
Hengartner, W.
1. Propriétés des restrictions d'une fonction plurisousharmonique ou entière dans $\mathbb{C}^n$ d'ordre fini aux droites complexes $C^1(zu)$. C.R. Acad. Sci. Paris Sér. A–B, 266 (1968), 649–651
2. Famille des traces sur les droites complexes d'une fonction plurishousharmonique ou entière dans $\mathbb{C}^n$. Comment. Math. Helv. 43 (1968), 358–377
Hörmander, L.
1. L^2 estimates and existence theorems for the $\bar{\partial}$ operator, Acta Math. 113 (1965), 89–152
Ibragimov, I.I.
1. Some inequalities for entire functions of finite degree in several variables (Russian), Dokl. Akad. Nauk. SSSR 128 (1959), 1114–1117
2. A bound for the norm of a linear operator in the class of entire functions of finite degree (Russian), Dokl. Akad. Nauk. SSSR. 152 (1963), 1054–1103.
3. Inequalities fo entire functions of finite degree in the metric of a generalized Lebesgue space (Russian), Akad. Nauk. Azerbaidzan. SSR, Dukl. 20 (1944), N° 4, 13–18
4. Mean values of entire functions of two complex variables that are represented by Dirichlet series (Russian), Izv. Vyss. Ucebn. Zaved. Matematika (1972), N° 6 (121)
Ibragimov, I.I., Džafarov, A.S.
1. Some inequalities for an entire function of finite degree and its derivatives (Russian), Dokl. Akad. Nauk. SSSR, 138 (1941), 755–758
2. Some inequalities for entire functions of finite degree in the norm of a generalized Lebesgue class (Russian), Izv. Akad. Nauk. Azerbaidzan SSR, Ser. Fiz-Mat. Tehn. Nauk (1962), N° 5, 17–28
Ibragimov, I.I., Nasibov, F.G.
1. Extremal problems for certain linear operators in the class of entire functions of finite degree (Russian), Leningrad Meh. Inst. Sb. Naucn. Trudov. N° 50 (1965), 116–125
Imatoshi, Y.
1. A theorem on uniformity of prime surfaces of an entire function of two complex variables. Tôkoku Math. J. (2), 27 (1975), N° 2, 285–290
Ivanov, V.K.
1. Relation between the growth of an entire function of several variables and the distribution of singularities of a function associated with it (Russian), Mat. Sb. N.S., 43 (85) (1957), 367–378
2. The growth characteristic of entire functions of several complex variables (Russian), Gosudarstv. Izdat. Fiz.-Mat. Lit., Moscow (1960). Mat. Sb. (N.S.) 47, (89) (1959), 3–16
3. The growth indicatirx of an entire function of two complex variables (Russian), Izv. Vyss Ucebn Zaved Matematika (1961), N° 2, (21), 24–31
4. A characterization of the growth of an entire function of two variables and its application to the summation of double power series (Russian), Mat. Sb. (N.S.) 47 (89) (1959), 3–16. English translation Amer. Math. Soc. Transl. (2), 19 (1962), 179–192
Jain, P.K.
1. On the means of an entire function of several complex variables. Yokohama Math. J. 20 (1972), 125–129
Jain, P.K., Gupta, V.P.
1. On the means of entire functions of several complex variables of small order. Kyungpook Math. J. 14 (1974), 185–194
Kamthan, P.K., Gupta, M.
1. Space of entire functions of several complex variables having finite order point. Math. Japan 20 (1975), N° 1, 7–19
2. Expansion of entire functions of several complex variables having finite growth. Trans. Amer. Math. Soc. 192 (1974), 371–382

Kamthan, P.K., Jain, P.K.
1. Remarks on the geometric means of entire functions of several complex variables, Riv. Mat. Univ. Parma (3) (1972), 113–117
Kardăs, A.I., Culik, I.I.
1. Majorant properties and Newton diagrams of entire functions of two complex variables (Ukranian; English and Russian summaries), Dopovidi Akad. Nauk. Ukrain. RSR, Ser. A (1969), 583–586, 665
Kioustelidis, J.
1. Eine einheitliche Methode zur Herleitung von Reihenentwicklungen für ganze Funktionen vom Exponential-typ, Composito Math. 26 (1973), 203–232
Kiselman, C.O.
1. On unique supports of analytic functionals, Ark. Math. 6 (1966), 307–318
2. On entire functions of exponential type and indicators of analytic functionals, Acta Math. 117 (1967), 1–35
3. The growth of restrictions of plurisubharmonic functions, Mathematical analysis and applications, Part. B, Adv. in Math. Suppl. Stud. 76, Academic Press, New York (1981), 438–454
Kneser, H.
1. Zur Theorie der gebrochenen Funktionen mehrerer Veränderlichen, Jahresbericht der Deutschen Math. Vereinigung, t. 48 (1948), 1–28
Kobeleva, N.L.
1. The relation between the growth of an entire function of two complex variables and the distribution of singularities of its associated functions (Russian), Izv. Vyss. Zaved Matematika (1962), N° 3, (28), 59–66
Korevaar, J., Hellerstein, S.
1. Discrete sets of uniqueness for bounded holomorphic functions $f(z, w)$. Entire functions and Related Parts of Analysis (Proc. Sympos. Pure Math., La Jolla, Calif., 1966), 273–284. Amer. Math. Soc., Providence, R.I. (1968)
Korobeĭnik, J.F., Moržakov, V.V.
1. A general form of the isomorphisms that commute with a differentiation operator in spaces of entire functions of slow growth (Russian), Mat. Sb. (N.S.), 91, (133) (1973), 475–487, 629. English translation: Math. Notes 20 (1973), 493–505
Kozmanova, A.A.
1. Polya's theorem for integral functions of two complex variables (Russian), Dokl. Akad. Nauk. SSSR (N.S.) 113 (1957), 1203–1205
Kramer, R.A.
1. Zeros of entire functions in several complex variables. Trans. Amer. Math. Soc. 176 (1973), 253–261
Kravčenko, F.G.
1. Analytic functions of roots of polynomials (Russian; English summary), Vycisl. Prikl. Mat. (Kiev), typ. 7 (1969), 77–93
Kujula, R.O.
1. Functions of finite λ-type in several complex variables, Bull. Amer. Math. Soc. 75 (1969), 104–107
2. Functions of finite λ-type in several complex variables, Trans. Amer. Math. Soc. 161 (1971), 327–358
Kurita, M.
1. A theorem on the value-distribution of a complex analytic mapping (Japanese). Sugahu 16 (1945), 195–202
Lal, J., Dikshit, G.P.
1. The Phragmen-Lindelöf principle for functions of several complex variables, Riv. Mat. Univ. Parma, (2), 6 (1945), 283–286
Lang, S.
1. *Introduction to transcendental numbers*, Addison-Wellesley (1966)
Lelong, P.
1. Sur l'ordre d'une fonction entière de deux variables, C.R. Acad. Sci. Paris 210 (1940), 470

2. Sur quelques problèmes de la théorie des fonctions de deux variables complexes. Ann. Ec. Norm. 58 (1941), 83–177
3. Sur les valeurs lacunaires d'une relation à deux variables complexes, Bull. Sci. Math. 56 (1942), 103–112
4. Sur la capacité de certains ensembles de valeurs exceptionnelles. C.R. Acad. Sci. Paris t. 214 (1942), 992
5. Définition des fonctions plurisousharmoniques, C.R. Ac. Sci., t. 215, 398–400 (1942)
6. Sur les suites de fonctions plurisousharmoniques, C.R. Ac. Sci., t. 215, 454–456 (1942)
7. Les fonctions plurisousharmoniques, Ann. Ec. Norm., t. 62, 301–338 (1950)
8. Propriétés métriques des variétés analytiques complexes définies par une équation, Ann. Ec. Norm., t. 67, 393–419 (1950)
9a. Sur la représentation d'une fonction plurisousharmonique à partir d'un potentiel, C.R. Ac. Sci., t. 237, 691–693 (1953)
9b. Sur l'extension aux fonctions entières de n variables, d'ordre fini, d'un développement canonique de Weierstrass, C.R. Ac. Sci., t. 237, 865–867 (1953)
9c. Sur l'étude des noyaux primaires et un théorème de divisibilité des fonctions entières de n variables, C.R. Ac. Sci., t. 237, 1379–1381 (1953)
10. Integration of a differential form on an analytic complex subvariety, Proc. Nat. Ac. of Sciences, 43, 246–248 (1957)
11. Fonctions entières (n variables) et fonctions plurisouharmoniques d'ordre fini dans $\mathbb{C}^n$, Journal d'Analyse, Jérusalem, t. 12, 365–406 (1964)
12a. Fonctions entières de type exponentiel dans $\mathbb{C}^n$, Ann. Inst. Fourier, t. 16, 269–318 (1966)
12b. Sur la structure des courants positifs fermés. Lect. Notes N° 578, 136–158 (1977)
13. Non continuous indicators for entire functions of $n \geqq 2$ variables and of finite order, Proceedings of Symposia in Pure Mathematics, t. 2, 285–297 (1966)
14. Fonctions entières et fonctionnelles analytiques, Cours professé à Montréal, Presse de Montréal (1968)
15. Un théorème de fonctions inverses dans les espaces vectoriels topologiques complexes, Lect. Notes N° 694, 172–195 (1978)
16. Potentiels canoniques et comparaison de deux méthodes pour la résolution du $\partial\bar{\partial}$ à croissance. Lect. Notes N° 822, 144–168 (1980)
17. Ensembles analytiques définis comme ensemble de densité. Inv. Math. 72 465–489 (1983)

Leontiev, A.F.
1. The representation of entire functions of several variables by Dirichlet series (Russian), Mat. Sb. (N.S.), 89, (131) (1972), 586–598

Levin, B.J.
1. Some extremal properties of entire functions of several variables (Russian), Doklady Akad. Nauk. SSSR (N.S.), 78 (1951), 861–864
2. Distribution of seros of entire functions, Translations of Math. Mono., Vol. 5, AMS Providence, R.I. (1964)

Litvinčuk, G.S., Haplanov, M.G.
1. On bases and complete systems in a space of analytic functions of two variables, Uspehi Mat. Nauk. (N.S.), 12 (1957), N° 4, 1976, 319–325

Logvinenko, V.N.
1. Theorems of the type M. Cartwright's theorems and real sets of uniqueness for entire functions of several complex variables (Russian), Teor. Funkcii Funkcional Anal. i Prilozen Vyp. 22 (1975), 85–100, 162
2. A certain multidimentional generalization of a theorem of M. Cartwright (Russian), Dokl. Akad. Nauk. SSSR, 219 (1974), 546–549. English translation: Soviet Math. Dokl. 15 (1974), 1617–1620 (1975)

Lok̆sin, B.I.
1. The sharpness of certain theorems on the growth of entire functions of several variables (Russian), Teor. Funkcii Funkcional. Anal. i. Prilozen. Vyp. 18 (1973), 81–90, ii
2. The growth with respect to one of thc variables of an entire function of finite order of two variables (Russian), Funkcional Anal. i. Prilozen 10 (1976), N° 2, 79

Lozinski, S.
1. A generalization of a theorem of S. Bernstein (Russian), Doklady Akad. Nauk SSSR (N.S.), 55 (1947), 9–12
Lunc, G.L.
1. The convergence of certain general series in the space of several complex variables (Russian), Sibirsk. Mat. Z. 13 (1972), 467–472
Maergoiz, L.S.
1. A property of the indicator of an entire function of several variables (Russian), Izv. Vyss. Uiebn Zaved Matematika (1964), N° 6 (43), 104–115
2. On the question of the relation between various definitions of orders of entire functions of several variables (Russian), Sibirsk. Mat. Z. 7 (1966), 1268–1292
3. Some properties of convex sets and their applications to the theory of the growth of convex and entire functions (Russian), Sibirsk. Mat. Z. 9 (1968), 577–591
4. Scales of growth of entire functions of several variables (Russian), Dokl. Akad. Nauk. SSSR 192 (1970), 495–498. English translation: Soviet Math. Dokl. 11 1970), 662–666
5. A function of the orders and scale of growth of entire functions of several variables (Russian), Sibirsk. Mat. Z. 13 (1972), 118–132. English translation: Siberian Math. J. 13 (1972), 83–93 (1973)
6. Types and their associated order of growth for entire functions of several variables (Russian), Dokl. Akad. Nauk. SSSR, 213 (1973), 1025–1028. English translation: Soviet Math. Dokl. 14 (1973), 1846–1850 (1974)
7. Functions having the types of an entire function of several variables with regard to its directions of growth (Russian), Sibirsk. Math. Z. 14 (1973), 1037–1056, 1157. English translation: Siberian Math. J. 14 (1973) 723–736 (1974)
8. The multidimensional analogue of the type of an entire function (Russian), Uspehi Mat. Nauk. 30 (1975), N° 5, (185), 215–216
Maergoiz, L.S., Yakolev, E.I.
1. Growth of convex and entire functions of infinite order with respect to the totality of variables, Some problems of multidimensional complex analysis (Russian), Akad. Nauk. SSSR, Sibirsk Otdel, Inst. Fiz. Krasnagarsk (1980), 79–83
Magnus, A.
1. On polynomial solutions of a differential equation, Math. Scand. 3 (1955), 255–260
Malgrange, B.
1. Existence et approximation des solutions des équations aux dérivées partielles et équation de convolution, Ann. Inst. Fourier 6 (1956), 271–355
Mamedhanov, D.I.
1. Some properties of an entire function of finite degree in a generalized Lebesgue space (Russian), Functional Anal. Certain Problems Theory of Difference. Equations and Theory of Functions (Russian), 150–160, Izdat. Akad. Nauk. Azerbaidzan SSR, Baku (1967)
2. Inequalities for positive entire functions in a generalized Lebesgue space (Russian), Dokl. Akad. Nauk. SSSR, 157 (1944), 526–528
Martineau, A.
1. Indicatrices des fonctions analytiques et inversion de la transformation de Fourier-Borel par la transformation de Laplace, S.C. Acad. Sci. Paris, 255 (1962), 2888–2890
2. Indicatrices des fonctionnelles analytiques et inversion de la transformée de Fourier-Borel par la transformation de Laplace, C.R. Acad. Sci. Paris, 255 (1962), 1845–1847
3. Sur les fonctionnelles analytiques et la transformation de Fourier-Borel, Jour. Analyse Math (Jerusalem) XI (1963), 1–164
4. Indicatrices des croissances des fonctions entières de N-variables, Invent. Math. 2 (1966), 81–86
5. Indicatrices de croissance des fonctions entières de N-variables; Corrections et compléments, Invent. Math. 3 (1967), 16–19
6. Unicité du support d'une fonctionnelle analytique: un théorème de C.O. Kiselman, Bull. Sci. Math. (2), 91 (1967), 131–141
7. Fonctionnelles analytiques non-linéaires et représentation de Polya pour une fonction entière

de n-variables de type exponentiel, Séminaire P. Lelong (Analyse), Année 1970, 129–165, Lecture Notes in Math., Vol. 205, Springer (1971)
8. Equations différentielles d'ordre infini, Soc. Math. France 95 (1967), 109–154
Meteger, J.
1. Local ideals in a topological algebra of entire functions characterized by non-radial rate of growth, Pacific J. Math., 51 (1974), 251–256
Maude, R.
1. Exceptional sets with respect to order of integral functions of two variables, Proc. Cambridge Philos. Soc. 53 (1957), 323–342
Molzon, R.E.
1. Capacity and equidistribution for holomorphic maps from $\mathbb{C}^2$ to $\mathbb{C}^2$, Proc. Amer. Math. Soc. 71 (1978), N° 1, 46–48
2. Sets omitted by equidimensional holomorphic mappings, Amer. J. Math. 101 (1979), N° 6, 1271–1283
3. The Bezout problem for a special class of functions, Michigan Math. J. 26 (1979), N° 1, 71–79
4. Potential theory on complex projective space: application to characterization of pluripolar sets and growth of analytic varieties, Illinois J. Math. 28, N° 1 (1984), 103–119
Molzon, R.E., Shiffman, B., Sibony, N.
1. Average growth estimates for hyperplane sections of entire analytic sets, Math. Ann. 257 (1981), N° 1, 43–59
Moržakov, V.V.
1. Convolution equations in spaces of functions that are holomorphic in convex domains and on convex compacta in $\mathbb{C}^n$ (Russian), Mat. Zametki 16 (1974), 431–440. English translation: Math. Notes 16 (1974), N° 3, 846–851 (1975)
Michiwaki, Y.
1. Several complex variables and Picard's theorem, Sci. Rep. Tokyo Kysiku Daigaku sect. A 5 (1955), 77–81
Motzkin, T.S., Schoenberg, I.J.
1. On lineal entire functions of n complex variables, Proc. Amer. Math. Soc. 3 (1952), 517–526
Muhtarov, A.Š.
1. The growth of entire functions of two complex variables (Russian), Akad. Nauk. Azerbaidzan, SSR., Dokl. 27 (1971), N° 3, 6–9
2. The characterization of the growth of functions (Russian), Akad. Nauk. Azerbaidzan, SSR, Dokl. 28 (1972), N° 4, 13–15
Napalkov, V.V.
1. Subspaces of entire functions of exponential type that are translation invariant (Russian), Sibirisk. Mat. Z. 14 (1973), 427–426, 463. English translation: Siberian Math. J. 14 (1973), 294–300
Newman, D.J., Shapiro, H.S.
1. Fischer spaces of entire functions. Entire Functions and Related Parts of Analysis (Proc. Sympos. Pure Math., La Jolla, Calif., 1966), 340–349, Amer. Math. Soc. Providence, R.I. (1968)
Nigram, H.N.
1. Use of the generalized Laplace transform to integral functions of several complex variables, Riv. Mat. Univ. Parma, (2), 7 (1966), 137–144
2. Some uses of the basic properties of Meijer transform to integral functions of two complex variables, Riv. Mat. Univ. Parma, (2), 7 (1966), 193–202
3. On "Borel-Laplace" transforms and integral functions of two complex variables, Istambul Univ. Feu. Fak. Mecm. Sér. A 33 (1968), 51–62 (1971)
Nikol'skii, S.M.
1. Some inequalities for entire functions of finite degree of several variables and their application (Russian), Doklady Akad. Nauk. SSSR (N.S.), 76 (1951), 785–788
Nishino, T.
1. Sur les valeurs exceptionnelles au sens de Picard d'une fonction entière de deux variables, J. Math. Kyoto Univ. 2 (1962–1963), 365–372

2. Nouvelles recherches sur les fonctions entières de plusieurs variables complexes I, J. Math. Kyoto Univ. 8 (1968), 49–100
3. Nouvelles recherches sur les fonctions entière de plusieurs variables complexes. II. Fonctions entières qui se réduisent à celles d'une variable, J. Math. Kyoto, Univ. 9 (1969), 221–274
4. Nouvelles recherches sur les fonctions entières de plusieurs variables complexes. III. Sur quelques propriétés topologiques des surfaces premières, J. Math. Kyoto Univ. 10 (1970), 245–271
5. Nouvelles recherches sur les fonctions entières de plusieurs variables complexes. IV. Types de surface première, J. Math. Kyoto Univ. 13 (1973), 217–272
6. Nouvelles recherches sur les fonctions entières de plusieurs variables complexes. V. Fonctions qui se réduisent aux polynômes, J. Math. Kyoto Univ. 15 (1975), N° 3, 527–553

Nishino, T., Yoshioka, T.
1. Sur l'itération des transformations rationnelles entière de l'espace de deux variables complexes, C.R. Acad. Sci. Paris, 240 (1965), 3835–3837

Noverraz, Ph.
1. Comparaison d'indicatrices de croissance pour des fonctions plurisousharmoniques ou entières d'ordre fini, J. Analyse Math. Jerusalem 12 (1964), 409–418
2. Fonctions entières ou plurisousharmoniques de type exponentiel, Ann. Soc. Sci. Bruxelles, Sér. I, 75 (1961), 113–122

Okada, M.
1. Un théorème de Bezout transcendant sur $\mathbb{C}^n$, J. Funct. Anal. 45 (1982), N° 2, 236–244

Paris, J.
1. Croissance des fonctions de plusieurs variables et domaines d'holomorphie associés, Acad. Roy. Bull. Cl. Sci. (5), 48 (1962), 29–36

Perami, H.
1. Sur le problème d'Abel-Gontcharoff pour les fonctions entières de deux variables, C.R. Acad. Sic. Paris, Sér. A–B (1966), 556–569

Petrenko, V.P.
1. The growth of entire curves and entire functions of two complex variables (Ukranian; English and Russian summaries), Dopovidi Akad. Nauk. Ukrain, RSR, Ser. A (1974), 792–794, 861

Plancherel, M., Polya, G.
1. Fonctions entières et intégrales de Fourier multiples, Comment. Math. Helv., Vol. 9 (1936–1937), 224–248; Vol. 10 (1937–1938), 110–163

Poincaré, H.
1. Sur les fonctions de deux variables, Acta Math. t. 2 (1883), 97–113
2. Sur les propriétés du potentiel et les fonctions abéliennes, Acta Math. t. 22 (1899), 89–180

Renyi, C.
1. On some questions concerning lacunary power series of two variables, Colloq. Math. 11 (1963–1964), 145–171

Ronkin, L.I.
1. On types of entire functions of two complex variables (Russian), Mat. Sb. N.S., 39, (81) (1956), 253–266
2. Integral functions of finite degree and functions of completely regular growth (of several variables) (Russian), Dokl. Akad. Nauk. SSSR (N.S.), 119 (1958), 211–214
3. A property of the distribution of singularities on the boundary of a polycylinder and its application to entire functions of several variables (Russian), Dokl. Akad. Nauk. SSR, 153 (1963), 278–281
4. On the conjugate orders and types of entire functions of several variables (Russian), Ukrain Mat. Z. 16 (1964), 408–413
5. Growth of entire functions of several complex variables (Russian), Mat. Sb. (N.S.), 71, (113) (1966), 337–356
6. Growth of entire functions of several complex variables (Russian), Dokl. Akad. Nauk. SSSR, 169 (1966), 509–532. English translation: Soviet Math. Dokl. 7 (1966), 974–977
7. An analogue of the canonical product of Weirstrass for entire functions of several complex variables (Russian), Dokl. Akad. Nauk. SSSR, 175 (1967), 767–769

8. The growth of plurisubharmonic functions and the distribution of values of entire functions of several variables (Russian), Dokl. Akad. Nauk. SSSR, 179 (1968), 290–292
9. The analog of the canonical product for entire functions of several complex variables (Russian), Trudy Moskov. Mat. Obsc. 18 (1968), 105–146
10. Characterizations of the distribution of seros of entire functions of several variables (Russian), Teor. Finkeiĭ Funkcional. Anal. i. Priložen typ 12 (1970), 111–116
11. The completeness of function system $e^{i\langle\lambda,x\rangle}$ and real uniqueness sets of entire functions of several variables (Russian), Funkcional Anal. i. Priložen, 5 (1971), N° 4, 86
12. Certain questions of the distribution of the zeros of entire functions of several variables (Russian), Mat. Sb. (N.S.), 87, (129) (1972), 351–368
13. Real sets of uniqueness for entire functions of several variables and the completeness of systems of functions $e^{i\langle\lambda,x\rangle}$ (Russian), Sibirsk. Mat. Z. 13 (1972), 638–644. English translation: Siberian Math. 13 (1972). 439–443
14. *Introduction to the theory of entire functions of several variables*, Translations of Mathematical Monographs, Vol. 44, American Mathematical Society, Providence, R.I. (1974)
15. Discrete uniqueness sets for entire functions of exponential type in several variables (Russian), Sibirsk. Mat. Z. 19 (1978), N° 1, 142–152

Rosenfeld, M.
1. On polynomials with related level sets, Canad. Math. Bull. 13 (1970), 137–138

Rubel, L.A., Taylor, B.A.
1. Uniqueness theorems for analytic functions of one and of several somplex variables, Proc. Cambridge Philos. Soc. 64 (1968), 71–82

Rudin, W.
1. A geometric criterium for algebraic varieties, J. Math. Mech. 17 (1967–1968), 671–683

Rütishauser, H.
1. Über Folgen und Scharen von analytischen und meromorphen Funktionen mehrerer Variablen, sowie von analytischen Abbildungen, Acta Math. 83 (1950), 249–325

Sadullaev, A.
1. Fatou's example, Math. Zametki 6, N° 4 (1969), 717–719

Saito, H.
1. Fonctions entières qui se réduisent à certains polynômes. I, Osaka J. Math. 9 (1972), 293–332

Salimov, F.G.
1. The order of entire functions of several complex variables that are defined by Dirichlet series (Russian), Izu. Vysš. Učebin. Zaved Matematika (1972), N° 5, (120), 74–79

Schneider, D.
1. Sufficient sets for some spaces of entire functions, Trans. Amer. Math. Soc. 197 (1974), 161–180

Schwartz, L.
1. Généralisation de la notion de fonction. Ann. Inst. Fourier (1945), 57–74
2. Courant associé à une forme différentielle méromorphe sur une variété analytique complexe, Colloque de Géométrie Différentielle CNRS, Strasbourg (1953). 185–195

Servien, Cl.
1. Espaces de fonctions entières et fonctionnelles analytiques, Séminaire P. Lelong (Analyse). Année 1967–1968, 57–71, Lecture Notes in Math., Vol. 71, Springer, Berlin (1968)

Shaw, J.K.
1. Whittaker constants for entire functions of several complex variables, Pacific J. Math. 38 (1971), 239–250

Sibony, N., Wong, P.M.
1. Some results on global analytic sets, Séminaire P. Lelong-H. Skoda (1978), Lecture Notes in Math. N° 822, Springer-Verlag (Berlin), 221–237

Siciak, J.
1. A note on functions of several complex variables, Proc. Amer. Math. Soc. 13 (1962), 686–689

Singh, J.P.
1. On the order and type of entire functions of several complex variables, Riu. Mat. Univ. Parma (2), 10 (1969), 111–121

Sire, J.

1. Sur les fonctions de deux variables d'ordre apparent total fini, Rend. Circ. Palermo 31 (1911), 1-91

Siu, Y.T. 1. Analyticity of sets associated to Lelong numbers and the extension of closed positive currents, Invent. Math. 27 (1974), 53-156

Skoda, H.

1. Solution à croissance du second problème de Cousin dans $\mathbb{C}^n$, Ann. Inst. Fourier (Grenoble) 21 (1971), 11-23
2. Croissance des fonctions entières s'annulant sur une hypersurface donnée de $\mathbb{C}^n$, Séminaire P. Lelong (1970-71), Lecture Notes in Math., N° 332, Springer-Verlag
3. Sous-ensembles analytiques d'ordre fini ou infini dans $\mathbb{C}^n$, Bull. Soc. Math. France 100 (1972), 353-408
4. Application des techniques L^2 à la théorie des idéaux d'une algèbre de fonctions holomorphes avec poids, Ann. Ec. Norm. Sup. 5 N° 4 (1972), 545-579
5. Nouvelle méthode pour l'étude des potentiels associés aux ensembles analytiques, Séminaire P. Lelong (1972-1973). Lecture Notes in Math. N° 410, Springer-Verlag
6. Estimations L^2 pour l'opérateur $\bar{\partial}$ et applications arithmétiques, Séminaire P. Lelong (1975-1976), Lecture Notes in Math. 578, Springer-Verlag (Berlin), 314-323

Šopf, G.

1. The dependence of the hypersurfaces of associated types for systems of associated orders on the choice of the associated orders (Russian), Izv. Vysš. Učebn. Zaved. Mat. (1974), N° 12, (151), 35-46
2. Construction of an entire function of several variables with a given asymptotic distribution of its zero points (Russian), Ukrain Mat. Z. (1981), N° 4, 476-481

Sreenivasulu, V.

1. A theorem on the order of an entire function of several complex variables, Indian J. Pure Appl. Math. 2 (1971), N° 2, 312-317

Srivastava, R.K.

1. On the derivatives if integral functions of several complex variables, J. Math. Tokushima Univ. 1 (1967), 51-56

Srivastava, R.K., Kumar, V.

1. On the order and type of integral functions of several complex variables, Compositio Math. 17 (1965), 161-166

Srivastava, R.K., Kumar, V.

1. On the order and type of integral functions of several complex variables, Compositio Math. 17 (1965), 161-166
2. On means of integral functions of two or more variables, Rev. Mat. Hisp.-Amer. (4), 29 (1969), 59-66

Srivastava, S.N.

1. On the mean values of an integral function of two complex variables, Ann. Polon. Math. 20 (1968), 57-60

Stavaskiĭ, M.Š.

1. The relation between the growth of an entire function of several complex variables and the set of singular points of its associated function, Izv. Vyss. Ucebn. Zaved. Mathematika (1959), N° 2, (9), 227-232

Stoll, W.

1. Mehrfache Integrale auf Komplexen Mannigfaltigkeiten, Math. Z. 57 (1953), 116-154
2. Ganze Funktionen endlicher Ordnung mit gegebenen Nullstellen Flächen, Math. Z. 57 (1953), 211-237
3. The growth of the area of a transcendental analytic set of dimension one, Math. Z. 81 (1963), 76-98
4. The growth of the area of a transcendental analytic set, I, II, Math. Ann. 156 (1964), 47-78 et Math. Ann. 156 (1964), 144-170
5. About entire and meromorphic functions of exponential type. Entire Functions and Related

Parts of Analysis (Proc. Sympos. Pure Math., La Jolla, Calif., 1966), 392–430, Amer. Math. Soc. Providence, R.I. 1968
6. About the value distribution of holomorphic maps into the projective space, Acta Math., 123 (1969), 83–114
7. Value distribution of holomorphic maps. Several Complex Variables, I (Proc. Conf. Univ. of Maryland, College Park, Md., 1970, 165–190, Springer, Berlin (1970)
8. A Bezout estimate for complete intersections, Ann. of Math. (2), 96 (1972), 361–401
9. Holomorphic functions of finite order on several complex variables. Conference Board of the Mathematical Sciences, Regional Conferences Series in Mathematics, N° 21, American Mathematical Society, Providence, R.I. (1974)

Strelic, Š.I.
1. The Wiman-Valiron theorem for entire functions of several variables, Dokl. Akad. Nauk. SSSR, 134 (1960), 286–288 (Russian), English translation: Soviet Math. Dokl. 1 (1961), 1075–1077
2. Generalization to entire functions of several complex variables of the theorem of Wiman and Valiron (Russian), Litovsk. Mat. Sb. 1 (1961), N° 1–2, 327–354
3. Relations for the derivatives of an entire transcendental function of several variables at points of maximum modulus (Russian), Dokl. Akad. Nauk. SSR, 145 (1962), 737–740
4. On the maximum modulus of analytic functions of several variables (Russian), Mat. Sb. (N.S.), 57, (99) (1962), 281, 296
5. Some questions of the growth and existence of entire transcendental solutions of partial differential equations (Russian), Litovsk. Mat. Sb. 2 (1962), N° 1, 167–178
6. Some properties of the maximum modulus of analytic functions of several variables (Russian), Litovsk. Mat. Sb.Z. (1962), N° 1, 153–166
7. The theorem of Wiman and Valiron for entire functions of several complex variables (Russian), Mat. Sb. (N.S.), 58, (100) (1962), 47–64
8. The growth of entire solutions of partial differential equations (Russian), Mat. Sb. (N.S.), 61, (103) (1963), 257–271
9. Behavior of an entire transcendental function of several complex variables for large values of its modulus (Russian), Litovsk. Mat. Sb. 4 (1964), 357–408

Suzuki, M.
1. Propriétés topologiques des polynômes de deux variables complexes et automorphismes algébriques de l'espace $\mathbb{C}^2$, J. Math. Soc. Japan 26 (1974), 241–257

Takijima, K.
1. The regularity of holomorphic mappings between analytic spaces, Sci. Rep. Tokyo Kyviku Daigaku Sect. A 10 (1969), 184–192

Taylor, B.A.
1. The fields of quotients of some entire functions. Entire functions and related Parts of Analysis (Proc. Sympos. Pure Math., La Jolla, Calif., 1966), 468–474, Amer. Math. Soc., Providence, R.I. (1968)

Temlyakov, A.A.
1. Entire functions of two complex variables (Russian), Moskov. Oblast. Pedagog. Inst. Uc. Zup. Trudy Kafedr. Mat. 20 (1954), 7–16

Trutnev, V.M.
1. A radial indicator in the theory of Borel summability and certain applications (Russian), Sibirsk. Mat. Z. 13 (1972), 659–664. English translation: Siberian Math. J. 13 (1972), 453–456 (1973)

Valiron, G.
1. Lectures on the general theory of integral functions. Privat. Toulouse (1923)

Vauthier, J.
1. Comportement asymptotique des fonctions entières de type exponential dans $\mathbb{C}^n$ et bornées dans le domaine réel, J. Functional Analysis, 12 (1973), 290–306

Vladimirov, V.S.
1. A generalization of Liouville's theorem (Russian), Trudy Mat. Inst. Stcklov. 64 (1961), 9–27

Waldschmidt, M.

1. Nombres transcendants et groupes algébriques, Astérisque 69–70, Soc. Math. France (1969)

Wang, S.P.

1. On difference equations of entire functions, Chinese J. Math. 2 (1974), N° 2, 291–306

Wiegerinck, J.

1. Growth properties of functions of Paley-Wiener class on $\mathbb{C}^n$, Indagationes Math. 46, N° 1 (1984)
2. Paley-Wiener functions with prescribed indicator, Thesis, Univ. Amsterdam

Winiarski, T.D.

1. Approximation and interpolation methods in the theory of entire functions of several variables (Polish and Russiam summaries). Proceedings of the Fifth Conference on Analytic Functions (Univ. Mariae Curie-Skłodowska, Lublin, 1970). Ann. Univ. Mariae Curie-Skłodowska, Sect. A, 22–24, 1968–1970, 189, 191 (1972)
2. Applications of approximation and interpolation methods to the examination of entire functions of n complex variables, Ann. Polon. Math. 28 (1973), 97–121

Wirtinger, W.

1. Eine Determinatenidentität und ihre Anwendung auf analystische Gebilde, Monatsh.Math. und Pysik 441 (1936), 343–365

Wu, H.

1. Normal families of holomorphic mappings. Acta. Math. 119 (1967), 193–233
2. An n-dimensional extension of Picard's theorem, Bull. Amer. Math. Soc. 75 (1969), 1357–1361

Yamaguchi, H.

1. Sur une uniformité des surfaces constantes d'une fonction entière de deux variables complexes, J. Math. Kyoto Univ. 13 (1973), 417–433
2. Sur le mouvement des constantes de Robin, J. Math. Kyoto Univ. 15 (1975), 53–71
3. Parabolicité d'une fonction entière, J. Math. Kyoto Univ. 16 (1976), N° 1, 71–92

Index

Algebraic dependence 160
- independence 160
- integer 155
- number 155
- - size 156
- variety 7
Analytic functional 177
- variety 46

Bounded family of polynomials 82

Carrier 177
Cauchy-Fantappié Formula 41
Compactly contained in 106
Complete intersection 47
- left stability 126
Complex dimension 47
- homogeneous function 5
- submanifold 46
Convergence exponent 64
Convolution operator 207
Cousin data 59
- area of 62
- current of integration 62
- multiplicity 60
Current 34
- closed 37
- continuous of order zero 36
- dominates 36
- positive 34
- - degree 34
- push forward 120

Denominator 156
Division Theorem 208

Entire function 2
Extension, finite type 155
- simple 155

Form modulus 35
- norm 35
- positive 30
- - decomposable 31
- pull back 120
Fourier-Borel transform 178/204
𝔉-support 186

Grassmannian 143
Genus 64

Harmonic function 230
Hartog's Lemma 22
Holomorphic function 2

Indicator of growth function, circled 21
- - - Cousin data 63
- - - positive current 37
- - - projective 179
- - - radial 21
- - - with respect to one variable 11
Inverse Function Theorem for Plurisubharmonic Functions 34, 240

Laplace transform, generalized 186
- - projective 183
Lelong number 37
Linearly separates 195

Maximum Principle 234
Minimal growth class 134

Order 8
- proximate 14
- conjugate 205
- strong proximate 16
- total 11
- with respect to one variable 12

Pluriharmonic function 231
Pluripolar set 24, 234
Plurisubharmonic function 3, 230
- - locally 237
Polar set 234
Polynomial domination 159
- size 159

Positively homogeneous function 5
Pseudo-algebraic 136

Regular growth 96
– system 32
Regularization 19

Slowly increasing function 14, 79
Subadditive function 5
Subharmonic function 3, 230
– – locally 237
Supporting function 178, 185
– hyperplane 197

Transcendence basis 160
– degree 160
Transcendental number 155
Type 8, 14
– maximal 8
– minimal 8
– normal 8

Weierstrass Preparation Theorem 47
– pseudo-polynomial 47

Grundlehren der mathematischen Wissenschaften

A Series of Comprehensive Studies in Mathematics

A Selection

190. Faith: Algebra: Rings, Modules, and Categories I
191. Faith: Algebra II, Ring Theory
192. Mal'cev: Algebraic Systems
193. Pólya/Szegö: Problems and Theorems in Analysis I
194. Igusa: Theta Functions
195. Berberian: Baer*-Rings
196. Athreya/Ney: Branching Processes
197. Benz: Vorlesungen über Geometrie der Algebren
198. Gaal: Linear Analysis and Representation Theory
199. Nitsche: Vorlesungen über Minimalflächen
200. Dold: Lectures on Algebraic Topology
201. Beck: Continuous Flows in the Plane
202. Schmetterer: Introduction to Mathematical Statistics
203. Schoeneberg: Elliptic Modular Functions
204. Popov: Hyperstability of Control Systems
205. Nikol'skiĭ: Approximation of Functions of Several Variables and Imbedding Theorems
206. André: Homologie des Algébres Commutatives
207. Donoghue: Monotone Matrix Functions and Analytic Continuation
208. Lacey: The Isometric Theory of Classical Banach Spaces
209. Ringel: Map Color Theorem
210. Gihman/Skorohod: The Theory of Stochastic Processes I
211. Comfort/Negrepontis: The Theory of Ultrafilters
212. Switzer: Algebraic Topology - Homotopy and Homology
215. Schaefer: Banach Lattices and Positive Operators
217. Stenström: Rings of Quotients
218. Gihman/Skorohod: The Theory of Stochastic Processes II
219. Duvant/Lions: Inequalities in Mechanics and Physics
220. Kirillov: Elements of the Theory of Representations
221. Mumford: Algebraic Geometry I: Complex Projective Varieties
222. Lang: Introduction to Modular Forms
223. Bergh/Löfström: Interpolation Spaces. An Introduction
224. Gilbarg/Trudinger: Elliptic Partial Differential Equations of Second Order
225. Schütte: Proof Theory
226. Karoubi: K-Theory. An Introduction
227. Grauert/Remmert: Theorie der Steinschen Räume
228. Segal/Kunze: Integrals and Operators
229. Hasse: Number Theory
230. Klingenberg: Lectures on Closed Geodesics
231. Lang: Elliptic Curves: Diophantine Analysis
232. Gihman/Skorohod: The Theory of Stochastic Processes III
233. Stroock/Varadhan: Multidimensional Diffusion Processes
234. Aigner: Combinatorial Theory
235. Dynkin/Yushkevich: Controlled Markov Processes
236. Grauert/Remmert: Theory of Stein Spaces
237. Köthe: Topological Vector Spaces II

238. Graham/McGehee: Essays in Commutative Harmonic Analysis
239. Elliott: Probabilistic Number Theory I
240. Elliott: Probabilistic Number Theory II
241. Rudin: Function Theory in the Unit Ball of C^n
242. Huppert/Blackburn: Finite Groups II
243. Huppert/Blackburn: Finite Groups III
244. Kubert/Lang: Modular Units
245. Cornfeld/Fomin/Sinai: Ergodic Theory
246. Naimark/Štern: Theory of Group Representations
247. Suzuki: Group Theory I
248. Suzuki: Group Theory II
249. Chung: Lectures from Markov Processes to Brownian Motion
250. Arnold: Geometrical Methods in the Theory of Ordinary Differential Equations
251. Chow/Hale: Methods of Bifurcation Theory
252. Aubin: Nonlinear Analysis on Manifolds. Monge-Ampère Equations
253. Dwork: Lectures on *p*-adic Differential Equations
254. Freitag: Siegelsche Modulfunktionen
255. Lang: Complex Multiplication
256. Hörmander: The Analysis of Linear Partial Differential Operators I
257. Hörmander: The Analysis of Linear Partial Differential Operators II
258. Smoller: Shock Waves and Reaction-Diffusion Equations
259. Duren: Univalent Functions
260. Freidlin/Wentzell: Random Perturbations of Dynamical Systems
261. Bosch/Güntzer/Remmert: Non Archimedian Analysis - A Systematic Approach to Rigid Geometry
262. Doob: Classical Potential Theory and Its Probabilistic Counterpart
263. Krasnosel'skiĭ/Zabreĭko: Geometrical Methods of Nonlinear Analysis
264. Aubin/Cellina: Differential Inclusions
265. Grauert/Remmert: Coherent Analytic Sheaves
266. de Rham: Differentiable Manifolds
267. Arbarello/Cornalba/Griffiths/Harris: Geometry of Algebraic Curves, Vol. I
268. Arbarello/Cornalba/Griffiths/Harris: Geometry of Algebraic Curves, Vol. II
269. Schapira: Microdifferential Systems in the Complex Domain
270. Scharlau: Quadratic and Hermitian Forms
271. Ellis: Entropy, Large Deviations, and Statistical Mechanics
272. Elliott: Arithmetic Functions and Integer Products
274. Hörmander: The Analysis of Linear Partial Differential Operators III
275. Hörmander: The Analysis of Linear Partial Differential Operators IV
276. Liggett: Interacting Particle Systems
277. Fulton/Lang: Riemann-Roch Algebra
278. Barr/Wells: Toposes, Triples and Theories
279. Bishop/Bridges: Constructive Analysis
281. Chandrasekharan: Elliptic Functions

Springer-Verlag Berlin Heidelberg New York Tokyo